Security and Reliability of Damaged Structures and Defective Materials

NATO Science for Peace and Security Series

This Series presents the results of scientific meetings supported under the NATO Programme: Science for Peace and Security (SPS).

The NATO SPS Programme supports meetings in the following Key Priority areas: (1) Defence Against Terrorism; (2) Countering other Threats to Security and (3) NATO, Partner and Mediterranean Dialogue Country Priorities. The types of meeting supported are generally "Advanced Study Institutes" and "Advanced Research Workshops". The NATO SPS Series collects together the results of these meetings. The meetings are co-organized by scientists from NATO countries and scientists from NATO's "Partner" or "Mediterranean Dialogue" countries. The observations and recommendations made at the meetings, as well as the contents of the volumes in the Series, reflect those of participants and contributors only; they should not necessarily be regarded as reflecting NATO views or policy.

Advanced Study Institutes (ASI) are high-level tutorial courses intended to convey the latest developments in a subject to an advanced-level audience

Advanced Research Workshops (ARW) are expert meetings where an intense but informal exchange of views at the frontiers of a subject aims at identifying directions for future action

Following a transformation of the programme in 2006 the Series has been re-named and re-organised. Recent volumes on topics not related to security, which result from meetings supported under the programme earlier, may be found in the NATO Science Series.

The Series is published by IOS Press, Amsterdam, and Springer, Dordrecht, in conjunction with the NATO Public Diplomacy Division.

Sub-Series

A.	Chemistry and Biology	Springer
B.	Physics and Biophysics	Springer
C.	Environmental Security	Springer
D.	Information and Communication Security	IOS Press
E.	Human and Societal Dynamics	IOS Press

http://www.nato.int/science
http://www.springer.com
http://www.iospress.nl

Series C: Environmental Security

Security and Reliability of Damaged Structures and Defective Materials

edited by

Guy Pluvinage
University Paul Verlaine - Metz
Metz, France

and

Aleksandar Sedmak
University of Belgrade
Belgrade, Serbia

Published in cooperation with NATO Public Diplomacy Division

Proceedings of the NATO Advanced Research Workshop on
Security and Reliability of Damaged Structures and Defective Materials
Portoroz, Slovenia
19–22 October 2008

Library of Congress Control Number: 2009928837

ISBN 978-90-481-2791-7 (PB)
ISBN 978-90-481-2790-0 (HB)
ISBN 978-90-481-2792-4 (e-book)

Published by Springer,
P.O. Box 17, 3300 AA Dordrecht, The Netherlands.

www.springer.com

Printed on acid-free paper

All Rights Reserved
© Springer Science + Business Media B.V. 2009
No part of this work may be reproduced, stored in a retrieval system, or transmitted
in any form or by any means, electronic, mechanical, photocopying, microfilming,
recording or otherwise, without written permission from the Publisher, with the exception
of any material supplied specifically for the purpose of being entered and executed on
a computer system, for exclusive use by the purchaser of the work.

PREFACE

The proposed workshop "Security and Reliability of Damaged Structures and Defective Materials" was programmed to treat the effect of damage and defects of different origins in structures and in materials, starting with crack as the most significant form of defects.

The problem of disaster failures of structures is still current, in spite of an important achievement in the forecast and prevention of failures. Let us mention just the bridge over Mississippi river in Minneapolis, which recently collapsed on 1 August 2007. It is to remind that this bridge was completed in 1967, when the achievements of Fracture Mechanics were well known and available.

Many unexpected failures occurred in the past. The analysis of such failures had been basically performed by experiments. Serious analysis started with the ship "Schenectady" fractured into two parts on 16 January 1943. Significant number of experiments had been performed during following years in investigating this case, the cause being found as the combined effect of nil-ductility-transition (NDT) temperature and presence of stress raisers. This was the initial stimulus to find a theoretical explanation. Fracture Mechanics (initially called Crack Mechanics) offered the answer about 10 years later, in the middle of 1970s. In the early stage only Linear Elastic Fracture Mechanics (LEFM) was available, able to explain only elastic behavior and brittle fracture by a fundamental crack parameter, the stress intensity factor K_I, and its critical value, plane-strain fracture toughness K_{Ic}. The help of K_{Ic} in forecasting and predicting of failures is important, but not sufficient. The next step was to define Elastic-Plastic Fracture Mechanics (EPFM), with crack parameters as crack-opening-displacement (COD) and J integral, enabling to also deal with cracks in plastic range. The great achievement followed by extension of the same principle to other loading and environment condition, i.e. variable loading (fatigue), elevated temperature effect (creep) and aggressive environment (corrosion), and also their simultaneous effects. Anyhow, in all these situations the material is considered as homogeneous and isotropic, which is far from its actual behavior. The first problem of that kind in actual structures, mainly when they are manufactured from steel by welding, is the behavior of welded joints, heterogeneous in microstructure and in properties, on one hand, and sensitive to crack occurrence, on the other. For that the most developed procedures for structural integrity assessment, such as PD 6493 (published in 1996), "Guidelines on some methods for the derivation of acceptance levels for defects in fusion welded joints" or SINTAP (Structural Integrity Assessment Procedure), accomplished recently, also include the problem of weld material heterogeneity.

This retrospective of development in crack analysis is given here with the aim to show that the problem of cracks in materials and structures has been continuously considered. Of course, the problem of crack, theoretically introduced by Griffith in early 1920s, had been reconsidered by Irwin and many other scientists in the 1950s and 1960s, but there is a general impression that it is not yet closed. New materials require a new or at least modified approach. Typical cases are pipelines and pressure vessels made of plastics and composites, which potential fracture can cause contamination of environment. Other cases are reinforced concrete and other composite materials. The failure modes, crack initiation and propagation behavior in composite structures can substantially differ from that classically treated in structures. Are the parameters defined by Fracture Mechanics applicable for these materials and structures, and how to do that, if applicable? Further development is required to forecasting and preventing failure in anisotropic materials and other new generation materials.

However, the list of crack problems is not yet closed. Continuous scientific research and investigation enabled developing nano-materials and micro- and nano-structures. They already are considered as convenient for operational control of several structures such as process equipment, automotive industry or airplanes. The failure of such structure in computer networking and electronic communications can cause disaster and must be taken into consideration. Although problems are different from LEFM, there is at present tendency to adopt or directly apply parameters as stress intensity factor or J integral. It is probably possible to do this formally, but understanding of obtained results should be required. The scale effect must be analyzed, as it was done at the beginning of fracture mechanics development, but also now as structures at a nano-scale are a reality. The crack development is considered in fact as the separation on atomic level, but it is not easy to transfer the same analysis at micro- and nano-scale, as it is necessary for micro- and nano-structures. Another question is: how to treat imperfections in nano-materials, as atomic distance is a very important parameter? And, at last, what is the meaning of a crack in nano-structures?

In any case, the crack problem as substantial for security of structures at macro, meso, micro and nano levels is very complex and its solution is far from being closed.

PROFESSORS G. PLUVINAGE AND ALEKSANDAR SEDMAK
NATO ARW Co-directors
March 2009

CONTENTS

PREFACE v

EXPERIMENTAL AND NUMERICAL ASPECTS OF STRUCTURE INTEGRITY 1
Petar Agatonović

FATIGUE FAILURE RISK ASSESSMENT IN LOAD CARRYING COMPONENTS 25
Ing. Dragos D. Cioclov

ASSESSMENT OF LOCALISED CORROSION DAMAGING OF WELDED DISSIMILAR PIPES 75
Ihor Dmytrakh and Volodymyr Panasyuk

PROCEDURES FOR STRUCTURE INTEGRITY ASSESSMENT 91
Nenad Gubeljak and Jožef Predan

EXPERIENCE IN NON DESTRUCTIVE TESTING OF PROCESS EQUIPMENT 119
Jano Kurai and Miodrag Kirić

CRACK–INTERFACE INTERACTION IN COMPOSITE MATERIALS 139
Liviu Marsavina and Tomasz Sadowski

MEASUREMENT OF THE RESISTANCE TO FRACTURE EMANATING FROM SCRATCHES IN GAS PIPES USING NON-STANDARD CURVED SPECIMENS 157
J. Capelle, J. Gilgert, Yu.G. Matvienko and G. Pluvinage

SAFE AND RELIABLE DESIGN METHODS FOR METALLIC COMPONENTS AND STRUCTURES DESIGN METHODS 175
G. Pluvinage

DEVELOPMENT AND APPLICATION OF CRACK PARAMETERS 209
Aleksandar Sedmak, Ljubica Milović and Jasmina Lozanović

CONTENTS

WELDED JOINTS BEHAVIOUR IN SERVICE WITH SPECIAL
REFERENCE TO PRESSURE EQUIPMENT 231
*Stojan Sedmak, Katarina Gerić, Zijah Burzić, Vencislav Grabulov
and Radomir Jovičić*

SECURITY OF GAS PIPELINES 253
Vladimir Stevanović

SAFETY, RELIABILITY AND RISK. ENGINEERING
AND ECONOMICAL ASPECTS 267
L. Tóth and Lenkey Biro Gy

ASSESSING THE DEVELOPMENT OF FATIGUE CRACKS:
FROM GRIFFITH FUNDAMENTALS TO THE LATEST
APPLICATIONS IN FRACTURE MECHANICS 281
Donka Angelova

EVALUATION OF SERVICE SECURITY OF STEEL
STRUCTURES 301
Edward Petzek and Radu Băncila

FATIGUE DESIGN OF NOTCHED COMPONENTS
BY A MULTISCALE APPROACH BASED ON SHAKEDOWN 325
K. Dang Van, H.M. Maitournan and J.F. Flavenot

MEASUREMENTS FOR MECHANICAL RELIABILITY
OF THIN FILMS 337
David T. Read and Alex A. Volinsky

FROM MACRO TO MESO AND NANO MATERIAL FAILURE.
QUANTIZED COHESIVE MODEL FOR FRACTAL CRACKS 359
Michael P. Wnuk

REPARABILITY OF DAMAGED FLUID TRANSPORT PIPES 387
Philippe Jodin

DAMAGE CONTROL AND REPAIR FOR SECURITY
OF BUILDINGS 399
Dragoslav Šumarac and Zoran Petrašković

INDEX 417

EXPERIMENTAL AND NUMERICAL ASPECTS OF STRUCTURE INTEGRITY

PETAR AGATONOVIĆ
Pappelweg 11, 85244 Röhrmoos, Germany

Abstract The paper presents an approach in solving the problem of fracture mechanics treatment of surface cracks, which are a typical source for the structural failure. Difficulties associated with predicting structural integrity is that the surface cracks are three dimensional, whereas fracture mechanics methods using characterization parameters as K_{Ic}, J_{Ic}, COD are derived from two-dimensional assumptions. The method proposed for the solution is based on consideration of the elasto-plastic stress–strain behavior independent on the collapse conditions. For the investigation and development of the method, systematic numerical and experimental investigations have been performed. The main goal of the verification part of the tests' program was to justify the proposed method and to demonstrate the transferability of data, i.e. use of measurements taken from simple test specimens for a prediction of failure in large and/or complex structural components. The agreement between the prediction and test results using proposed procedure is very good. It has been shown that the proposed method gives more accurate prediction of verification test results and is less conservative than the current methods. In all cases where unnecessary conservatism is undesirable this method is more advantageous.

Keywords: Safety assessment, residual strength, surface crack, elasto-plastic fracture mechanics

1. Introduction

Systematic approach as a tool for easier, quicker, more innovative and more competitive development has developed over the last few decades into a well established strategic tool in solving engineering problems. It focused attention on the need for an integrated process of evaluating and solving problems based on scientific and engineering effort. Within of the integrated approach the main aspects of numerical and experimental investigations are focused on their planning and execution so that they support each other in a most efficient way.

Consequently, the paper presents an approach based on management of evidences systematically considering results of literature recherché, numerical calculation, material performances and testing of labor specimens and components within of the integrated evaluation concept in solving the problem of fracture mechanics treatment of surface cracks.

Failures of engineering structural components and structures have been mostly traced to surface cracks. This can be shown on latest sad example. On 3 June 1998, the *ICE* 884 Munich-to-Hamburg train derailed near Eschede, north of Hannover, at a speed of 200 km/h. Most of the trains crashed into the pillar of a concrete bridge over the track, the remains of two cars were buried under the ruins of the bridge. The accident caused 100 deaths and 88 injuries.

The accident was caused by a broken wheel tyre on the third axle of the first middle car. On the *ICE 1*, it was chosen to install new wheels of the type Bochum 84", manufactured by VSG, with a layer of rubber between the body and the tyre. Such wheels are common for light rail vehicles, but not for high speed trains. An undiscovered surface crack on the inside of the tyre (Figure 1) became longer under the fatigue loading of the rotation and caused the tyre to break in the end – something which never occurred before with this wheel type.

Figure 1. Fracture surface of broken ICE train wheel tire.[9]

The wheels of the *ICE 1* middle cars were originally of the "monobloc" type – made of one piece of steel with no separate tyre. However, these wheels became unround with time and caused vibrations which the steel springs of the bogies transmitted into the car body. The new rim, dubbed a "wheel-tire" design, consisted of a wheel body surrounded by a 20 mm thick rubber damper and then a relatively thin metal wear rim. The new design was not tested at high speed before it was commissioned and brought into service, but proved successful at resolving the issue of vibration at cruising speeds. The wheel "Bochum 84" was designed for speeds of up to 284 km/h.

There are other large number of similar design examples, concerning such components as engine crankshafts, turbine disks and blades, pressure vessels, tanks, pipes and similar. Generally, for practical application surface cracks with their limited dimensions dominate real structures. Especially for a surface crack containing structure that is thin, the limit collapse of a ligament is usually the main cause of structure rupture. This means, in spite of small dimensions, although approaching so-called collapse conditions the critical section is influenced (reduced) by the crack presence.

A further difficulty associated with predicting structural integrity is that the surface cracks are three dimensional, whereas fracture mechanics methods using characterization parameters as K_{Ic}, J_{Ic}, COD are derived from two-dimensional assumptions. Therefore, there is a relative uncertainty concerning the application of a critical value K_{Ic} to predict failure of surface-cracked components. Because of this, the K_{Ie} has been proposed, determined on surface crack specimens and limited to the LEFM application. However, for the material characterization purposes in general, the specimen dimensions must in all direction be sufficiently large to isolate the crack tip stress distribution from the borders of the specimen. Otherwise, the measured critical values would not solely be a characterization of the material, but would also depend on the size and geometry employed in the test or in actual structure. These dependencies, caused by the yielding conditions, represent a main problem in fracture mechanics, being still not entirely resolved – that of transferability of results from laboratory testing to real assessment case especially if non-linear effects are prominent.

Study of the formula relating to the critical load of the surface crack ligament is, therefore, an important project in the area of elasto-plastic fracture assessment.

Investigations presented here are mostly based on space structure experience. Accordingly, the problem of lightweight, thin wall structure appears as dominating subject and particularly limits the verification of the applicability of the results to the relevant circumstances examined [3,4,6].

2. Energy concern for fracture

Modern fracture mechanics traces its beginning back to Griffith, who in 1920 used a simple energy balance to predict the onset of fracture in brittle materials. Thus, fracture is associated with the consumption of energy. The crack will grow when the energy available for crack extension is greater than or equal to the work required for crack growth (i.e. material resistance). The conventional Linear Elastic Fracture Mechanics (LEFM) uses single parameter to represent material resistance (R) against failure, as K_{IC}. K_c, COD and J_c, being developed based on the Irwin solution for *strain energy' release rate (G)*, necessary for crack extension. These methods assume the crack tip volume wholly confined with the elastic deformations, so that the above defined parameters are interdependent and exchangeable. With other words, corresponding critical values should be a material characteristic and independent on size or geometry, but this is true only for pure elasticity. The increase in fracture resistance under elasto-plastic and stabile crack extension conditions shows that the distinct limiting parameters of the material are not valid in this case and the solution based on their application could lead to large scatter.

The failure criterion for plane strain fracture $K_{Ic} \geq K$, is supposed to give a conservative failure load prediction since the fracture toughness in plane strain (K_{Ic}) is less than the true fracture toughness of the material. However, for many materials and loading conditions, and or specimen geometry, the assessment based on this simple solution could be very pessimistic and can lead to unnecessary replacement and shut-downs, with larger cost and inconveniency. In the real structure a sudden failure, as assumed in LEFM regime, occurred by fast unstable crack propagation, this means in a most dangerous way, is a rare event. More often, the breakdown of the structure is accompanied with plastic deformations and slow crack growth accompanied with increase in material fracture resistance, characterized as ductile failure.

Accordingly, an important aspect of the behavior of structural metals is that crack-tip yielding precedes fracture. Therefore, the analysis of failure in structural components depends on two inputs, the fracture behavior and the deformation behavior, and they, furthermore, both depend on the constraint conditions in the section with the crack. However, all of the previous work on constraint has emphasized the fracture events itself and not the deformation process (represented by the stress and strain state in the components), responsible for the continuous constraint conditions changes.

Nevertheless, the Griffith model, with some modifications, is still applied today. However, this model is only applicable to materials that fail in a completely brittle manner. When the plastic deformations are restricted, due to the high constraint, the resistance increase could not act and the material fails at critical stress intensity K_{IC}, corresponding to plane strain conditions.

The physical mechanisms involved in the volume energy necessary to cause crack extension or failure are illustrated in Figure 2. In a most usual case the failure is preceded by the significant plastic deformation. After the onset of plasticity the total energy spent, shown by the surface under the curved line, is partly dissipated mostly by plastic deformation. The remain stored or recoverable part of energy becomes visible only after unloading. It corresponds to the triangle UPS (*). The dissipated portion, or plasticity energy, (dW/dV)p is varying depending on actual conditions controlling plastic deformations. Note that, in the case of crack extension the unloading path slope is different to the initial one. Unloading from a point P leads to a permanent strain ε_p that remains in the specimen. Furthermore, re-loading traced the curve which is coincident with the unloading path PU and the new yield stress is the same as the stress from which we unloaded (P). This

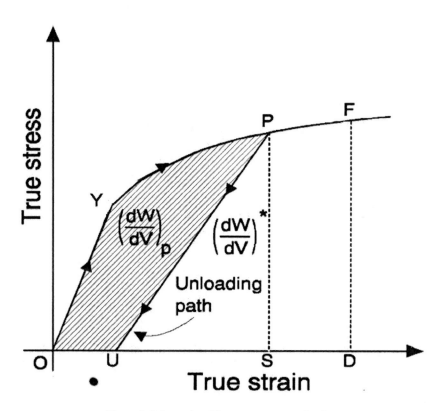

Figure 2. Schematic of fracture energy evaluation.

increase in yield stress as loading proceeds is known as work or strain hardening of material. The yield stress increase also shows that the development of plastic strains is always accompanied with changes in microstructure of material. To this end, dislocations are the discrete entities that carry plastic deformation (measured by "Burgers vector"). As a result, all material parameters change including also the fracture mechanics parameters. For ductile fracture in metals, plastic zone size increases with crack growth and hence R-curve rises.

At failure, assumed in F, only the recoverable energy is available and, therefore, relevant as the measure of the failure resistance. It is directly related to the J-integral (G, etc..). This is the reason that the failure energy under these conditions is not easy to determine.

Under the elasto-plastic regime stress fields at the proximity of the crack tip can be divided into hydrostatic and shear components. Behind strain hardening, one of the most important features concerning plastic deformation in metals, that influence fracture behavior, is the observation that the hydrostatic pressure on an element of material does not cause any plastic strain. Yielding of the material and the crack-tip blunting that plastic load conditions can produce are governed by the shear component of the stress field. Tensile hydrostatic stresses contribute directly to the opening-mode tensile stresses, but do not influence yielding and crack-tip blunting. Higher crack tip stresses occur when the material near the crack tip is highly "constrained", such as under plane strain conditions, where no contraction in the direction parallel to the crack line is allowed. It follows that fracture toughness will be directly influenced by an increase in the hydrostatic component of the crack-tip stress field, because of reduced crack-tip blunting, which increases the crack-tip strain concentration. Accordingly, *crack-tip constraint is* the term used to describe conditions influencing the performance of the hydrostatic component of the crack-tip stress field, which exhibits strong effect on apparent fracture toughness, in dependence on flaw depth and geometry, material properties and loading conditions.

On the other hand, the reduction in constraint results in a free plastic deformations and a large reduction in the tensile and hydrostatic stress ahead of the crack. Consequently, Figure 3 shows that thinner specimens produces higher R-curve than thicker ones and that bending loads produce lower R-curves than tension, because greater thickness and bending mode loading result in higher constraints near the crack tip.[7]

Figure 4 illustrate application of the R-curve for the assessment of stable and unstable behavior for a structure with an initial crack of length a.

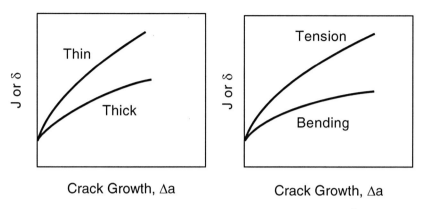

Figure 3. Effect of thickness and loading mode on R-curve.[8]

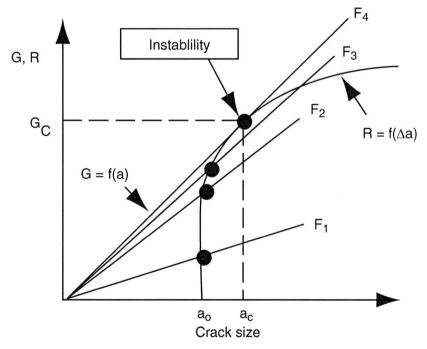

Figure 4. Driving force analysis for instability.

The driving force is represented by a series of G = f(a) curves (representing driving force) for increasing (F_1 to F_4) constant load levels. Up to the load F_2, the crack does not grow from its initial value because G < R (material's resistance). For materials, which have a rising R-curve, when the load reaches F_3 the crack grows a small amount, but further crack growth at this load is impossible, as the driving force is less than the material resistance. When the load reaches F_4 the structure becomes unstable because the rate of increase in G with crack extension starts to exceed the material's resistance. Therefore, instability occurs behind the point of tangency between the driving force and the R curve.

3. Proposed elasto-plastic surface crack solution

Although some of the aspects of the behavior under elasto-plastic conditions discussed in this chapter are complex, they still do not consider all features of the surfaces crack problem. The analytical fracture mechanics solutions assume two-dimensional conditions and the simple representation of overall conditions by unique driving force parameter (K, J, or CTOD). This means the analyses do not account for the distribution of the driving force along the crack front, which varies considerably along the contours of surface flaws.

On the other hand, in spite of the complex local deformation behavior, due to the redistribution, an adaptation take place and, for practical fracture control methods, the requirements concerning theoretical soundness of the solution could be negotiated. With other words, it is not practical to strive toward precise analytical solutions, anyway based on assumptions limiting the accuracy, because such solutions may require unnecessary large calculation time. Accordingly, practicable, robust solutions that are based on simplified estimation, if possible, may be more appropriate. However, robustness in this contest, does not implies ability to provide usable results on the basis of less reliable input but the simplicity based on unified solution applicable in the broad range, covering different kinds of the failure, and difference in geometry.

Because the failure is energy controlled, for a proper evaluation, consideration is necessary of two parameters representing under general conditions the stored energy leading to the failure and the deformation behavior, i.e. stress and strain in the critical (net) section with the crack.

The adequate method has been developed and firstly used for ARIANE 5 structure evaluation[1,5].

The well-known LEFM solution for J-Integral

$$J = \frac{K^2}{E} \quad (1)$$

can be written as the product between two K-solutions: one for stress and another for strain:

$$J = K_\sigma K_\varepsilon \quad \text{(where } K_\sigma = \frac{K}{E}\text{)} \quad (2)$$

The similarity to the Neuber solution for the case of stress concentration cannot be prized coincidental.

Constitutive laws based on classical plasticity generally define total strain as the sum of its elastic and inelastic components, with independent constitutive relationships describing each.

$$\varepsilon = \varepsilon_{el} + \varepsilon_{pl} \quad (3)$$

Therefore

$$J = K_\sigma \left(K_{\varepsilon_{el}} + K_{\varepsilon_{pl}}\right) = \frac{K_\sigma^2}{E} + K_\sigma \cdot K_{\varepsilon_{pl}} = J_{el} + J_{pl} \quad (4)$$

As usual, the actual strain is evaluated according to the stress–strain curve of the material. For this purpose use of well known Ramberg–Osgood approximation is typical

$$\sigma = B.\varepsilon_{pl}^{\,n} \quad (5)$$

Thus, for given stress

$$\varepsilon = \frac{\sigma}{E} + \left(\frac{\sigma}{B}\right)^{1/n} \quad (3')$$

In addition, the presence of a crack reduces the cross section in the structure. Net-section yielding refers to the point when the plastic zone spreads throughout the net cross section. With deep cracks, gross section yielding probably could not occur and become insignificant for this situation. Therefore, in case of limited dimensions, as for example for the plate with surface crack, relevant strain evaluation is only appropriate if the net section is considered, similar to the notch factor consideration in fatigue. Elastic-plastic fracture analyses such as the EFRI and R-6 methods assume net-section yielding in the structure.

To allow this, the linear elastic solution for K, should be scaled by the net per remote area ratio:

$$K_\sigma = \sigma_n \frac{A_N}{A} \sqrt{\frac{\pi.a}{Q}} .F \qquad (6)$$

$$K_\varepsilon = (\varepsilon_{el} + \varepsilon_{pl})_n \frac{A_N}{A} \sqrt{\frac{\pi.a}{Q}} \cdot F \qquad (7)$$

with, for the plate

$$A_N = A - \frac{\pi}{2}.a.c \qquad (8)$$

When ε_{pl} is negligible, the solution (4) reduces to the simple LEFM case. This means, the overall procedure results in a continuous linear–non-linear solutions' array for the fracture caused by the crack, independent of the fracture mechanics criteria limitations or collapse. Complicated collapse analysis is not necessary. The evaluation is based on net-section stress, i.e. section with the crack.

Using the relationships for the J-integral (4), the crack driving forces (CDF) can be calculated and corresponding method applied for growing crack, so that also standard crack resistance method may be used[2]. The procedure requires, for the material consideration, the actual engineering stress–strain curve and fracture toughness value and, for the consideration of the crack effect, the LEFM crack stress-intensity solution for given geometry.

Due to the net-section dependence on the crack size, for the evaluation of the critical crack size for given stress level using Eq. (7) the iteration is necessary.

4. Numerical verification of proposed solution

At the beginning of the investigation of the proposed procedure, for the purposes of verification and to examine conditions for the reliable data transfer from small specimens to the real structures i.e., to close possible gaps appearing due to the difference in the geometry and the loading conditions of the laboratory specimens and larger structures, a series of numerical calculations has been performed. For each specimen simulation, a referred cylinder segment, with the diameter that corresponds to the test pressure vessel foreseen for the verification testing, has been calculated. This allows a direct comparison of identical wall thickness and crack geometry parameters, by varying only the loading conditions and the curvature effect of the cylinder. Detail description of the used calculation procedure and software is given.[1,2] Here only results interesting for further investigation will be presented.

In Figure 5 a typical change of a plastic part of the J-integral for different levels of the section stress is shown. It is obvious that the distribution along of the crack contour change depending on the load level. This confirms that the prediction of the fracture by a simple scaling of the LEFM solution can lead to inadequate results.

The constraint effect, which is based on hydrostatic portion of the global stress state can be evaluated based on the ration of the hydrostatic pressure and equivalent stress (von Mises) or $h = \sigma_m/\sigma_e$. Typical results shown in Figure 6 demonstrate the complexities of constraint. Near to the surface deformation is free and h takes minimal values. However, inside of the crack, as expected, there is fast increase in all values and with the load increase the maximum values for cylinder appears already after 20°. At the same time, the crack tip constraint relaxes at the point of maximum depth, which is close to the back wall surface. Because of this, the local highest true driving force and the apparent maximum could be at two different locations.

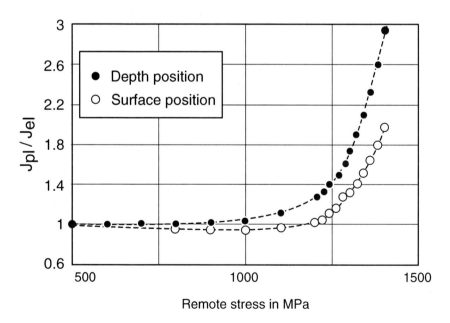

Figure 5. Change of the plastic portion of J-integral with load increase.

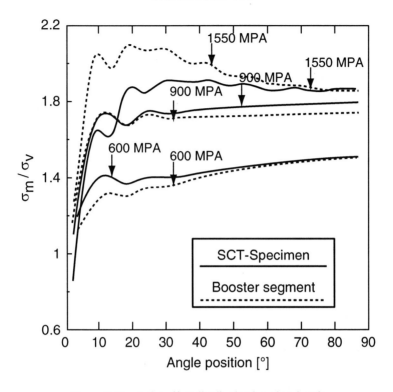

Figure 6. Constraint effect distribution based on h ratio.

In spite of the complex local behavior along of the surface crack contour, the stress redistribution may take place initiated by the local crack instability growth and for the fracture behavior; the significance of stress redistribution effects may be meaningful.

To the end, probably the most important results, for the goals of this investigation, are shown in Figure 7. The agreement between the FE-results and the solution based on proposed stress–strain approximation (2) at the place of crack contour with the maximum J values is excellent. Therefore, based on this very good agreement with the FE results concerning J-integral evaluation, the stress–strain approximation appears to be not only a very simple but also accurate method for the assessment of the plasticity effects and the prediction of the failure behavior of structures with the surface crack under non-linear conditions.

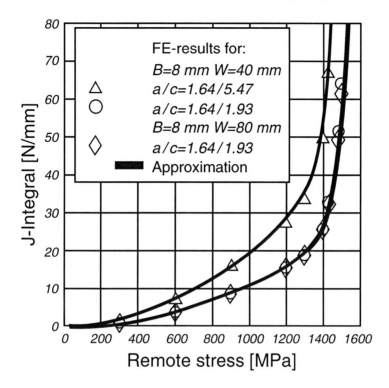

Figure 7. Comparison between FE-Results and approximation.

5. Experimental evaluation

Finally, for the investigation and verification of the method, systematic experimental tests have been carried out. Possible variation in the behavior, based on material differences is considered using two typical very different space structure materials: high strength low alloy steel D6AC and low strength aluminium alloy AL 2219 T62. To ensure proper verification, test matrix, which includes the variation of crack size and shape has been established.

5.1. SPECIMEN TESTING

Testing of the surface crack specimens requires careful preparation of the artificial defect. To this end, as a crack starter a semi circular notch is machined by spark erosion and all specimens were fatigue loaded to produce sharp crack. However, using tension loading leads to semi circular fatigue cracks (a/c = 1). Therefore, in order to get a/c < 1 the specimen must be additionally fatigue loaded in a four point bend fixture.

As known, representation of results considering typically only one crack dimension, is not conform with the analytical relationship for elliptical crack where the crack size is represented both by the crack dimensions and crack shape factor Q. Considering the ARIANE 5 experience it has been found that the useful and simple normalization can be achieved using "effective" crack size derived from crack area. The representation based on crack area has practical advantages:

- Crack size is represented by a value which directly corresponds to the NDI signal.
- The imperfection of the crack shape compared to the theoretical elliptical form can also be considered.

Analysis confirmed that between both normalization, one based on a/Q ratio and another on area, the differences in the most important a/c range (0.2 to 1) are within 10%.

The results of investigation for D6AC specimens are shown in Figure 8. According to the results, there is the trend for larger sections (based on thickness) to produce higher failure stresses. This effect can be only partly

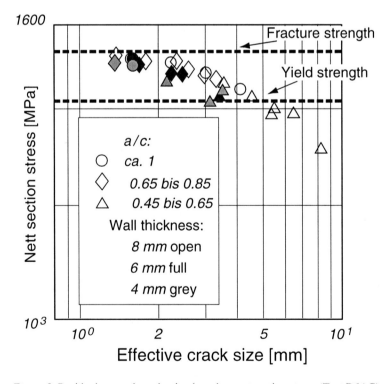

Figure 8. Residual strength evaluation based on net section stress (Test D6AC).

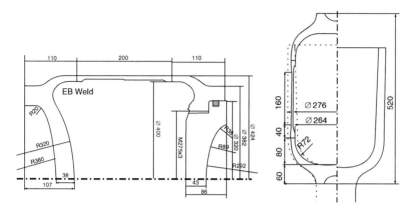

Figure 9. Two vessel design for testing.

compensated if instead of the remote stress the net section stress (this means section reduced by the area of the crack) is used. Figure 9 shows that percentile difference in results for different thickness is very small. Nevertheless, this kind of size effect must be considered based on experimental results.

5.2. COMPONENT TESTING

The effort to create the procedure adequate for the assessment of the integrity of the defect-containing components and structures was completed by the verification testing. The objectives of the verification part of the test programme was to justify the proposed method and to demonstrate the transferability of data, i.e. use of measurements taken from simple test specimens for a prediction of failure in large and/or complex structural components. To this end, the geometry and the loading conditions for testing should be representative to the structures containing surface defects (thin wall pressure vessels).

The used forms of the test pressure vessels based on available preforms were different for two materials (Figure 9). It has been found by numerical calculation (MARC) that the form of the specimen must be very accurately defined to control the loading in the areas outside of the cracked wall (welding and cylinder cover). The selected geometries fulfilled very successfully all requirements for reliable testing and can also be recommended for future testing.

In all test coupons, surface defect starters have been introduced in the centre position on the outside of the vessel wall at three equidistant positions at circumference. The introduction of more than one crack has been successfully proved in test programme. In this way more than one usable result has been achieved, even though the final fracture was always initiated from only one of the defects. However, data concerning crack growth and the significance of the defect size variation could be investigated additionally. The crack starters are cut using spark erosion. Before burst testing, the initial artificial crack has been fatigued to produce natural crack sharpness. The instrumentation of the specimen included the measurement of the load controlling parameter, surface crack extension measurements and the strain measurements. The crack extension at the surface was measured during the tests' duration at both sides of the crack using special crack extension sensors. All data from testing has been stored on disks.

The fracture appearance of all burst vessels is similar. Figure 10 shows one of the steel barrels with the wall thickness 8 mm after test. Dynamic crack is retarded at covers at both sides at the place of the significant increase in wall thickness. More detailed picture (Figure 11) shows the position and the appearance of defect leading to the fracture.

Burst aluminum vessels with 3 mm wall thickness are shown in Figure 12. Again cracks arrest at the section with increased thickness. In case of 3 mm wall vessels the general figure is the same.

5.3. ANALYSIS AND DISCUSSION OF RESULTS

Predictions were based on the plane strain fracture toughness K_{Ic} and the engineering stress–strain curve for the pressure vessel and surface crack specimens. The necessary critical fracture toughness values were derived from test results. At lower stress level the results approach the value 140 MPa$\sqrt{m}$ for steel material and 60 MPa$\sqrt{m}$ for the aluminium alloy. For all calculation engineering stress–strain curve has been used. Figure 13 shows the results for D6AC materials. As can be seen, the agreement for steel material is nearly perfect.

For comparison purposes the same results has been also predicted based on R6 method (Open points in Figure 13). The lower accuracy, which appears on the safe side, is in most application cases capable to fulfill the safety requirements.

EXPERIMENTAL AND NUMERICAL ASPECTS 17

Figure 10. Outlook of the typical vessel fracture mode.

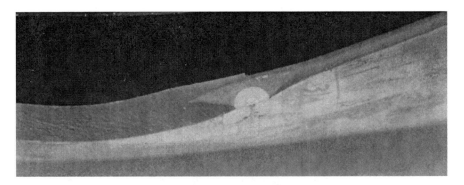

Figure 11. Detail outlook of the initial crack.

Figure 12. Outlook of the fracture mode of Al-specimens (3 mm wall thickness).

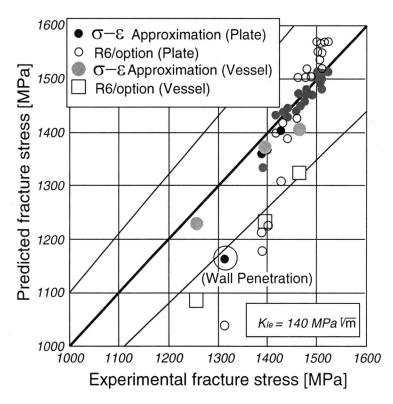

Figure 13. Results of prediction for D6AC considering plate and vessel results.

Accuracy of the prediction, based on proposed method, can also be demonstrated based on prediction of strain at fracture which was evaluated on plate specimen based on strain gage measurements (Figure 14). Under the yielding conditions the stress change is low compared to the strain change and, therefore, the accuracy of the evaluation can be better controlled based on strain measurements.

The prediction for the Al2219 material (Figure 15) suffers from the larger scatter of the specimen results. Again, the prediction for verification pressure vessel coupons is very good. However, the predictions for plate specimens approaching material strength are increasingly conservative. It appears that for the deviation, stress–strain properties of the material might be responsible. This could be explained, for example, by the material (Al 2219) softening during crack starter fatigue or by different sizes of tensile test specimen and specimen with the crack. The sensitivity of the aluminium alloys in this respect is well known.

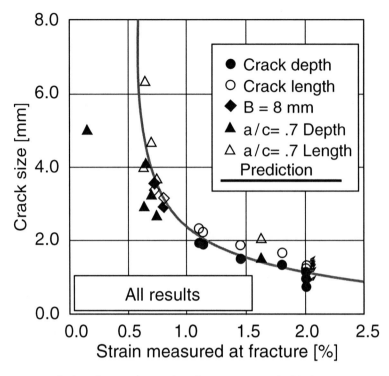

Figure 14. Prediction of net section strain at fracture compared with the measurements.

Non-conservatism of the prediction for specimens with thin wall (3 mm) is also, based on similar results with the plate specimens with the steel materials in some way expected. There is the trend to lower toughness values for thin wall vessels (3 mm). This is in full agreement with the general findings and with the plate tests for the steel material. On the other hand, in both cases the dimension of the ligaments of the surface crack was very low (lower than 0.5 mm). Similar results has been achieved in the case of two plate specimens with surface crack approaching back wall their results in Figure 15 are on the same line defining deviation of 10% against prefect prediction. For the decision in this respect additional testing would be necessary. Nevertheless, the predictions are still very good and generally within of the margin of 10%.

It is shown again that the size effect is a very important consideration within fracture mechanics. The effect is, however, reduced to the wall thickness only. The global size of the structure is, based on comparison of the plate and pressure vessel results, not significant. Because the results of the verification tests were in full agreement with the component-like test results,

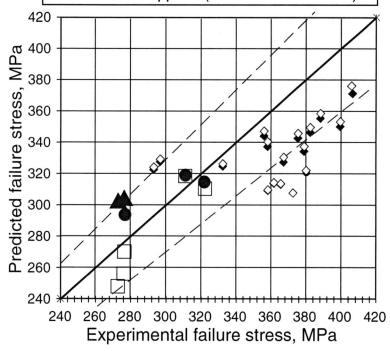

Figure 15. Final Prediction (Al2219).

the transferability of specimen with surface crack results to large structures is fully justified for the considered pressure vessel geometry. This is also in agreement with the numerical results, as shown before.

Finally, important possible extensions of the method for the evaluation of FAD (Fracture Assessment Diagram) should be mentioned. FAD is the interaction curve between the proximities to LEFM failure and plastic yielding which include the factors influencing the fracture behavior of structural components:

- Geometry and size of the component
- Properties of the material based on stress–strain and toughness behavior
- Location, type, shape and size of crack

Based on
$$\sigma \cdot \varepsilon = const.$$
$$\sigma \cdot (\varepsilon_{el} + \varepsilon_{pl}) = \sigma_e \cdot \frac{\sigma_e}{E}$$

Reference stress values can be determined
$$\sigma_e = \sqrt{\sigma \cdot E \cdot (\varepsilon_{el} + \varepsilon_{pl})}$$

After substitution $\sigma = L_r \cdot \sigma_Y$ and based on (5')

$$K_r = \frac{\sigma}{\sqrt{\sigma \cdot E \cdot (\varepsilon_{el} + \varepsilon_{pl})}} = \frac{L_r \cdot \sigma_Y}{\sqrt{L_r \cdot \sigma_Y \cdot E \cdot \left[\left(\frac{L_r \cdot \sigma_Y}{E}\right) + \left(\frac{L_r \cdot \sigma_Y}{B}\right)^{1/n}\right]}}$$

and finally the formula for the construction of FAD can be evaluated.

$$K_r = \sqrt{\frac{L_r \cdot \sigma_Y}{E \cdot \left[\frac{L_r \cdot \sigma_Y}{E} + \left(\frac{L_r \cdot \sigma_Y}{B}\right)^{1/n}\right]}}$$

Comparison with the R-6 Option 2 solution for the same steel materials (Figure 16) shows significant differences in the direction of less conservative use of material in the area of elsto-plasticity. This is full agreement with the experimental results (compare Figure 13) and can be recommended for further use.

6. Conclusions

The main subject of the presented work was the use of combined numerical and experimental investigation on adequate specimens and components with the surface crack for the development and verification of new fracture mechanics methods for reliable evaluation of the residual strength of components.

Numerical FE-investigations for the case of surface cracks showed that the effect of plasticity is twofold: an increase of the J integral due to the plasticity and a loss of the plane strain constraint at the free surface. This

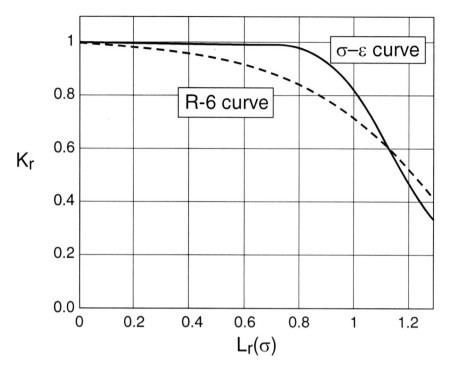

Figure 16. Comparison of FAD diagrams based on proposed method and R-6/Option 2.

leads to the redistribution of the J-integral compared to the linear K-solution. As a consequence, linear elastic solutions cannot be simply scaled to a yielding situation and a failure criterion, based on this could not be sufficient.

For the purposes of residual strength calculation, however, the redistribution within of the surface crack contour allows accurate and sufficient crack severity evaluation based on crack area. Using this kind of crack representation and considering elasto-plasticity effect, based on approximation of the stress and strains in the net section area, very accurate results have been achieved. It has been shown that the stress–strain approximation gives more accurate prediction of verification test results and is less conservative than the current methods.

In summary, the proposed accurate structure integrity assessment method provides the technology for future designs to obtain an optimization in weight while not incurring higher risk and cost. The presented very robust new analysis method (applicable in all range from LIEFM to net section yielding conditions) should significantly contribute to the state-of-the-art EPFM methodology. Being directly applied to the design of the lightweight structures it wills easy the effort of the design engineer to develop the successful and reliable hardware and keep in place with other advancing technologies.

ACKNOWLEDGEMENT Presented results were obtained within the space structure Programs supported and managed by ESA (ARIANE 5, Structure Integrity of Launchers).

References

1. Agatonovic, P.: Development of Residual Strength Evaluation Tool Based on Stress–Strain Approximation, *International Journal of Fracture* **88**: 129–152, 1997.
2. Agatonovic, P. and T.K. Henriksen: Development of Residual Strength Prediction Tools for the Structure Integrity of Launchers Based on Elasto-Plastic Fracture Mechanics, in *Proceedings of the Conference on Spacecraft Structures, Materials and Mechanical Testing*, Noordwijk, The Netherlands 1996 (ESA SP-386, June 1996) pp. 389–398.
3. Agatonovic, P. and M. Windisch: Role of Combined Numerical and Experimental Investigation in the Justification of the Structural Integrity and Damage Tolerance of Space Structures, *Proceedings of an International Conference Spacecraft Structures and Mechanical Testing*, ESA SP-321, Noordwijk, The Netherlands, 1991, pp. 679–685.
4. Agatonovic, P. and M. Windisch: Non-linear Fracture Analysis of Specimens and Components with Surface Cracks, in *Numerical Methods in Fracture Mechanics*, Ed. by A.R. Luxmoore, Peneridge Press, Swansea, 1990, pp. 597–610.
5. Anderson, T. L.: Elastic-plastic fracture mechanics: A critical review, (Part 1) SSC-345, Ship Structure Committee, April 1990.
6. ESA Contract 9934/92: *Final Report*, MAN Technologie AG (1995).
7. *ICE 1* crash at Eschede, 3rd June 1998; www.railfaneurope.net.
8. Schwalbe, K.-H.: Ductile Crack Growth Under Plane Stress Conditions: Size Effects and Structural Assessment, *Engineering Fracture Mechanics* **42**(2): 211–219, 1992.
9. Zehnder, A.: *Lecture Notes on Fracture Mechanics*, Department of Theoretical and Applied Mechanics, Cornell University, Ithaca, NY, January 7, 2008.

FATIGUE FAILURE RISK ASSESSMENT IN LOAD CARRYING COMPONENTS

ING. DRAGOS D. CIOCLOV
Consultant, CEC. Saarbrücken, Germany

Abstract A methodology is presented for integrating probabilistic fracture mechanics (PFM) with quantitative non-destructive inspection (NDI) for the purpose of failure risk assessment in load-carrying elements, mainly, under cycling loading. The definition of the failure risk in structural components is made in the context of the general approach of structural reliability with highlights on the sources of uncertainties and variability encountered in the analysis. The quality of NDI is accounted by the probability of detection (POD), as function of the flaw size. The main focus in the presentation is placed on the fatigue failure risk assessment in conjunction with the quality and timing of the envisaged NDI. The management of failure risk in aerospace technology is exemplified in the framework of established philosophies known as fail-safe and damage tolerance approach. Fail-safe or total life (TL) approach is outlined by the probabilistic assessment of the fatigue life of landing gears components under realistic loading spectra encountered by a combat aircraft. A parametric analysis of the interplay between the iso-probable fatigue life and deterministic safety factors, as applied to the mean life, is outlined. Damage-tolerance (DT) approach to structural safety, having at the core fracture mechanics technology, is presented in the framework of probabilistic paradigm with its inter-relation with variability and uncertainty associated with non-destructive inspection practice. On the base of Monte Carlo computer simulation, a rationale have been developed encompassing the fatigue crack growth (FCG) both under short- and long-crack regime, thus addressing the entire fatigue life. For this purpose, the concepts of "initial fictitious crack" size and "equivalent initial flaw" size are discussed for the purpose of the implementation in the FCG analysis. By computer simulation, it is exemplified the FCG in an aluminum alloy of class 2024 T, for a coupon with central hole. The scatter of the total fatigue life and the crack size at a given life in the simulation follows from the probabilistic input of material strength characteristics (ultimate tensile strength, yield point and fracture toughness) as well as from probabilistic FCG parameters entering a Paris–Klensil type relationship. Further, by fitting the simulated data into continuous statistical distributions enabled to model the key distributions involved in the fatigue failure risk assessment. Finally, the probability of failure, at a specific timing during the fatigue life, is estimated by Monte Carlo massive simulation of the final failure on the base of Failure Analysis Diagram (FAD) approach. It is demonstrated the benefit of applying a non-destructive inspection technique qualified by a specific probability of detection (POD) dependence on the crack size. Quantitatively, the decrease of failure risk is evinced in terms of failure probability. Computer re-sampling simulation known as "bootstrap" technique has been applied for constructing confidence intervals on POD vs. crack size in order to use in the risk analysis safe bounds of POD vs. crack size which, otherwise, are established, merely, on expert bases. It is also discussed the integration of quantitative NDI with probabilistic fracture mechanics, from the perspective of assessing a better timing and capability ranking of non-destructive inspection procedures.

Keywords: Failure risk, probabilistic fracture mechanics, uncertainty, quantitative NDI, POD, probability of failure, Monte Carlo simulation, fatigue crack growth, equivalent initial flaw, fatigue life distribution, fatigue crack size distribution, failure assessment diagram, bootstrap method, aluminum alloy

1. Failure risk concept and its assessment

Operational reliability is, in many industrial fields, a key issue on the ever-challenging international markets. Not long ago, quantitative assessing of operational reliability was restrained only to certain privileged industrial fields, such as nuclear and aerospace industries. Nowadays, operational reliability concerns large industrial companies as well as small enterprises.

The broad concept of operational reliability is primarily connected to the question of what defines the *failure risk*, in a way that enables to evaluate it quantitatively for the purpose of forthcoming decisions.

One possible definition of risk is:

$$RISK\left(\frac{Consequence}{UnitTime}\right) = FREQUENCY\left(\frac{Event}{UnitTime}\right) \times SEVERITY\left(\frac{Consequence}{Event}\right) \quad (1)$$

Risk frequency assessment may be made by statistical inference on past events (*a posteriori* analysis) or by probabilistic prediction (*a priori* analysis). The assessment of severity is a matter of economic, social, environment or political nature.

The general context of failure risk in load-carrying structures and machine elements with various contributions, concurrently implied in its assessment, are summarized in Figure 1. Factors which decide the acceptable failure risk level pertain to the specificity of the product, as related to the engineering design and manufacture, operational circumstances, together with in-service inspection (ISI), and maintenance policy. The safety philosophy of the society in conjunction with the past experience (i.e. proved sound engineering incorporating lessons drawn from past-accidents) is the general framework of the failure risk assessment and the process of decision-making.

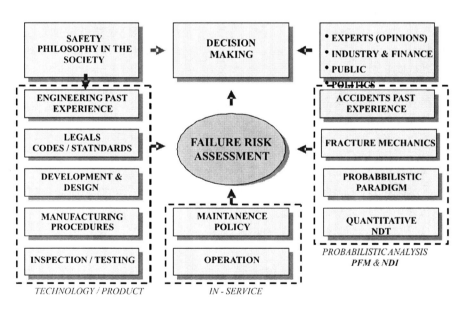

Figure 1. The context and concurrent contributions implied in the quantitative failure risk assessment.

2. On the approach to failure risk assessment

2.1. A-POSTERIORI EXPERIENCE

The past experience represents, since the maturing of technical activity of the mankind,[1] the most obvious background for new developments. Usually, the quantitative perception of the failure risk is achieved in terms of failure frequencies or failure rates defined as the number of failures (#) in an unit of time (#/time) for the entire set of products under consideration or normalized to the entire number of items (#/time × nb. items).

Failures due to design, manufacturing or unknown causes imply high rates of failure and rupture. These types of accidents are not related with physical degradation mechanisms of materials, hence, there is only a weak correlation with ISI strategy but correlates with ISI strategy and practice. Service data analysis has shown no correlation between design stresses (safety factors) and failure probability assessed on *post-factum* considerations.

When standard statistics inference is applied to past-experience of failure and rupture data, the point values correspond to maximum likelihood estimates. However, there are some problems with estimating failure or rupture rates in terms of point estimates frequencies. For instance, only upper bounds on the frequencies can be obtained if zero failure or rupture events are observed. These problems, arising from frequency-based inference, can be circumvented by the application of Bayes statistics. This rationale treats failure and rupture in terms of uncertainty distributions inferred from the state of knowledge *a-priori* to the collection of data and derive, on this basis, the likelihood function related to *a-posteriori* observed data (e.g. Box and Tiao 1973, Gelman et al. 1995).

2.2. PROSPECTIVE APPROACH

The scientific background of failure risk assessment encompasses and synthesizes essential procedures such as:

- Non-destructive inspection (NDI) techniques
- Fracture mechanics (FM)
- Mathematical theory of reliability
- Theory of probability and applied statistics
- Computerized methods for data processing
- Damage tolerance concepts, as "fitness for purpose"

---

[1] Hammurabi Code (*Codex Hammurabi*) 1758 B.C. (average) gives the first known written reference on building safety and legal consequences in the case of failure of a newly constructed building (e.g. The Code of Hammurabi, translated by L.W. King, 2007, Yale University).

- Human reliability quantitative assessment related to specified tasks
- Failure consequences on human life, economy, social or natural environment

Industrial inspection with non-destructive techniques has the objective to detect and locate defects during manufacturing and, further, during operation. It plays an ever-increasing role in the early, fast and reliable detection of failure potential in order to maintain the productivity and operation reliability of plants. In addition, industrial inspection is a major factor of economic protection by excluding, by all means failures that, nevertheless, nowadays, typically implies total costs at the level of approximately 4% of GDP in industrialized countries, as well as virtually unpredictable cost associated with the whole range of possible negative consequences on human life and natural environment.

The modern approach to failure risk resulted from the material damage or rupture has been matured in a distinct field of engineering science known as Fracture Mechanics. Combined with quantitative assessment of NDI it offers a powerful tool for mechanical failures prediction, far beyond the classical approach based only on safety factors.

As far as mechanical products are the result of conscious engineering and matured technology, the sources of failure risk reside, mainly, in uncertainty and variability (Cioclov and Kröning 1998, Cioclov 2002a, 2003).

Objective *uncertainties* are involved in manufacturing and during the operation of load-carrying structures. V*ariability* in material properties and loading circumstances is also encountered, as results of the stochastic nature of underlying physical phenomena and natural actions. Uncertainty is the assessor's lack of knowledge about physical laws and parameters which characterize physical and technical systems. Uncertainty is reducible by further experiments and study. Variability is the effect of *chance* and is function of system. Variability is *objective* since it resides in the nature of the involved physical mechanisms related, in our case, with the material strength. The encountered random loading pattern and environment actions are, equally, sources of variability. Basically, variability is not reducible by either study or further testing and measurements. Variability may be reduced only by changing the system. Uncertainty and variability (U&V) act conjointly to erode our ability to predict the future behavior of a system. Quantitative characterization of U&V is at the base of failure risk assessment. The most common methods available for this purpose are: the probability theory, applied statistics, fuzzy logic, neural networks, and elicitation procedures. In the present study only methods pertaining to the theory of probability and applied statistics will be considered.

It should, however, be emphasized that no matter how advanced are the theoretical backgrounds and the computation algorithms or how sound is the experimental support of failure risk assessment methodology, the inevitable involvement of U&V brings into play the purely non-algorithmic experts opinions (see Figure 1).

3. Fatigue failure risk assessment and management

3.1. FRACTURE CONTROL AND FAILURE RISK MANAGEMENT

Fracture control and failure risk management is a systematic approach aiming to prevent failure by fracture during operation. The extent of fracture control depends on: criticality of components in a load-carrying structure, the extent of the damage that might be incurred to the structure by the failure of one or several components, the economic consequences of the failure and the impact on society and environment, to cite the most salient issues. Fracture management strategy includes:

- Testing on specimens, joints and components under conditions of static, cyclic and dynamic loading at ambient and operating temperature (creep, thermal fatigue, alone or in combination) environmental actions (corrosion, neutron irradiation, hydrogen embrittlement, etc.).
- Inspection procedures
- Repair/replacement management

An example of systematic fracture control and failure risk management, from design, through manufacture and in operation is offered by aerospace industry where fatigue damage structure is one of the main concerns.

3.2. FAILURE RISK MANAGEMENT IN AEROSPACE INDUSTRY

There are two lines of thinking in approaching fracture control and risk of failure assessment in load-carrying components operating under cyclic loadings which induce cumulative fatigue damage of the material. One, is based on the total fatigue life assessment based on empirical test data organized in a representation: applied cyclic stress as function of number of applied cycles, $\sigma(N)$, the Wöhler curves. This is known as total life (TL) or safe-life approach to fatigue behavior evaluation. The alternate way of approach resorts to fracture mechanics methodology for assessing, explicitly, the propagation of the dominant crack leading to fracture. Because this methodology is constructed with the aim to explicitly quantify the fatigue damage of the component in terms of crack-size growth during operation it is termed as damage tolerance (DT) method. A former variant of this methodology, applied to aircraft structures, is known under the denomination of "fail-safe" approach. Let's give an outline of these two basic philosophies of approaching failure risk under material fatigue circumstances (e.g. UASF (1984) and Welch et al. (2006) for application to fuel tanks of F-16 combat aircraft fleet)

3.3. TOTAL-LIFE APPROACH (E.g. Cioclov 1975, 1998b)

TL method applies to structuresof single load path, within structures with load-carrying components arranged in series. This way of thinking stems from the early fatigue tests performed by Wöhler in the mid of the nineteenth century. In a series-organized multi-component structure, the failure of one component implies the failure of the entire structure. In aircraft industry the application of TL method is practically limited to landing gears although elements of TL approach are sometimes inserted into DT assessments. According to TL method, fatigue life calculation is performed for every component in a load carrying structure with components in series arrangement. The fatigue life assessment rationale attempts to reflect the material response to external cyclic loading by a cumulative fatigue damage rule. The material response to cyclic loading is evaluated on the base of experimental, $\sigma - N$, Woehler endurance curves, usually evaluated under constant amplitude cyclic loading on standard specimens or "coupons" which reflects the dominant feature of the component (e.g. thin-walled rectangular coupons of Al-alloys with a central hole simulating a rivet hole in an aircraft fuselage – see further Figure 13).

Figure 2 illustrates, in principles, the TL approach to the assessment of the fatigue life in service of a component from landing gear assembly.

Probabilistic nature of the fatigue damage is accounted for on the base of experimental scatter of the fatigue life at various loading levels which enables to fit probabilistic distributions of the fatigue life at a constant level of cyclic loading intensity (amplitude or maximum stress) or, alternately, the fatigue strength (expressed in stress amplitude or maximum stress) distribution at a specified fatigue life. On the base of these two complementary distributions the iso-probable endurance curves, P, are used, further, in a cumulative fatigue damage rule in order to assess the endurance – at P probability of occurrence – under the specific pattern of loading intensity variation from cycle-to-cycle (e.g. Cioclov 1975, 1999).

Applied loading action, essentially of random nature, is synthesize in loading spectra and the associated probability distributions of instantaneous loading intensity of occurrence, $p(\sigma)$. The loading synthesis is based on the recordings during operation on similar products or prototypes in the product development stage.

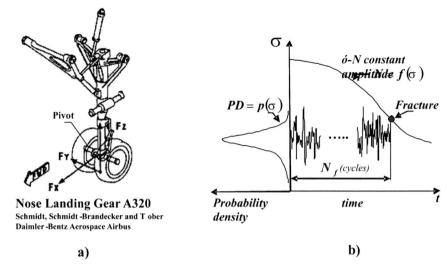

Figure 2. Schematics of safe-life life approach to landing gears. (a) Sketch of nose landing gear assembly (after Schmidt et al. 1999); (b) schematics of the random loading stress-spectrum in a critical cross-section, as in the wheel pivot, and interaction loading-fatigue life in Wöhler durability curves representation.

The interaction, cyclic loading vs. material response, is modeled by a cumulative fatigue damage rule which quantifies the damage per cycle by a phenomenologic damage parameter D. In the simplest formulation (Palmgren-Langer-Miner) it is hypothesize that fatigue damage cumulates linearly during the fatigue life, N, at a constant rate (damage per cycle) of $D = 1/N$. Obviously, in the initial undamaged state we have $D = 0$ and, at fracture, under constant amplitude loading the limit state, $D = 1$, corresponds. It follows that under irregular, n_i, loading cycles between i loading levels of constant amplitudes, σ_{ai}, to which i fatigue lives correspond, N_i (on the Wöhler curve), the condition of fracture occurrence is:

$$\sum_{i=1}^{N_r} \frac{n_i}{N_i(\sigma_{ai})} = 1 \qquad (2)$$

where: N_r is the fatigue life under irregular cyclic loading. When irregular cyclic random loading, stationary around a mean stress, σ_m, is systemized in loading spectra of probability density of the loading intensity, $p(\sigma_a)$, as shown in Figure 2b, then the integral form of linear cumulative fatigue damage rule enables to solve Eq. (2) in terms of the fatigue life under irregular cyclic loading, N_r:

FATIGUE FAILURE RISK ASSESSMENT 33

$$N_r = \int_{\sigma_{inf}}^{\sigma_{sup}} \frac{p(\sigma)d\sigma}{N(\sigma)}, \ \sigma \in (\sigma_{inf}, \sigma_{sup}) \quad (3)$$

In practice, the computation of the fatigue life under irregular loading is validated by full-scale fatigue tests.

The operational (allowable) fatigue life is established by amending the computed fatigue life by a safety factor, usually between 3 and 10. The higher value of safety factors applied to fatigue life in comparison with those practiced to establish allowable loads reflects the wider statistical scatter around mean value associated with the fatigue life.

A probabilistic TL (safe-life) approach to the estimation of failure risk of landing gear components, as results of fatigue damage accumulation, has been developed by the author (Cioclov 1999). Figure 3 exemplifies the estimation of failure risk of a landing gear component (pivot) as results of fatigue damage accumulation (Cioclov 1999, Iordache and Cioclov 2002).

In this exercise, the operational random variable loading, forces and moments, monitored experimentally, have been reduced to stresses in the critical cross-sections of the analyzed components. Stress spectra are organized in deterministic blocks of constant amplitude stresses. For a specific cross-section, the constant amplitude loading stress spectrum consists of five blocks. Each block corresponds to the main stages of operation:

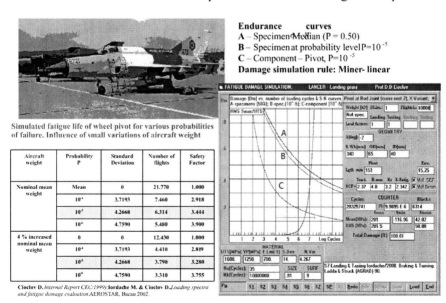

Figure 3. Safe-life method for reliability assessment as applied to landing gears of LANCER aircraft (Cioclov 1999, Iordache and Cioclov 2002). Table gives probability based safety factors.

touch-down contact and ground rolling at landing, taxiing, braking, towing and ground rolling at take-off. Due account has been given to the mean stress levels in every operational stage. Computation strategy consisted in computer simulation of block-by-block fatigue damage accumulation according to the linear Miner rule applied to constant amplitude endurance (Wöhler) curves of the landing gears components, expressed in number of flights. Ultimately, the computational simulation yields the endurance of the component.

The constant amplitude endurance curves of the components have been derived on the base of endurance curve assessed experimentally on standard specimens. In order to account on the real geometry of the component, endurance curves have been amended for size and notch effects according to well established models (e.g. Cioclov 1975). Probabilistic aspects resulting from estimated variability in the material strength (UTS, YP) and fatigue characteristics (fatigue limit) have been implemented in the simulation algorithms. Figure 3 illustrates the iso-probable endurance curves A and B determined on standard smooth specimens, for median and $P = 10^{-5}$ probability of occurrence, respectively. The iso-probable endurance curve of the wheel pivot component, for the probability of occurrence of $P = 10^{-5}$ is represented by the curve C in Figure 3. Details on the analytical form of iso-probable endurance curves can be found elsewhere (Cioclov 1975, 1998b).

The total median value of pivot endurance under assumed operational loading spectrum has been estimated at 21.770 flights. At an acceptable probability of failure occurrence, say, 10^{-5}, the maximum number of flights must be lower than 6.314 flights which correspond to a safety factor related to the median endurance of 3.444.

It should be emphasized that the outlined probabilistic TL methodology based on cycle-by-cycle fatigue damage accumulation simulation evinces, explicitly, the correlation between the failure risk, expressed by the probability of failure event occurrence, and the usual engineering safety factor applied to median value of the fatigue life.

3.4. FATIGUE DAMAGE TOLERANCE APPROACH

Fracture control and failure risk management in aerospace structures, when the initiation and growth of fatigue crack during operation is not excluded, is systemized in a strategy denominated as damage tolerance approach. DT aims to prove that catastrophic failure of a structure of particular design, owing to fatigue crack growth (FCG), is avoided throughout the operational life. If fatigue cracks subsist in a load carrying component and their growth cannot be avoided, the structure must be capable to sustain the damage without catastrophic failure in emergency circumstances or until the next prescribed inspection, when repairs or replacements are made. For this purpose of analysis fracture mechanics methods are applied.

In DT approach the pre-existence of an initial crack or a flaw of crack acuity is implied. Defects of crack-nature may be introduced in a structural component either during manufacturing or in operation. In the latter case, cracks may be induced at unavoidable notches, corrosion pits or microstructural defects in components which, nevertheless, are validated as defect-free at inspections. Initial crack-like defects are not uncommon in welded structures being related with joints geometry, welding procedure and execution circumstances (e.g. welding position, environment, workmanship skill, etc.)

Forcefully, common sense dictate that a pristine load-carrying component, with smooth geometry design which has passed all inspections during manufacturing processes is, supposedly, defect free. However, for FCG assessment by fracture mechanics methods some sort of initial crack size must be considered. The various interpretations of the initial crack size, a key issue in DT methodology, will be taken up in more details in Section 5. For now being, in order to expose the principles of DT approach, it suffices to consider that an initial crack, of known size and location, pre-exists in the structural component under analysis.

The objectives of DT analysis are to determine the effect of the presence of cracks on the structural strength and assess if during the period between inspections for crack detection, the cracks growth remains within tolerable limits, i.e. do not provoke failure. Formalized prescriptions of DT approach to airplanes design have been published by UASF (1984) as Damage Tolerant Design Handbook.

Figure 4 illustrates the steps in applying DT procedure to a load-carrying element (Broeck 1986, 1993).

First step consists in assessing the critical crack size, a_{cr}, on the base of the knowledge (by testing or fracture mechanics modeling) of static residual strength of the component, as function of crack size, $P_{res}(a)$. If the component is crack-free, then its strength is P_{ult}. A safety factor, j, is set to P_{ult} in order to specify the maximum permissible load in service, P_{max}, i.e., $P_{ult} = jP_{max}$. Typically, $j = 3$ for stationary load-carrying structures and $j = 1.5$ for aircraft structures. If it is admitted that a crack subsists in the component and the minimum acceptable residual strength, P_{res}, is set, via the safety factor $g < j$, at $P_{res} = gP_{max}$. Then, the critical crack size, a_{cr}, can be ascertained on the base of $P_{res}(a)$ relationship as shown in Figure 4a.

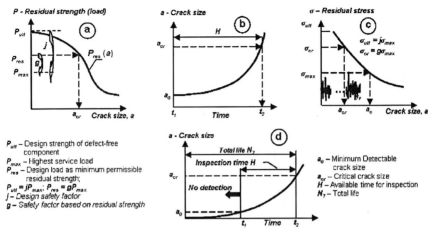

Figure 4. The steps in applying DT rationale (Broeck 1993). (a) Assessing critical residual strength and associated critical crack size, a_{cr}; (b) assessing the time, H, for fatigue crack growth from the initial crack size, a_0, to critical crack size, a_{cr}; (c) strength-loading interaction in terms of applied cyclic stresses; (d) assessment of inspection intervals.

The critical crack size, a_{cr}, is an "ultimate defense" against failure conceived to incorporate, by the intermediate of the safety factor, g, unknowns and uncertainties which are not apprehensible by analysis or other *a priori* knowledge. Safety factors j and g assure the conveyance of the past experience accumulated in structural design. j is a conservative safety factor and in the past five or six decades remained, virtually, unchanged proving to be realistic in the context of ever increasing operational loading, on the one hand, but acting on aircraft of refined design and manufacturing methods. More design-specific safety factor, g, is related with the interplay between the strength, deformation capability of the material and maximum loading which is expected to be encountered in service.

To assess the time H for the dominant fatigue crack to growth from the initial size a_0 to the critical size a_{cr} (Figure 4b), material-specific fatigue crack propagation rules are used (e.g. Paris, Forman, Walker rules, etc.). In the present state of art, the parameters entering in the analytical form of fatigue crack growth rules are derived on the base of fatigue testing on specimens containing pre-existent cracks. In practical computation, stress rather than force representation is preferred (Figure 4c).

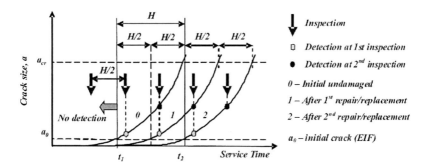

Figure 5. Setting damage tolerance inspection intervals in service. (After Broeck 1993).

If initial crack size, a_0, is identified with the minimum crack size detectable with the NDI technique in use then H is the available time for the application of NDI (Figure 4d). Before the fatigue crack attains the size, a_0, crack detection is not possible with the considered NDI technique. The aim of the outlined strategy, merely, to establish inspection intervals in relationship with the capability of the NDI technique to detect small cracks and to introduce safety provisions for the circumstance when a crack is missed during inspection than to assess the total life. Figure 5 portrays how applies a DT-based NDI strategy.

Obviously, an inspection interval ought be smaller than, H. Hence, the available time for NDI is H/i. The factor i has a typical value of 2 which implies that if a defect is missed in a scheduled NDI session, it remains, nevertheless, one possibility in the next session to detect the defect before it reaches the critical size. If a defect is found, replacement or repair at the initial quality of the component is made and the FCG curve, from this moment on, is considered to correspond to undamaged material. The new position on time axis in Figure 5 is shifted to the left with the amount of time elapsed until crack detection.

To implement DT strategy in practice, in a larger context of design and operation, supplementary supporting engineering methods are available (see UASF 1984, Schmidt et al. 1998). Besides fracture control of FCG by conjoint consideration of fracture mechanics and quantitative non-destructive inspection one can also cite:

1. Redundancy by design. It means: introducing additional load paths with elements designed to carry the load in the case of failure along the main load path; specific design to confer capability to arrest fatigue crack propagation.

2. Hazard analysis to identify potentially critical items prone to crack initiation, growth and fracture.
3. Design to enable and facilitate NDI.
4. Failure risk assessment in quantitative terms (failure probabilities).

On the deterministic DT methodology, as described above, probabilistic pattern related with material strength and FCG statistical variability, and probabilistic uncertainties assessment arising from the application of NDI can be superposed. It is achieved by integrating probabilistic fracture mechanics and probabilistic models of NDI reliability, the latter having at the core the concept of probability of detection (POD). This novel development in DT methodology will be fully presented in the following sections.

Finally, it should be emphasized that in the outlined rationale, it remains open to interpretation the central issue of how to define and assess, quantitatively, the initial crack size under the circumstance when there is no physical basis to accept the pre-existence in the structural element of initial flaws of a crack nature.

4. Probabilistic paradigm in mechanical failure risk assessment

Probabilistic aspects are encountered, virtually, in all steps of failure risk assessment in load-carrying structures. The source resides in the variability of materials properties and cyclic loading intensity, together with various uncertainties which cannot be avoided in the process of design, manufacturing and operation. In refining both TL and DT approaches to structural reliability two probabilistic aspects are of relevance: one related with the process of FCG, the other related with NDI process. An outline of these aspects will be given in this section.

4.1. PROBABILISTIC DESCRIPTION OF THE FATIGUE CRACK GROWTH

It is now accepted that the observed scatter of the experimental data on fatigue crack growth evinces two components: one type of scatter, the "collective scatter" is observed across the sample of specimens; the other type of scatter, the "individual scatter" is due to the randomness observed during the FCG within an individual specimen. The former scatter is related with the variability of material properties at macroscopic scale. It may be described by a multi-dimensional random variable (vector). The latter type of scatter is conveniently described, in first approximation, by a random process with independent increments. It is worth to mention that, from engineering standpoint, the knowledge of collective scatter, rather than individual scatter, is of more significance since it relates with the fatigue behavior of a set of items, such as an aircraft fleet or the set of load-carrying elements of the same.

Both approaches imply the use of an explicit FCG rule validated, in the present state of art, on purely empirical bases or semi-empirical mechanistic models.

FCG assessment by models stemming from primary physical mechanisms of the evolution of fatigue damage is still lacking. The basic mechanisms of fatigue damage occur at mesoscale where the material behavior is governed by dislocations dynamics. The entire assessment construct relies, nowadays, on coupled multi-scale simulation with the means of massive parallel computing. Multi-scale simulation encompasses wide size- and time-scales in conjunction with hierarchical physical theories ranging from the first quantum mechanics principles, through molecular dynamics, coupled with continuum mechanics, up to macroscopic level. Such an approach becomes possible by the tremendous development of computing power in the peta-flop domain. On this line, mutations in nowadays engineering thinking are to be expected, even in the near future. Research projects have been recently envisaged to develop predictive, experimentally verified advanced computation capabilities for materials fatigue damage evolution mased on fundamental materials physics at nano- and microscale with the aim at development of dislocations dynamics based multi-scale models of fatigue crack nucleation and cracks growth in the early stage (Ghoniem 2006, Brinckmann and Giessen 2003, Brinckmann and Siegmund 2008). A general review of the state-of-art in this novel field of materials science research can be found in a recent publication of the author (Cioclov 2008).

The concepts and methods presented in this study remain, however, in the realm of advanced but well established continuum mechanics engineering. Particularly, the FCG will be assessed by a Paris–Klensil type equation (Paris et al. 1961, Klensil and Lucás 1972) which is one of the most frequently used in engineering practice. This rule will be developed into a probabilistic model and used, consistently, in the description of scatter and uncertainties in FCG. Obviously, many other existing models of FCG, apart of Paris–Klensil one, can be probabilistically adapted.

As outlined, one way of approach to probabilistic description of the FCG is based on the association of random variables (RV) to a FCG rule, which is defined, usually, in a deterministic average sense. The procedure of randomization of the empirical parameters of a FCG rule, which also makes easy the application of the Monte Carlo technique, falls into this category. On this way, one can model successfully the collective scatter encountered in the FCG (e.g. Lidiard 1979, Varanasi and Whittaker 1976, Engeswik 1981, Shaw and LeMay 1981, Yang et al. 1983, Ditlevsen and Olsen 1986, Annis 2003, Cioclov 2002–2004, Cioclov et al. 2003). However, it is worth to mention the main limitations of this approach. First, randomization technique has developed, to date, only for the parameters of the Paris-Klensil type FCG rule[2]: C, m and less ΔK_{th}. Secondly, the parameters of the fitted distribution for these parameters have no clear-cut physical meaning and no link with material micro-structural properties has been attempted, so far. Thirdly, though probabilistic models offer a fair prediction of the scatter encountered in the FCG process they do not provide an insight into the nature of fatigue damage process in order to enable further generalizations (see Ghonem and Dore 1987). However, probabilistic FCG models have gained popularity among practitioners and many applications in fatigue reliability management have been developed on these bases.

Another way of approach is to associate deterministic FCG rules, which describes the mean or median trend of FCG, with a time-dependent random process defined by a probabilistic distribution of unity mean value and an auto-covariance function which describes probabilistic dependence between two point realizations along the random process path. In principle, by this approach, both the collective and the individual scatter displayed in the FCG can be addressed. On this line of thinking, one can cite the works of Ghonem and Provan (1980), Lin et al. (1984), Ghonem and Dore (1985), Sobczyk (1986), Ortiz and Kiremidjian (1988), Ihara and Misawa (1988), Tang and Spencer (1989), Tsurui et al. (1989) and Wu and Ni (2003).

With the advent of massive automatic computing power, the direct application of Monte Carlo random simulation technique enables to model directly both the collective and individual scatter in the process of the fatigue crack growth. Monte Carlo modeling of FCG is implemented into *pFATRISK*® methodology which will be further described together with relevant engineering fatigue case studies (Section 6).

Much theoretical material on probabilistic fatigue crack growth modeling, with references to test data, can be found in the monographs of Bolotin (1965), Bogdanoff and Kozin (1985) and Sobczyk and Spencer (1992).

A less studied aspect of fatigue cracks statistics in DT approach is that related with the initial crack size and the various expedients referred under the collective heading of "equivalent initial flaw". This issue will be taken up in more details in Section 5.

---

[2] FCG rule: $da/dn = C\left(\Delta K^m - \Delta K^m\right)$, with $\Delta K, \Delta K_{th}, C, m$ the range of applied cyclic stress intensity factor, the inferior threshold range, and two empirical parameters, respectively.

4.2. PROBABILISTIC DESCRIPTION OF QUANTITATIVE NON-DESTRUCTIVE INSPECTION

Quantitative non-destructive inspection (QNDI) encompasses flaws detection and their quantitative evaluation, together with assessing uncertainties and variability (U&V) associated with a specific NDI technique in terms of probability of detection (*POD*).

NDI systems are driven to their extreme capability to find small flaws. To the edge of extreme capability of NDI, not all small flaws may be detected. Because of U&V, NDI capability (here meant as reliability of the flaws detection process) is characterized in terms of POD as function of the flaw size, *a*. *POD(a)* function is defined as the proportion of all flaws of size, *a*, which will be detected by a given NDI system and associated prescribed procedures. Probability of non-detection (*PND*) is simply the probabilistic complementary of POD, i.e. *PND = 1 − POD*.

Probability of detection can be estimated only by statistically planned non-destructive testing (NDT) experiments on specimens containing flaws of known size. A large experimental effort has been made in this field in the past decades and extended literature is available on this subject (e.g. Berens and Hovey 1981, Berens 1988, Nichols and Crutzen 1988). Such experiments have enabled to derive various models of POD variation as against the flaw size, *a*. For the purpose of failure risk analysis, as incorporated into *pFATRISK®* methodology, following POD models will be considered:

- Asymptotic-exponential (Marshall 1982)

$$POD(a) = A\left[1 - \exp\left(-\frac{a}{a_1}\right)\right] \quad (4)$$

- Asymptotic-exponential with lower threshold

$$POD(a) = A\left\{1 - \exp\left[-\frac{(a-a_o)}{(a_1-a_o)}\right]\right\} \quad (5)$$

- Log-logistic or log-odds (Berens 1988)

$$POD(a) = \frac{A\exp(a_o + a_1 \ln(a))}{1 + \exp(a_o + a_1 \ln(a))} \quad (6)$$

- Asymptotic of power-law type (Cioclov 1998a)

$$POD(a) = A\left(\frac{A_1}{A}\right)^{\ln(a_1/a_o)/\ln(a/a_o)} \quad (7)$$

Parameters A, A_1, a_0 and a_1 in Eqs. (4) to (7) are fitting parameters. It is worth mentioning that the parameter A is an upper asymptote to POD. It reflects the possibility to miss a flaw irrespective of its size, hence its value is lower than unity. This parameter has been proposed by Marshall (1982) in the early stages of PISC research program with the aim to process the POD data available at the stand of technique of that time, coupled with the reliability of human factor implied in the NDI. Nowadays, owing to the refinement of NDI hard- and software, together with improved management and execution of the inspection, the parameter A may be related merely to the gross human errors.

4.3. PROBABILISTIC RE-SAMPLING SIMULATION IN A SET OF POD VERSUS CRACK-SIZE DATA

New powerful statistical inference methods and probabilistic algorithms have been developed in the last decades with the purpose of extracting more information from small sample test data such as POD vs. crack size experiments. Computer simulated re-sampling in the parent (real) test sample – the "bootstrap" method (Efron and Tibishrani 1993, Cioclov 2002b) – is appropriate to set non-parametric confidence intervals to POD vs. crack size correlation.

Figure 6 illustrates a bootstrap simulation with *pFATRISK*® computer program. For instance, it is common in fracture mechanics reliability assessments of aerospace structural components to use a conceptual or "equivalent" initial crack (see Section 5) which is related with the capability of applied NDI to assess a crack size $a_{90/95}$, at $POD = 90\%$ on 97.5% conservative bound (95% confidence interval-CI). In the case documented in Figure 6, $a_{POD/CI} = a_{90/95} = 3.042$ mm. The illustrated example, which refers to ultrasonic inspection of samples of aluminum alloys, will be used in Section 6 to outline the integration of QNDI with probabilistic simulation of failure by fatigue crack growth.

5. On the evaluation of initial (equivalent) fatigue crack size and its statistics

Fracture mechanics formalism becomes inconsistent for vanishingly small crack size (e.g. Smith (1977) for a comprehensive discussion). In order to extend, in a formal manner, the linear-elastic fracture mechanics (LEFM) methods to fatigue behavior description, dominated by crack growth to failure and to integrate this methodology into Damage Tolerance philosophy it is necessary to hypothesize some sort of initial crack. A key concept of DT methodology is the equivalent initial flaw (EIF) size which ought to be specified in order to make possible fracture mechanics description of the dominant fatigue crack growth. In practice, load carrying components in pristine state are considered to be free from of initial defects. This owes to the high-quality manufacturing methods and the reliability of the nowadays NDI techniques. EIF concept enables to link the traditional total life approach (safe-life) with explicit assessment of the fatigue crack growth by fracture mechanics methods. To make more explicit the EIF concept, a closer insight into FCG process is appropriate.

In the case when well defined pre-existent cracks in a structural element with smooth geometry cannot be invoked, in order to make the fracture mechanics formalism workable, some sort of plausible equivalent initial flaw (EIF) must be considered conceptually. EIF can be related with material microstructure and/or defects at this size-level, with in-service inspection crack-sizing capability or even with an EIF derived as a purely computational parameter. In the latter case, EIF is assessed from the condition of durability

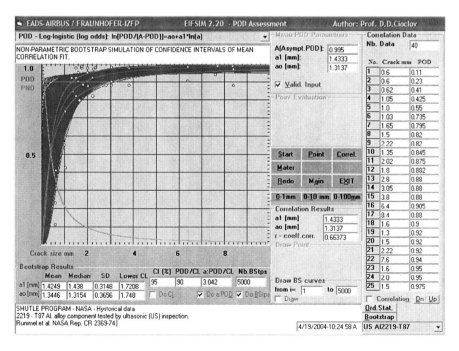

Figure 6. Two-parameter bootstrap simulation of ultrasonic inspection POD vs. crack size data. Components of Al 2219-T87 alloy. Rummel et al. (1974), NASA Rep. CR 2369-74. 5000 re-samplings on linear transformed of Log-odds rule of POD vs. crack size (a). Construction of non-parametric confidence intervals of 95%. Crack size at POD=90%, on 97.5% conservative bound is $a_{90/95}$ = 3.042 mm. Simulation performed with *pFATRISK*[R] software.

correspondence when the same problem of durability is approached, comparatively, by TL (Wöhler) and DT (FCG) methodologies.

Fatigue damage by FCG in metallic materials implies, as a rule, five phases: (1) nucleation of many shear micro-cracks, typically in the localized slip bands emerging at surface; (2) as cyclic loading proceeds, micro-cracks with orientation tendency perpendicular to maximum applied tensile stress coalesce into one dominant fatigue crack which has preeminence over the other small cracks and propagates, further, to fracture. Once a well defined fatigue crack is generated – it behaves in first instance as short fatigue crack – phase; (3) In this phase, the crack growth mechanism is dominated by the interplay between microstructure and the local plastic deformation at the crack tip. Further, in the phase (4), the dominant fatigue crack propagates by stable growth as macroscopic long crack when the dominant factor is the size of the crack. In the final phase (5) of propagation, the dominant fatigue crack growths unstably at rapid increasing rate until, finally, failure occurs by fracture. Figure 7 illustrates the stages of the fatigue damage in terms of fatigue crack growth.

Stages of fatigue cracks growth

1. **Nucleation** of many micro-cracks
2. **Coalescence** of some micro-cracks into one dominant crack
3. Growth of the dominant crack as a **short crack**
4. Stable growth of the dominant crack as a macroscopic **long crack**, until,
5. Failure by **fracture** of the remnant ligament ahead the crack tip (by plastic tear or brittle fracture)

• The limits between phases are not always well defined.

• A controversial issue in fatigue design is the setting of crack initiation point (defined in crack size and time).

t_{inc} incubation time,
t_{ini} macroscopic crack initiation,
t_f – time to fracture

ULT – ultimate static strength

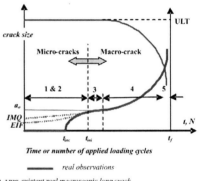

a_o : pre-existent real macroscopic long-crack
IMQ – initial material quality (notional long-crack)
EIF – equivalent initial flaw (notional short and long-crack)

Figure 7. Stages of the fatigue crack growth.

From the perspective of DT analysis, crack initiation can be placed at the onset of phase 4 when crack growths in long-crack regime is detectable and observable with non-destructive techniques, visual inspection included.

In the context of the outlined fatigue damage phenomenology various interpretations of the initial flaw size have been attempted for the purpose of DT analysis. When a crack of size a_0 pre-exists in the structural element the interpretation of EIF is unambiguous. When the crack pre-existence cannot be invoked various interpretations of the meaning of EIF are in use. The most publicized ones are known under various headings such as: constant-size EIF, initial fictitious crack (IFC) size or, in a meaning of a crack-nature generic parameter termed as "initial material quality" – IMQ. All these interpretations are associated with macroscopic long cracks propagating under cyclic loading in a solid continuum. Occasionally, constant-size EIF may encompass both short- and long-crack stages of FCG. The physical meaning of these parameters is more or less obvious and taking into account the various conventional premises used for their derivation they may be regarded in engineering applications, merely, as computation parameters. However, their justification for application resides only in the concordance they assure between fatigue life prediction and experimental evidence. This concordance can be ascertained, computationally, by "backward" FCG simulation from the end number of cycles at failure, N, towards the initial state, just before the first loading cycle.

Another approach to EIF size is by correlations with key microstructure size characteristics which have been proved to have a key role in the fatigue crack initiation as is the size of inclusion particles in aluminum alloys. These aspects are now under research and will be discussed in more details in the followings in the context of statistics aspects of the short-fatigue cracks and their growth simulation.

The concept of EIF has nowadays matured in some well established design procedures. For instance, constant EIF value used in the design of aluminum thin-wall aircraft structures amounts $a_0 = 1.27$ mm according to design recommendations in USA while the trend in Europe is towards the value of $a_0 = 1$ mm. IMQ global value of 0.127 mm has been proposed for Al alloys used currently for fuselage alloys (Brooks et al. 1999). It has been suggested that for more refined assessments by FCG simulation, when both short and long-crack regime are taken into account, to consider EIF in the range of 5 to 25 µm, as suggests the results of backward FCG simulations (Cioclov 2000–2004). This range identifies with the range of micro-structural defects size, as evince the quantitative analysis of micro-structural observations illustrated, further, in Figures 9 and 10.

Tests made at AIRBUS revealed that artificial defects introduced by saw cuts in hidden locations in Al longitudinal lap joint were detected with POD of 90% for defects size greater than 2 mm while at the same POD performance natural cracks only greater than 5 mm have been detected (Tober and Klement 2000, Schmidt et al. 1999). This obvious evidence suggests that EIF formulation in quantitative terms is related with circumstances of crack location in corroboration with the specific shape design of the component.

According to the experimental evidence (Bokalrud and Karlsen 1982, DNV 2008) it is accepted that in welded joints pre-existent cracks follow an exponential distribution with mean EIF of 0.11 mm. In welded jackets of off-shore North See platforms, exponential distribution of initial cracks has been evinced in tubular joints in more than 4,000 in-service NDIs by means of magnetic particles and eddy currents (Moan et al. 2000). A mean initial crack of 0.38 mm per "hot spot" in each welded joint has been ascertained. In butt-welded joints other experiments revealed Log-Normal distribution with mean initial crack size of 0.78 mm (Kontouris and Backer 1989). A recent review on the experimental data on initial cracks subsisting in various types of load carrying structures has been published by ISSC (2006), Committee III.2.

In the followings, two closely related interpretations of EIF will be detailed, namely the initial fictitious crack size and equivalent initial flaw.

5.1. THE INITIAL FICTITIOUS CRACK SIZE CONCEPT

As outlined, fracture mechanics formalism becomes inconsistent for vanishingly small crack size. In order to circumvent this pitfall and in view of extending in a formal manner the applications of LEFM to the incipient stage of short fatigue cracks El Haddad et al. (1979) proposed a simple modified general expression of the stress intensity factor (SIF) range expression in the form:

$$\Delta K = Y \Delta \sigma \sqrt{\pi (a + a_0)} \qquad (8)$$

where a_0 is a fictitious crack length, a, the physical crack size, $\Delta\sigma$, the applied cyclic stress range and $Y(a, L)$ is the geometry correction factor dependent on the crack and structural element size (L). Formally, when the physical crack vanishes, $a \rightarrow 0$ Eq. (8) still predicts a finite value of the SIF range associated with fictitious crack length, a_0. The quantity a_0, represents a limit circumstance and it seems reasonable to relate it with two correlated fatigue limit states perceived from the standpoint of the classical total life, on the one hand, and fracture mechanics approach, on the other. In TL perspective a_0 is related with $\Delta\sigma_d$, the fatigue limit while from fracture mechanics perspective it is related with ΔK_{th}, the lower threshold to FCG. Both $\Delta\sigma_d$ and ΔK_{th} are lower thresholds under which no fatigue damage occurs. They depict, via two different formalisms, the same physical situation. Thus from the definition of SIF, with ΔK identified with ΔK_{th} and $\Delta\sigma$ with $\Delta\sigma_d$, when $a \rightarrow 0$, it results:

$$a_0 = \frac{1}{\pi}\left(\frac{\Delta K_{th}}{Y\Delta\sigma_d}\right)^2 \qquad (9)$$

Taking into account that incipient short cracks are surface cracks, the geometry correction factor assumes the limit value $Y = 1.12$ (e.g. Murakami 1987). It is worth to note that owing to the dependence on the element geometry by intermediate of Y correction, a_0, cannot be regarded, in principle, as a material property (see DuQuesnay et al. 1988, Yu et al. 1988 and Megiollaro 2007 for a detailed discussion) though, it was related by some authors with slip bands length (Davidson 1983, Tanaka et al. 1981, Gall et al. 1997).

5.2. THE EQUIVALENT INITIAL FLAW CONCEPT

The basic idea underlying the initial fictitious crack size concept has gain popularity among the practitioners in aerospace industry under the denomination of "equivalent initial flaw" (EIF). It was conjectured that with EIF values of some tens of micrometre it is possible to predict, on the base of Paris rule, fatigue lives which comply with S-N fatigue test data.

It is worth to recognize that despite the effort made to substantiate this approach, theoretically (e.g. Taylor 1999), or by fractographic analyses (e.g. Wang 1980, Fawaz 2003), it is merely a computation expedient introduced for the needs of the applications when fatigue damage tolerance philosophy is applied, rather than a concept with sound experimental support.

The justification of application of EIF concept in the damage tolerance analysis resides in the proper "calibration" of a_0, at a constant value, for a specific cyclic loading (say $\sigma_{max} / \sigma_{min} = 80 / 8$ MPa for fuselage structures of Al 2024 alloy), regardless of the component geometry, and using this value, as conceptual initial crack size, in a fracture mechanics computation of the fatigue lives (e.g. Kurth et al. 2002, Cioclov 2003).

The concept of EIF has been used originally in aircraft structural design to assess the effects of initial flaws, particularly, the machining marks on the surface of riveted holes supposed, as a worst case scenario, to exist prior to cyclic loading in service (Rudd and Gray 1978, Wang 1980, Yang and Manning 1980, Gallagher et al. 1984, Yang et al. 1990, Grandt 2000). The use of EIF concept in managing the fatigue damage of aircraft structures has been extensively discussed by Manning and Yang (1987). This approach is well suited to the safety management of aircraft structures by inspection when EIF size identifies with the minimum crack size which is detectable with a high probability of detection (POD) by the currently used NDI technique. For this purpose, as outlined in Section 4.3 it is common to define the index, $POD_{90/95}$, the probability of detection of 90% on the bound 97.5% confidence curve (95% confidence level interval).

Originally, (Wang 1980, Gallagher et al 1984), EIF size has been assessed after the tear down of test specimens of retired aircraft components which have been subjected to a known number of loading cycles, N. Computational back-extrapolation to initial time ($N = 0$), enables to evaluate an EIF size. Probabilistic interpretation of this procedure is illustrated, schematically, Figure 8. It consists in assessing, by statistical fractography, a crack size distribution at the time of observation considered as 'time to crack initiation' – t_{TTCI}, It is the time for a crack to reach a size which is well observable by metallographic techniques. Typically, a mean or median value of, $a_{TTCI} = 0.75$ mm is considered (e.g. Yang and Manning 1980 and for experimental support, Norohna et al. 1978). Then, the distribution of crack size at t_{TTCI} is constructed (usually a three-parameter Weibull distribution), by order statistics, on the base of the experimental set $\{(a_{TTCI})_{P_i}\}$. Further, the P_i-quantils of the fitted crack size distribution, $(a_{TTCI})_{P_i}$, are back extrapolated to the initial time by a simple crack growth rule (typically $da/dt = Qa^m(t)$, with Q and m as empirical parameters depending on material, structural geometry and loading spectra. A set of EIF size, $\{(a_0)_{P_i}\}$, is obtain at $t_{TTCI} = 0$ which enables to construct the distribution of EIF size, $p(a_0)$. Obviously, $p(a_0)$ distribution is not unique and, as underlying premises suggest, it is material, loading pattern and, presumably, component geometry dependent.

On the base of this data processing methodology, Yang and Manning (1980) used the cited metallographic data of Norohna et al (1978), and derived a two-parameter Weibull distribution of the EIF in Al 2024-T3 alloy (see also Maymon (2005) for statistical re-assessing of this set of data). The resulted cumulative probability function and the associated statistical parameters are:

$$P(a_0) = 1 - \exp\left[-\left(\frac{a_0}{\beta}\right)^\alpha\right] \text{ with } \beta = 18.594\,\mu m; \text{ and } \alpha = 0.918 \quad (10)$$

and the mean value and standard deviation (SD) results as:

$$M[a_0] = \beta\,\Gamma\left(\frac{1}{\alpha}+1\right) = 19.36\,\mu m$$

and

$$SD[a_0] = \beta^2 \left[\Gamma\left(\frac{2}{\alpha}+1\right) - \Gamma^2\left(\frac{1}{\alpha}+1\right) \right] = 21.1 \mu m$$

with a coefficient of variation of variation of 1.09.

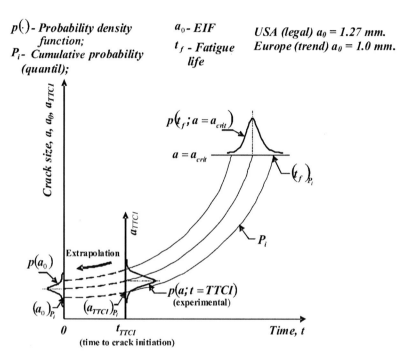

Figure 8. Schematic of equivalent initial flaw (EIF) size distribution evaluation by back-extrapolation from experimental crack size distribution assessed by quantitative fractography at conventional time t_{TTCI}.

Figure 9 shows a representation of the probabilistic distribution function given by Eq. (10) in comparison with the later studies made by Brooks et al. (1998) and an alternate fitting according to exponential distribution which appears a reasonable choice conferring analytical simplicity in DT analyses.

As discussed by Barter et al. (2001), the outlined method of EIF distribution assessment is sensitive to the accuracy of the fractographic results and the rule of crack size back extrapolation. It is not uncommon that, by back-extrapolation, negative EIF size results which, obviously, is unrealistic. These difficulties forced another ways of thinking. In one way of approach (Cioclov 2002–2004), it is assumed an initial crack size, a_0, and the crack growth is iteratively simulated with an accepted FCG rule until the predicted life corresponds to the experimental fatigue life of unflawed component. In fact, by this procedure, the crack growth model is back-projected from failure to the initial undamaged state. This way of approach is enabled by *pFATRISK*[R] rationale by direct back-simulation of the FCG by coupling short- and long-crack regime (see Section 6).

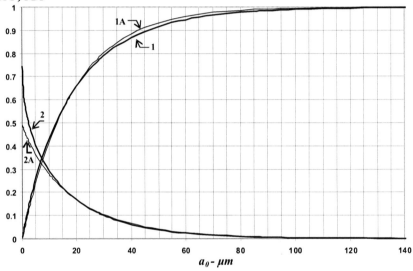

1- CPF, Al-2024 Norohna et al (1978); Yang and Manning (1980); 2P-Weibull $\beta = 18.594\ \mu m$ and $\alpha = 0.918\ \mu m$
1A – CPF, same data as 1. Exponential distribution.
2 – PDF, pooled 2024 and 7076 Al-alloys; Brooks et al. (1999); 2P- Weibull
2A – PDF, same data as 2. Exponential distribution.

Figure 9. Cumulative probability function (CPF) and probability density function (PDF) of the two-parameter Weibull distribution of the equivalent initial flaw (EIF) size a_0 derived by Yang and Manning (1980) on the base of experimental data of Norohna et al (1978) and Brooks et al. (1999) for the historical data base of FCG data on Al-alloys. Comparison with exponential distribution.

The concept of equivalent initial flaw is understood in a broader sense by some authors. Wang (1980) used the term of "initial fatigue quality" which, in principle, is claimed to be related, globally, with the material and manufacturing quality. In the case of fastener holes of riveted aluminum alloy joints of aircraft structures, initial fatigue quality is identified with EIF. EIF is assessed, statistically, on the base of combined fractographic and direct observation of crack size growth as function of the number of flights simulated in fatigue testing by representative service loading spectra. On the base of these data, EIF distribution is evaluated by backward extrapolation to the initial state at zero number of flights. This concept forms the basis of the UASF Airplane Damage Tolerance Requirements, MIL-A-83444 (UASF 1984).

It is worth to note that in aluminum alloys it was actually observed that the fracture of dispersed secondary particles (deliberately introduced to promote strengthening mechanisms) generate initial fatigue cracks which, subsequently, under cyclic loading, coalesce into a dominant fatigue crack which grows to complete failure. Bowels and Schijive (1973) showed that large secondary particles of size 1 to 10 μm act as crack nuclei in Al 2024-T3 alloy. The fracture incipience has been observed as interface failure between particles and metallic matrix and, occasionally, as the particles cracking. These data suggest that the adherence between matrix and particles is deteriorated during the incubation period owing to a mechanism of dislocations accumulation in the slip bands at the crack tip and, concurrently, with dislocations blocking on inclusions (basically a Zener-type dislocation mechanism – Zener 1948).

Half a century ago, Landes and Hardrath (1956), by way of FCG simulation with empirical rules, not derived from Fracture Mechanics principles, predicted endurance curves of 2024-T3 coupons with central hole radius of 1.6 mm. Compliance of simulation with test data has been achieved when EIF was 6 μm which is in the range of inclusions size. Newman and Edwards (1988) observed in thin coupons of Al-2024-T3 that fatigue cracks nucleated at particle inclusions and inclusion clusters with an average depth of some 12 μm develop quickly into semi-elliptic cracks emerging at the surface. Simulated FCG performed by Newman et al. (1992) with *FASTRAN* code comply with an EIF of approximately 10 μm.

Barter et al. (2001) evinced for a high-strength 7050-T7451 alloy wider range of inclusions size and from a large number of scanning electronic microscopy observations derived the histogram of the inclusions depth (measured from the specimen surface) which fits into a Log-Normal distribution. Figure 10 illustrates the data obtained by Barter and co-workers and the fitting of inclusions depth size into a Log-Normal distribution.

Laz et al. (2001) found that in Al 2024-T3 alloy, the mean particles cross section area in undamaged material is 5.4 μm^2 with a conventional diameter of d_p = 2.6 μm and a standard deviation of 9.2 μm. In fatigue fractured specimens it was evinced that fractured particles acts as fatigue crack nucleation sites. It should be noted that in the tests of Laz and co-workers an incubation period have been register until initial cracks are generated by particles fracturing. It is apparent that a fracture mechanics prediction based exclusively on the statistics of d_p neglects the incubation period in the fatigue life and this statistics cannot be adopted as physically substantiated EIF input.

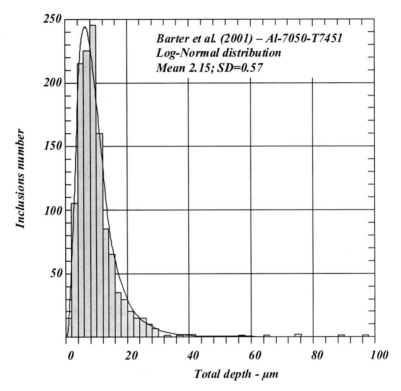

Figure 10. The histogram of the inclusions depth in a high-strength aluminum alloy. (After Barter et al. 2001.)

An EIF can be constructed on the base of an integrated Fracture Mechanics with quantitative NDI. For this purpose, conservative lower bounds curves of $POD(a)$ versus crack size may be used. A *bootstrap* procedure has been outlined in Section 4.3 which enables to draw from a small sample of experimental POD data a representative crack size, $a_{POD/CI}$, at specific POD and confidence interval (CI), (e.g. $a_{90/95}$). By identifying $a_{POD/CI}$ with the initial crack size that may subsists in the load carrying component, i.e. an EIF interpretation. This way of approach will be exemplified in Section 6.2, Module G in the context of fatigue failure risk assessment in the case when NDI inspection is performed at a specified life of the component.

Finally, it should be remarked that in the present state of art of the EIF assessment, neither the influence of cyclic loading intensity, which is clearly revealed in EIF size back-computations, nor the distinction between short- and long-crack stages of growth have been considered.

In order to cope with these neglected aspects in the EIF methodology in the next section, on the canvas of *pFATRISK*® code, refinements of EIF assessment will be presented. It is introduced the dependence of EIF size on the cyclic loading intensity and both short- and long-crack aspects will be considered in the back-projection from total fatigue life to the pristine initial state of the material.

6. Probabilistic fatigue damage simulation by crack growth – *pFATRISK*® method and software (Cioclov 1995–2007)

6.1. GENERAL DESCRIPTION

pFATRISK® rationale is based on cycle-by-cycle Monte Carlo simulation of FCG. The underlying model is constructed on the principles of fracture mechanics. The ultimate state criterion at failure (before the last supported loading cycle) is formulated in terms of Failure Assessment Diagram (FAD). Various formats of FAD are considered in order to comply with estabished national and international codes (R/H/R6-Rev. 3-1986, SINTAP-1999, BS 7910:2000). FAD methodology encompasses both elastic dominated (brittle) and elastic-plastic dominated (ductile) fracture. Under fatigue circumstances, there are implemented various models of fatigue crack growth (FCG) under long- and short-crack regime. Under long-crack regime, the implemented FCG rules are those proposed by Paris et al. (1961), Klensil and Lucás (1972), Forman et al. (1967) and Walker (1970). Crack closure effect is taken into account according to various models which have gained acceptance in DT practice. Under short-crack regime the FCG rule is described by a power law with upper conservative parameters or with parameters depending on the intensity of the cyclic loading (Cioclov 2002–2004). By considering in FCG models both stages of short- and, subsequently, long-crack growth, from EIF size to the crack size at failure, it is possible to predict the total endurance (Wöhler) curves. Stress intensity factors are calculated according to known analytical solutions having, nevertheless, the possibility to implement numerical data derived from finite element analysis.

Statistical pattern of FCG simulation results from the random variables associated with the governing parameters and the algorithms of the limit curve of FAD. A wide variety of statistical distributions of governing FCG and FAD parameters is implemented in the rationale.

Table 1 gives an account on the input variable and their enabled probabilistic distributions.

The logic of probabilistic failure risk simulation under both static and fatigue loading circumstances is illustrated in Figure 11 together with the contributions and interactions implemented in *pFATRISK*® rationale.

TABLE 1. Enabled probabilistic distributions for the input variables.

Distribution	Probabilistic input				
	Material characteristics UTS, YP, Kc^1	FCG characteristics C, m, Ktho, B, m_1^3	Crack size a, b	Loading static and cyclic	POD rule
Normal	●	●	●	●	○
Log-Normal	●	●	●	●	●
3P-Weibull	●	●	●	●	●
1/x 3P-Weibull^2	○	○	○	●	○
Extremal Type I	○	○	○	●	○
Extremal Type II	○	○	○	●	○
Extremal Type III	○	○	○	●	○
Logistic	○	○	○	○	○
Log-logistic	○	○	○	●	●
Rayleigh	○	○	○	●	○
Exponential	●	●	●	●	●
Uniform	○	○	○	●	○
Pareto	○	○	○	●	○

● – Implemented; ○ – not-implemented.

[1] *UTS*, ultimate tensile strength; *YP*, yield point; *Kc*, fracture toughness.

[2] Inverse of the random variable argument (1/x) follows 3P-Weibull distribution.

[3] Material parameters *C, m, Ktho* define Paris and Klensil-Lucas rule for FCG under long-crack regime: $da/dn = C(\Delta K^m - \Delta Ktho^m)$ where Δ means the range of SIF variation. $\Delta Ktho$ refers to a cycle ratio of $R = 0$. Material parameters *B* and m_1 define the rule of FCG under short-crack regime: $da/dn = B\Delta K^{m_1}$.

The loading considered in the static and fatigue failure analysis can be also modeled by statistical distributions which are most common in fracture reliability analysis. Under fatigue circumstance, the cyclic loading may be modeled with constant amplitudes, blocks of constant amplitudes or with stationary and quasi-stationary random variation.

Figure 12 shows the screen display of the probabilistic input of the conventional strength characteristics and FCG parameters according to Paris rule in the long-crack range and to a power-law in the short-crack growth regime. The considered material is an aluminum alloy of 2024-T3 class. Throughout in the example presented in Figure 12, Normal distribution is used to model the statistical variability of the input parameters pertaining to static strength characteristics (UTS, YP and Kc) and short-and long-crack growth characteristics (C, m K_{th0}, B and m_1). The graphical representation in Figure 12 illustrates the scatter in FCG curves (growth rate vs. SIF range) when the input parameters are Monte Carlo sampled. The mean curve is shown in dark-blue color.

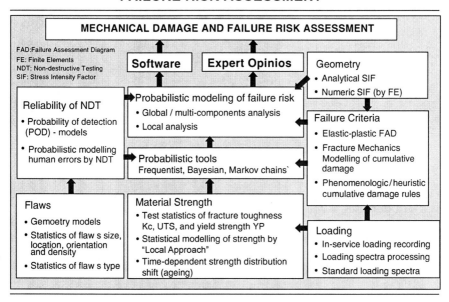

Figure 11. The logic of probabilistic models for mechanical failure risk assessment under static and fatigue loading circumstances.

6.2. THE OUTPUT OF *pFATRISK*®

In the various modules of the *pFATRISK*® rationale the following results may be obtained:

6.2.1. *Material module*

a. Visualization of the FCG curves with their scatter across the mean curve (e.g. Figure 12)
b. Crack growth rate (Vo) and SIF range ΔK_0 at $a = a_0$, the transition from short- to long crack regime (Figure 7 and Figure 12)

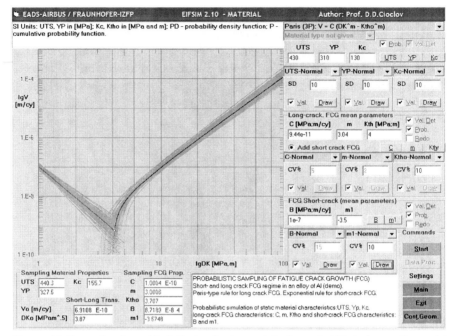

Figure 12. Example of graphical display in the module for probabilistic input of material characteristics (conventional static strength and short- and long-crack FCG characteristics). Data are representative for an aluminum alloy of 2024-T3 class.

6.2.2. *Geometry input module*

1. Stress intensity factor corrections $Y(a)$ as function of crack size and structural component dimensions (Figure 13 bottom left);
2. FE point data fitting into continuous $Y(a)$ curves by polynomial interpolation (Lagrange and spline functions) and higher order polynomial correlation.

6.2.3. *Loading input module*

1. Visualization of the loading spectra and computation of their characteristics (e.g. the root mean square – *RMS*).

6.2.4. *FCG simulation module* (Figure 13)

1. Fatigue life N_f to grow a macroscopic (long) crack from an initial size, until failure.
2. Total fatigue life N_{ft} encompassing both short- and long-crack growth range as implied by the application of EIF concept.
3. The remnant strength R of the component as function of applied loading cycles.

4. Fatigue life N_i at the transition from short- to long-crack growth regime.
5. Fatigue life $N(a)$ at the attainment of a given crack size, a.
6. Fatigue life $N(R)$ at the attainment of a given remnant strength, R.
7. Fatigue life $N(POD)$ at the attainment of a given POD.
8. Crack size $a(n)$ at the attainment of a given life, n.

Figure 13 shows the screen display when the probabilistic simulation of FCG is performed with the input material data shown in Figure 12. The component geometry is of a test coupon, i.e. a plate of finite width having a central hole of 2 mm radius and two opposite through thickness initial cracks, each of 1 mm, emanating from the hole. SIF analytical solution is according to Bowie (1956) and Newman (1976). Long-crack regime is assumed. Blue curves in the upper-left field of the screen portrays the simulates scatter of the fatigue life, N_f, necessary to grow a crack from an initial size until the failure (Item 1 in Section 6.2.4). The green curves in the same field represent the scatter in the remnant strength R, representation on ordinate axis, R/R_o, with R_o the initial strength of the component having an initial crack (Item 3 in Section 6.2.4). In the upper-right field is the FAD representation with the path visualizing succesive damage states as initial cracks growth under cyclic loading. FAD representation is made in the

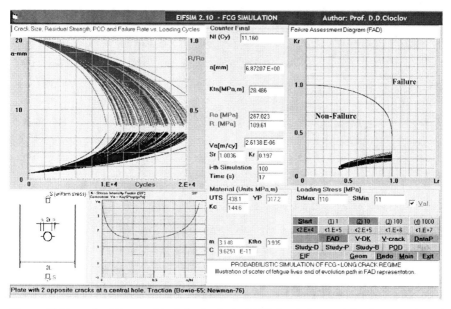

Figure 13. Example of the graphical display in the FCG simulation module after a session of probabilistic simulation of FCG until failure. Representation $a(n)$ – crack size (blue) and $R(n)/R_o$ (green) vs. the number of applied loading cycles n (upper left) and FCG path in FAD diagram (upper right).

coordinates $Kr = K/Kc$ and $Lr = $ *applied load/limit load at plastic collapse.* Dark red curve represent the FAD limit boundary for failure/non-failure, according to Dugdale model (Dugdale 1960)[3].

Figure 14 shows the screen display when the same probabilistic simulation of FCG is made, as illustrated in Figure 12, but with FCG until a preset number of loading cycles (here 10,000 cycles). In this case it is evinced the scatter in the crack size $a(n)$ and the remnant strength $R(n)/R_o$ at the preset number of loading cycles (items 3 and 8 in Section 6.2.4).

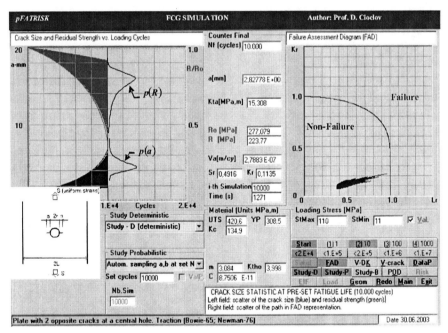

Figure 14. Probabilistic simulation of FCG until a pre-set number of loading cycles (10,000 cycles). Same input as in *Figure 12.* Statistics of the crack size and remnant strength at the pre-set number of loading cycles of $\sigma_{max}/\sigma_{min} = 110/11$ MPa.

6.2.5. *Module for statistic fitting of simulated data*

Probabilistic simulation of the fatigue crack growth with *pFATRISK*® method enables to evaluate the statistical distributions of many parameters of interest in fatigue reliability analysis: fatigue life to grow a macroscopic (long) crack from an initial size to critical size at ultimate fracture, N_f; fatigue life until transition from short- to long-crack stage, N_i; total fatigue

---

[3] Uncorrected Dugdale model has been used instead of large scale plasticity corrected variants, for $L_r > 1$, since under FCG, immediately prior to ultimate fracture, the plastic strained is still confined at the crack tip. This is a conservative assumption.

life, N_{ft}; fatigue life to propagate a crack to a pre-set size, $N(a)$; crack size at a number of applied loading cycles, $a(n)$; the instantaneous remnant strength during FCG, $R(n)$.

Data resulted from probabilistic simulation module (items 1–5, 8 in Section 6.2.4), can be fitted into various statistical distributions, as given in Table 2. Non-parametric order statistics methods are used for this purpose (e.g. Meeker and Escobar 1998).

TABLE 2. Allowed distributions for input parameters.

Fitting distribution	Simulated parameter			
	N_f, N_{ft} N_i	$N(a)$	$a(n)$	$R(n)$
Normal	●	●	●	●
Log-Normal	●	●	●	●
3P-Weibull	●	●	●	●
1/x 3P-Weibull	○	○	●	●
Extremal Type I	○	○	●	●
Extremal Type II	○	○	●	●
Extremal Type III	○	○	●	●
Logistic	○	○	●	○
Log-logistic	○	○	●	●
Exponential	●	●	●	●

● – Implemented; ○ – not-implemented.

Figure 15 shows the screen display in the module for statistical treatment of simulated fatigue data. The example, the same as that illustrated in Figure 14, refers to a sample of 2,000 simulated data of the crack size at 10,000 loading cycles.

Figure 16 illustrates the fitting of simulated crack size sample, of 10,000 items, according to various statistical distributions: Normal, Log-Normal, 3P-Weibull and 2P-Weibul of the inverse of the crack size. On the ordinate, $Log(1-P)$ representation has been used since this transforms enlarges the scale on probabilistic axis, enabling to better visualize the fitting goodness in the right tail of the distribution which corresponds to the large crack size values in the sample. The probability of occurrence of cracks having large size is determinant for the structural component reliability during FCG.

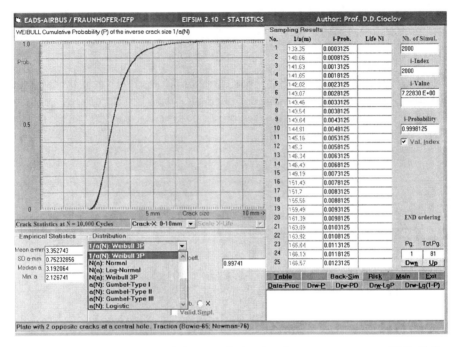

Figure 15. Fitting of simulated fatigue crack size data into a two-parameter Weibull distribution of the inverse value of the crack size. Ordered values and associated probabilities are shown in the table.

Comparatively, among the checked distributions, a very good fit has been obtained with 2P Weibull distribution applied to the inverse value of the crack size ($1/a$). To the best knowledge of the author, this statistical pattern of behavior of the fatigue cracks is a novel observation which may be of far reaching importance in structural reliability prediction. Moreover, the good fitting of the inverse of the crack size by a two-parameter Weibull distribution evinces that there is no limiting threshold effects in the range of large size cracks, in contrast with the range of small cracks range in the sample where trends towards an inferior threshold is clearly evinced both by test an simulated data.

It is worth to mention that the outlined conjecture on the distribution of the fatigue cracks has been obtained by computer simulation of the FCG in a large sample of structural components (e.g. 10,000 items) having at the base a well established fracture mechanics model. The availability of such a large statistical sample of the fatigue crack size, obtained directly by fatigue testing, is virtually impossible.

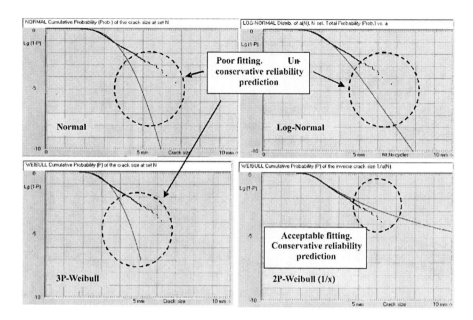

Figure 16. Simulated statistics of the fatigue crack size at 10,000 loading cycles of $\sigma_{max}/\sigma_{min} = 110/11$ MPa. Data fitting to Normal, Log-Normal, 3P-Weibull distribution and to 2P-Weibull distribution of the inverse of the crack size (1/a). On the probability ordinate representation of $Log(1-P)$ is used in order to enlarges the scale in the right tail of the distribution.

The 2P-Weibull distribution of the inverse of the crack size is further used to exemplify the simulation of the probabilities of failure in the next module of failure risk assessment.

For the case under analysis, Figure 17 shows how the simulated statistical scatter of the crack size changes with the number of applied loading cycles. The obvious trend of the scatter is to increase with the accumulation of the applied loading cycles, i.e. as the mean size of the fatigue crack increases. Also to the best knowledge of author, this effect, which might be intuitively conjectured, is explicitly evinced by computer simulation for the first time.

Figure 18 illustrates the shape of probability density (PD) function of the simulated data shown in Figure 17. Fitted data into Log-Normal distributions of the crack size in various stages of FCG evince that the coefficient of variation increases in concert with accentuation of the distribution skewed pattern.

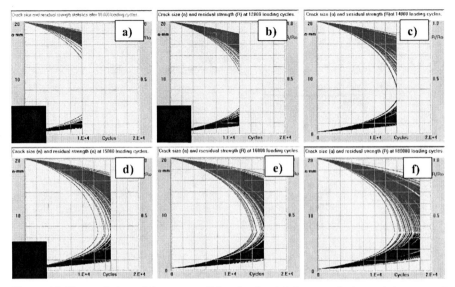

Figure 17. The evolution of the scatter of the simulated fatigue crack size and the remnant strength as the number of applied loading cycles is increased.

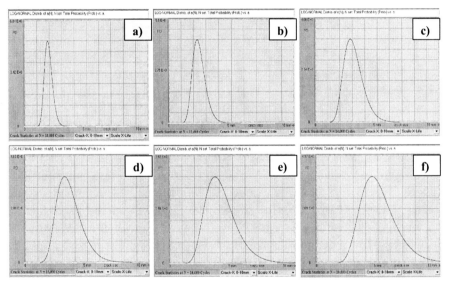

Figure 18. The variation of the shape of probability density function of the simulated fatigue crack in various stages of the FCG.

6.2.6. *Module for failure risk assessment (probability of failure P_f)*

In this module, having as input the fitted distribution of the simulated crack size $a(n)$ at the attainment of a given life, n (see Item 8), the FAD analysis is performed, probabilistically, by Monte Carlo simulation of crack size data, material characteristics scatter (according to the distributions input in item (a) in Material Module, Figure 12) and the loading statistics. In this module, for the convenience of DT analysis, the re-input of crack size and loading distributions may be performed. After a large number of iterations (usually greater than 10^9) the probability of failure is estimated by the ratio of the number of failure scenarios, resulting from FAD analysis, to the total number of iterations. Efficient programming algorithms in FAD analysis have been used which makes that 10^9 FAD simulation scenarios to be performed on commercial computers within a couple of minutes. With slight modifications of the *pFATRISK*® code for enabling portability on parallel computers, over 10^{15} FAD simulation can be attained in a reasonable time. It makes possible to estimate stable failure probabilities of as a low values as 10^{-9}. The nowadays petaflop computing technology brings in the realm of engineering assessment stable failure probabilities of values of 10^{-11} to 10^{-12}.

The interpretation of the probability of failure, P_f, resulted from FAD analysis, requires explanations. Since FAD analysis refers to singular loading, the probability of failure, P_f, inferred by Monte Carlo simulation of all assumed random variables in the FCG model represents, after n cycles of loading, an estimation of the momentary probability of failure in the next loading cycle, i.e. ($n + 1$)th cycle. In the illustrated analysis the distributions of materials strength characteristics ($\boldsymbol{UTS}$, $\boldsymbol{YP}$ and $\boldsymbol{K_c}$) and parameters entering in FCG rule ($\boldsymbol{C, m}$ and $\boldsymbol{K_{th}}$) are time-independent reflecting the overall inter-components variability of these parameters. The distribution of the size of the propagating fatigue crack is time dependent, as evinces FCG simulation, reflecting the statistics of the fatigue damage evolution within a single component (see Section 4.1).

FAD module enables also deterministic fracture analysis at a specified number of applied loading cycles when all involved parameters entering into computation assume mean values. The result of the analysis is of the type "fail/not-fail" and in the case of not-failure a safety index is computed. Under deterministic FAD analysis various parametric (sensitivity) fracture analyses may be performed at specified moments in the fatigue life. Parametric sensitivity analysis refers to the limit values at failure of one of the governing parameters such as crack size, fracture toughness or loading.

The graphical display in the module of failure risk simulation gives a "dynamic" view of repeated Monte Carlo scenarios in FAD representation of the damage state points associated with each of the computation iteration.

Figure 19 shows the screen display in the module for failure risk assessment of the component illustrated in Figure 13, after 10^6 iterations of FAD analyses. Simulation has been performed by Monte Carlo sampling of the material static strength characteristics according to Normal distributions, as shown in Figure 12, and the crack size according to 2P-Weibull distribution applied to the inverse value ($1/a$) of the crack size (Figure 15). The "cloud" of representative points marks the scenarios of non-failure, if they are represented inside FAD boundary, and the scenarios of failures, if represented points fall outside FAD boundary. The ratio of the number of scenarios of failure to the total number of repetitions yields, in this case, an estimation of the failure probability of $P_f = 0.005673$. The coefficient of variation (COV) when the FAD simulation session consists of 10^6 iterations is approximately 1.3% (Cioclov 2002–2004).

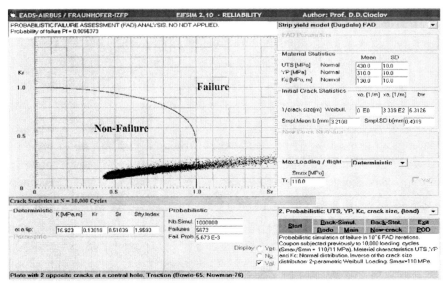

Figure 19. Example of graphical display in the module for failure risk assessment. The case when no NDI is applied.

6.2.7. *POD module*

A particular feature of the module of failure risk simulation is that it enables the estimation of the probability of failure, P_f, under the circumstances of application or non-application of NDI. When NDI is considered applied, an algorithm of crack detection is operating which, in each FAD iteration, simulates the crack detection/non-detection according to the POD rule introduced as input in the POD module.

POD module can be addressed in various stages of computer simulation of FCG with *pFATRISK*®. It enables to input *POD(a)* vs. crack size curves according to the models described in Sections 4.2 and 4.3, Eqs. 4 to 7. It is also possible to introduce experimental data, as point values (POD_i, a_i), and then to perform correlation analysis in transformed coordinates that make linear the POD rule.

A unique feature is implemented in this module, namely, the possibility to perform bootstrap computer re-sampling in the experimental (parent) data set (POD_i, a_i), in order to construct non-parametric confidence intervals over the mean values. This technique, already discussed in Section 4.3, is supplemented with algorithms pertaining to order statistics (Cioclov 1975, 1998a). In this way, it can be assessed for the purpose of further failure risk analysis, conservative lower bounds curves of $POD(a)$ versus crack size. A representative crack size, $a_{POD/CI}$, at specific POD and confidence interval (CI), (e.g. $a_{90/95}$), can be ascertained. By identifying $a_{POD/CI}$ with the initial crack size that may subsists in the load carrying component (similar with EIF interpretation) the simulation of the fatigue crack growth yields the mean fatigue life (deterministic simulation) or the distribution of fatigue life or the distribution of the crack size at pre-set fatigue life (probabilistic simulation as outlined in Figure 14) for the use in the module F of failure risk assessment.

Figure 6 shows the screen display of the POD module. The illustrated experimental POD data have been obtained by Rummel et al. (1974) on an aluminum alloy of type T2219-T87 components subjected to ultrasonic NDE. The experimental data have been fitted into a Log-logistic POD rule (Eq. (6)) and this particular rule transferred to failure risk assessment (module F) in order to assess the decrease of failure probability (or increase of reliability) when a specific NDI technique is applied. The quality/reliability of NDI is defined in this exercise by a Log-logistic model of $POD(a)$, as given by Eq. (6).

Figure 6 also gives an example of two-parameter bootstrap re-sampling simulation in the fitted $POD(a)$ Log-logistic curve. Bootstrap simulation enables to construct a non-parametric confidence interval $CI = 95\%$ and estimate, at 90% POD, on the conservative bound of 97.5% (computed by formula $0.5 + CI/2$), the crack size $a_{90/95} = 3.042$ mm. This parameter may be regarded as a global "one-point" characterization of the reliability of NDI technique.

Figure 20 shows the results of probability of failure, P_f, simulation as in the case illustrated in Figure 19 but under circumstance of NDI application on the structural component under analysis at the timing of 10,000 loading cycles, approximately 75% of the mean fatigue life of the component. When in an iteration the detection algorithm simulates crack detection, then the component is considered to be removed or repaired at the initial quality. Of course, as POD concept reflects also the still existing uncertainty and variability (see Section 2.2) involved in the NDI, there is a chance that critical defects which lead to failure be missed during inspection. This aspect is captured in the *pFATRISK* simulation rationale and is contained in the final result of the simulation, the value of the probability of failure, P_f. The benefit of applying NDI of a specific quality is thus quantified. The undertaking of integrating, on probabilistic bases, fracture mechanics and quantitative interpretation of NDI is, by these means, amenable into the realm of current engineering analysis.

Table 3 outlines, in the case under analysis, a comparison between the simulated probabilities of failure under circumstances of application and not application of NDI.

A decrease of 50 times in the probability of failure is achieved by applying the considered NDI characterized by a well defined POD vs. crack size relationship.

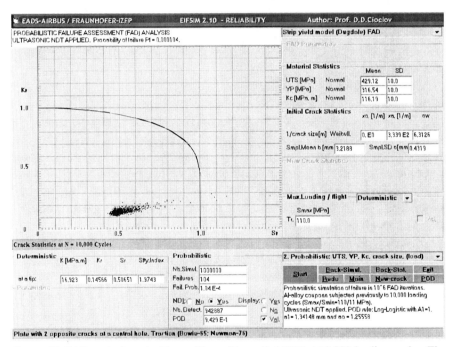

Figure 20. FAD analysis as in *Figure 6* but with applied NDI at 10,000 loading cycles. The POD rule as in *Figure 6*.

TABLE 3. Comparison between the simulated probabilities of failure under circumstances of application and not application of NDI.

FAD probabilistic	Nb. simulations	Nb. detections	POD	Nb. failures	P_f
No NDI	1,000,000	–	–	5,673	0.005673
With NDI	1,000,000	942,887	0.9429	104	0.000104

7. Discussion, conclusions and prospects

The study develops a methodology for integrating, on probabilistic bases, fracture mechanics with quantitative NDI for the purpose of failure risk assessment in load-carrying structures and machine elements, with special emphasize on fatigue damage circumstance.

The uncertainty in our knowledge and the intrinsic stochastic variability implied in the physical mechanisms confer a probabilistic character of fracture behavior under static and cyclic loading. Under these circumstances, the estimation of the probability of failure may be achieved by integrating probabilistic fracture mechanics with quantitative NDE. Heuristic probabilistic failure models may be constructed on the base of the principles of the probability theory which, in end result, imply the computation of multidimensional convolution integrals. This way of approach is tedious and necessitates the development of accurate numerical approximation methods which may by a difficult task in engineering practice. An alternate way of approach, outlined in this study, is to simulate, massively, on computer by Monte Carlo technique the governing random variables of the underlying physical models of material response to applied loading. Monte Carlo techniques, i.e. methods of direct simulation of variability and uncertainty offer, nowadays, a straightforward alternative owing to the powerful, commercially available, computational facilities.

In the study it is outlined that new statistical inference techniques related with computer re-sampling in the initial (parent) test data – *bootstrap* algorithms – enable a better substantiation of the confidence which can be placed on the test data used in the assessment of the failure risk by fracture.

Using Monte Carlo computer simulation techniques, a rationale have been developed encompassing the fatigue crack growth description by fracture mechanics formalism under short (local)- and long (macroscopic dominant)- crack, thus addressing the entire fatigue life. The application of fracture mechanics concepts at local level (material microstructure) or global level (material as a continuum body) necessitates the definition of the size of a relevant initial flaw size which pre-exists in the material. In the context of the assessment of evolutionary fatigue damage by simulation of the fatigue crack growth, various interpretation of what is a relevant initial flaw size has been discussed from the standpoint of the physical meaning, experimental support and engineering interpretation.

The implication of quantitative NDI into failure risk assessment is made via the concept of probability of detection, POD, as function of crack size. Some specific POD models are presented and log-logistic model is further used to illustrates how more refined statistic information can be extracted by bootstrap re-sampling simulation and the implication of this technique in the estimation of failure risk under fatigue damage circumstance.

A practical engineering tool emerged from these concepts and the ensuing algorithmic construct which is materialized in *pFATRISK*^{R} rationale and software. On this line of approach, a short presentation of *pFATRISK*^{R} is given in the context of an application pertaining to the fatigue damage of aerospace components by crack growth. The sequences of a simulation sessions with *pFATRISK*^{R} are documented: probabilistic input of material characteristics, the random generation of fatigue crack growth curves, computation of crack-associated stress intensity factors, probabilistic crack growth simulation when an equivalent initial flaw is specified, statistical treatment of simulated fatigue life and crack size scatter in various stages of fatigue life of a structural component and, finally, the probabilistic FAD analysis is presented as the analytical model that enables to estimate the time-dependent probabilities of structural failure.

On the base of integrated module of POD simulation, a comparative analysis outlines, quantitatively, in terms of decreasing probabilities of failure, the benefit that can be achieved by applying NDI of a specific quality.

Prospectively, much research and development remains to be accomplished on this line in order to consolidate and implement the modern probabilistic failure risk assessment into engineering practice. By gaining the acceptance of this approach a better substantiation of the information needed for risk-decision making can be achieved.

In a larger prospect of nowadays techno-centric society, the assessment made by professionals of the failure risk is transferred to decision making levels and projected into public conscientiousness by extremely complex and intricate ways and systems of communication. The transfer of information on the technical risk to all partners, involved or interested, implies the evaluation of a meta- or hierarchical risk, of interpretive nature, which may arise from distortion, misunderstanding and, not irrefutably excluded, from conscious manipulation. Mass-media plays a non-negligible role in interpreting, for society at large, the significance of technical risk in every aspects of human activity. Assertiveness in risk interpretation is unconditionally necessary. On this line of reasoning, a new paradigm emerged. It can be defined as the societal perception of man-made risk which may impair our lives, the environment and the wealth of society. On these existential issues on failure risk involvement in contemporary technology, the reader is referred to an extensive and ever increasing literature such as: Slovik et al. (1981), Kepllinger (1990), Krüger et al. (1991), Peters (1991), Kasperson and Stallen (1991), Sfercoci (2007) or, more comprehensively, in specialized periodicals such as Journal of Risk and Uncertainty (Springer) and Risk Analysis (Blackwell-Wiley Publishing).

References

Annis, C., 2003. Probabilistic life prediction "isn't" as easy as it looks. In: *Probabilistic Aspects of Life Prediction*, Eds. Johnson and, W.S., Hillbery, B.M., ASTM STP 1-1450. ASTM International, West Conshohocken, PA.

Barter, S.A., Sharp, P.K., Holden, G. and Clark, G., 2001. Initiation and early growth of fatigue cracks in an aerospace aluminum alloy. *Fatigue Fract. Eng. Mater. Struct.* **25**: 111–125.

Berens, A.P., 1988. NDE Reliability Data Analysis. In: *ASM Metals Handbook*. Vol. 17, 19th Edition: *Nondestructive Evaluation and Quality* Control: 689–701. ASM International, Materials Park, OH.

Berens, A.P. and Hovey, P.W., 1981. Statistical Methods for estimation the crack detection probabilities. In: ASMT Special Technical Publication, STP **798**: 79–94.

Bogdanoff, J.L. and Kozin, F., 1985. *Probabilistic Models of Cumulative Damage*. Wiley, New York.

Bokalrud, T. and Karlsen, A., 1988. A probabilistic fracture mechanics evaluation of fatigue failure from weld defects. In: *Proceedings of the Conference on Fitness for Purpose Validation of Welded Constructions*. Paper No. 8, London, UK.

Bolotin, V.V., 1965. *Statistical Methods in Structural Mechanics*. Holden Day Inc., San Francisco, CA.

Bowles, C.Q. and Schijve, J., 1973. The role of inclusions in fatigue crack initiation in aluminum alloy. *Int. J. Fract.* **9**(2): 171–179.

Bowie O.L., 1956. Analysis of an infinite plate containing radial cracks originated at the boundaries of an internal circular hole. *J. Mathematics Phys.* **35**: 56.

Box, G.E.P. and Tiao, G.C., 1973. *Bayesian Inference in Statistical Analysis*. John Wiley Classics, New York.

Brinckmann, S. and Van der Giessen, E., 2003. Towards understanding fatigue crack initiation: A discrete dislocation dynamics study. *ICM9, Proceedings of 9th International Conference on the Mechanical Behavior of Materials*, Geneva, Switzerland.

Brinckmann, S. and Siegmund, T., 2008. Computation of fatigue crack growth with strain gradient plasticity and irreversible cohesive model. *Eng. Fract. Mech.* **75**: 2276–2294.

Broeck, D., 1986. *Elementary Engineering Fracture Mechanics*. 4th edition, The Hague, Martinus Nijhoff.

Broeck, D., 1993. Concepts of fracture control and damage analysis. In: *ASM Handbook*, Vol 17: 410–419. ASM International, Materials Park, OH.

Brooks, C.L., Prost-Domansky, S. and Honeycutt, K., 1999. Predicting modeling for corrosion management: modeling fundamentals. *Third Joint NASA/FAA/DoD Conference on Aging Aircraft.* www.apesolutions.com.

BS – 7910, 2000. *Guide on Methods for Assessing the Acceptability of Flaws in Metallic Structures*. British Standard Institution.

Cioclov, D.D., 1975. *Strength and Reliability under Varying Loading* (in Romanian). Facla Publishing, Timisoara, Romania.

Cioclov, D.D., 1995–2007. *pFATRISK®, Probabilistic Fatigue and Damage Failure Risk Assessment*. Proprietary method and software. Prof. D.D.Cioclov, Lille, France; Saarbücken, Germany;New York, USA.

Cioclov, D.D., 1998a. *FATSIM – A Software for Fatigue Damage Simulation*. Internal Note, Fraunhofer Institute of Non-Destructive Testing – IZFP, Saarbrücken, Germany.

Cioclov, D.D., 1998b. Probabilistic modeling of structural fatigue. In: *Uncertainty Modeling and Analysis in Civil Engineering*, Ed. Ayyub, B.M.: 289–320. CRC Press, Boca Raton, FL.

Cioclov, D.D., 1999. Analiza durabilitatii unui tren de aterizaj (in Romanian), Durability analysis of a landing gear. Research Contract Report for AEROSTAR Bacau, Romania.

Cioclov, D.D. and Kröning, M., 1999. Probabilistic fracture mechanics approach to pressure vessel reliability evaluation. In: *Probabilistic and Environmental Aspects of Fracture and Fatigue*. Ed. Rahman, S. *The 1999 ASME Pressure Vessels and Piping (PV&P), Conference*: 115–125, Boston, MA, 1–5 August 1999.

Cioclov, D.D., 2002a. *Third American-European Workshop on Reliability of and Demining*, BAM-Berlin, Sept. 2002. www.9095.net

Cioclov, D.D., 2002b. *Bootstrap Analysis of the Burst Pressure of Composite Pressure Vessel*. Internal Report for SGS-TÜV Saarland, Germany, Project Vh 3217-CNG/4 and ULLIT S.A. La Châtre, France.

Cioclov, D.D., 2002–2004. FATSIM – A Rationale and Code for Probabilistic Modelling of Fatigue crack growth from holes. Internal Reports Fraunhofer IZFP Saarbrücken and AIRBUS Hamburg.

Cioclov, D.D., 2003. Failure risk simulation in steel pipes. Plenary Session of the Annual Assembly of the International Institute of Welding (IIW) – Bucharest, Romania, July 2003.

Cioclov, D.D., 2008. Nanomechanics of Materials. Reprint from *Welding and Material Testing* No. 2, 3, 4/2007 and 1/2008. National R&D Institute for Welding and Material Testing – ISIM Timisoara, Romania.

Cioclov, D.D., Escobedo Medina, M.C. and Schmidt H-J., 2003. On the probabilistic fracture mechanics simulation of the fatigue life and crack statistics for the purpose of damage tolerance assessment. *ICAF 2003, Fatigue of Aeronautical Structures as an Engineering Challenge*, Vol 1. Ed. Guillome, M.: 189–209. EMAS Publishers. Lucerne, Swiss.

Davidson, D.L., 1983. A model for fatigue crack advance based on crack tip metallurgical and mechanics parameters. *Acta Metall.* **32**: 707–714.

Ditlevsen, O. and Olsen, R., 1986. Statistical analysis of Virkler data on fatigue crack growth. *Eng. Fract. Mech.*, **25**(2); 177–195.

DNV, 2008. Det Norske Veritas (DET) Recommended Practice DNV-RP-C203. *Fatigue Design of Offshore Steel Structures*. DET, April 2008. Høvik, Norway.

Dugdale, D.S., 1960. Yielding of steel sheets containing slits. *J. Mech. Phys. Solids.* **8**: 100–104.

DuQuesnay, D.L., Yu, M.T. and Topper, T.H., 1988. An analysis of notch size effect on the fatigue limit. *J. Test. Eval.* **16**: 375–385.

Efron, B. and Tibishirani, R., 1993. *An Introduction to the Bootstrap*. Chapman & Hall, New York.

El Haddad, M.H., Smith, K.N. and Topper, T.H., 1979. Prediction of non-propagating cracks. *Eng. Fract. Mech.* **11**: 573–584.

Engesvik, K.M., 1981. *Analysis of uncertainties in the fatigue capacity of welded joints*. Ph.D. Thesis, Division of Marine Structures, The University of Trondheim and The Norwegian Institute of Technology, Dec. 1981, Norway.

Fawaz, S.A., 2003. Equivalent initial flaw testing and analysis of transport aircraft skin splices. *Fatigue Fract. Eng. Mater. Struct.* **26**: 279–290.

Forman, R.G., Kearney, V.E. and Engle R.M., 1967. Numerical analysis of crack propagation in cyclic loaded structures. *Transaction ASME, J. Basic Eng* **89**: 459.

Gall, K., Sehitlogu, H. and Kadioglu, Y., 1997. A methodology for predicting variability in microstructurally short fatigue cracks. *ASME J. Eng. Mater. Technol.* **119**: 171–179.

Gallagher, J.P., Giessler, F.G., Berens, A.P. and Engle, Jr. R.M., 1984. *USAF Damage Tolerant Design Handbook: Guidelines for the Analysis and Design of Damage Tolerant Aircraft Structures*. AFWAL-TR-82-3073, AD-A153-161.

Gelman, A., Carlin, J.B., Stern, H.S. and Rubin, D.B., 1995. *Bayesian Data Analysis*. Chapman & Hall, London.
Ghonem, H. and Provan, J.W., 1980. Micromechanics theory of fatigue crack initiation and propagation. *Eng. Fract. Mech.* **13**: 963–977.
Ghonem, H. and Dore, S., 1985. Probabilistic description of fatigue crack propagation in polycrystalline solids. *Eng. Fract. Mech.* **21**: 1151–1168.
Ghonem, H. and Dore, S., 1987. Experimental study of the constant probability crack growth curves under constant amplitude loading. *Eng. Fract. Mech.* **27**: 1–25.
Ghoniem, P.J., 2006. Atomistic-dislocation dynamics modeling of fatigue microstructure and crack initiation. Research Project Proposal to *The Air Force Office of Scientific Research – AFOSR*. Arlington, VA,. http://osiris.seas.uda.edu/projects/AFOSR-fatigue.pdf.
Grandt, Jr. A., 2000. Analysis of aiging aircraft scenarios and models for fracture mechanics research. In: *Proceeding 2000 USAF Aircraft Structural Integrity Program (ASIP) Conference*, San Antonio, TX, United Technologies Corp., Dayton, OH.
Iordache, M. and Cioclov, D. D., 2002. Loading spectra and fatigue damage evaluation Report AEROSTAR, Bacau, Romania.
Ihara, T. and Misawa, T., 1988. A stochastic model for fatigue crack propagation with random propagation resistance. *Eng. Fract. Mech.* **31**(1): 95–104.
ISSC, 2006. Fatigue and Fracture, Committeee III.2. *16th International Ship and Offshore Structures Congress*, **1**: 445–527. 20–25 August, 2006. Southampton, UK.
Kasperson, R.E. and Stallen, P.M. (Eds), 1992. *Communicating Risk to the Public: International Perspectives*. Kluwer, Dordrecht, The Netherlands.
Kepplinger, M., 1990. *Wertewandel: Technikdarstellung in der Medien und Technikverständnis der Bevölkerung*. Chemie-Ingenieur-Technik. **62**(6): 465–473.
Klensil, M. and Lukás, P., 1972. Influence of the strength and stress history on the growth and stabilization of failure cracks. *Eng. Fract. Mech.* **4**: 77–92.
Kountouris, I.S. and Backer, M.J., 1989. Defect assessment analysis of defects detected by MPI in offshore structure. *CESLIC Report*, No. OR6. Department of Civil Engineering, Imperial College, London, UK.
Kurth, R.E., Brust, F.W., Ghadiali, N.D., Backukas, J. and Tan, P., 2002. Damage initiation and residual strength modeling in aircraft structures and comparison to experimental test results from fastener. *6th Joint FAA/DoD/NASA Aging Aircraft Conference*. Sect. 9A, Prob. Methods and Stress analysis. Sept. 16–19, San Francisco, CA. www.galaxyscientific.com/agingaircraft2002.
Krüger, J. and Ruβ-Mohl, S. (Eds), 1991. Risikokommunikation und Kommunikationsrisiken. Technikakzeptanz – Medien – Öffentlichkeit. Sigma Velag, Berlin.
Landers, C.B. and Hrdrath, H.F., 1956. Results of axial-load fatigue tests with electropolish 2024-T3 and 7075 aluminum alloy sheet specimens with central holes. *NACA TN-3631*.
Laz, P.J., Craig, B.A. and Hillberry, B.M., 2001. A probabilistic total life model incorporating material inhomogeneities, stress level and fracture mechanics. *Int. J. Fatigue*, **23**: S119–S117.
Lidiard, A.B., 1979. Probabilistic fracture mechanic. In: *Fracture Mechanics, Current Status, Future Prospects*. Pergamon Press, Oxford, U.K.
Lin, Y.K., Wu, W.F. and Yang, J.N., 1984. Stochastic modelling of fatigue crack propagation. In: *Probabilistic Methods in the Mechanics of Solids and Structures. Proc. IUTAM Symp*. Stockholm: 103–110. Springer Verlag, Berlin.
Manning, S.D. and Yang, J.N., 1987. *Advanced Durability Analysis: Analytical Methods, Volume I*. Air Force Wright–Patterson Aeronautical Laboratories Report. AFWAL-TR-86-3017, Dayton, OH.

Marshall, W., 1982. *An Assessment of the Integrity of the PWR Pressure Vessels.* "Summary Report". June 1982, UEKA Authority, UK.

Maymon, G., 2005. Probabilistic crack growth behavior of aluminum 2024-T353 alloy using a "unified" approach. *Int. J. Fatigue*, **27**: 828–834.

Meeker, W.Q. and Escobar, L.A., 1998. *Statistical Methods for Reliability Data.* Wiley.

Meggiolaro, M.A., Miranda, A.C.O., Castro J.T.P. and Freire, J.L.F., 2007. A fracture mechanics based model for explaining notch sensitivity effects on fatigue crack initiation. In: *Mechanics of Solids in Brazil 2007*, Eds. Alves, M. and Mattos, C., Brazilian Society of Mechanical Sciences and Engineering.

Moan, T., Vårdal, O.T., Hellevig, N-C. and Skjodli, K., 2000. Initial crack depth and POD values inferred from in-service observations of cracks in North-Sea jackets. *J. Offshore Mech. Arct. Eng.* **122**(3), p.157.

Murakami, Y., 1987. *Stress Intensity Factors Handbook*, Vols. 1 and 2. Murakami Y. Ed.-in Chief. Pergamon Press, Oxford, UK.

Newman, Jr. J.C., 1976. Predicting Failure of Specimens with either Surface Cracks or Corner Cracks at Holes. NASA TN D-8244, National Aeronautics and Space Administration, June 1976.

Newman, Jr. J.C., 1992. *FASTRAN-II.* A fatigue crack growth structural analysis program. *NASA Technical Memorandum* TM-104159. Langley Research Center, Hampton, VA.

Newman, Jr. J.C. and Edwards, P.R., 1988. Short-crack behavior in an aluminum alloy. An AGARD Cooperative Test Programe. Advisory Group for Aerospace Research & Development, AGARD, R-732, Neuilly-sur-Seine, France.

Nichols, R.W. and Crutzen, S., 1988. *Ultrasonic Inspection of Heavy Section Steel Components: The PISC – II Final Report.* Elsevier Applied Science, London.

Norohna, P.J. et al., 1978. *Fastener Hole Quality, Volume 1.* AFFDL-TR-78-206, Wright Patterson Air Force Base, Dayton, OH.

Ortiz, K. and Kiremidjian, A., 1988. Stochastic modeling of fatigue crack growth. *Eng. Fract. Mech.*, **29**(3): 317–334.

Paris, P.C., Gomez, M.P. and Anderson, W.E., 1961. A rationale analytic theory of fatigue. *Trend Eng.* **13**: 9–14.

Peters, H.P., 1991. Durch Risikokommunikation zur Technikakzeptanz? Die Konstruktion von Risiko "Wirklichkeiten" durch Experten, Gegenexperten und Öffentlichkeit. In: *Risikokommunikation. Technikakzeptanz, Medien und Kommunikationsrisiken*, Eds Krüger, J. and Ruβ-Mohl, Edition Sigma, Berlin.

R/H/R6-Revision 3, 1986. *Assessment of the Integrity of Structures Containing Defects.* Authors: Milne I., Ainsworth, R.A. Dowling, A.R. and Stewart, A. T. British Energy (previously CEGB).

Rudd, J.L. and Gray, T.D., 1978. Quantification of fastner-hole quality. *J. Aircraft*, **15**: 143–147.

Rummel, W.D., Todd, P.H., Frecska, S.A. and Rathke, R.A., 1974. Report CR-2369, NASA Feb. 1974.

Schmidt, H.-J., Schmidt-Brandecker, B. and Tober, G., 1999. Design of modern aircraft structures and the role of NDI. NDT.net. AccessedJune 1999, **4**(6).

Sfercoci, L., 2007. Journalismus und Technik. Bericht, March.2007, Fraunhofer IZFP, Saarbrücken, Germany.

Shaw, W.J.D. and LeMay, I., 1981. Intrinsic material scatter in fatigue crack propagation. *Fatigue Fract. Eng. Mater. Struct.* **4**: 367–375.

SINTAP, 1999. *SINTAP: Entwurf einer vereinheitlichen europäischen Fehlerbewertungsprozedur – eine Einführung.* Authors: U. Zerbst, C. Wiesner, M. Koçak, L. Hodulak. GKSS-Forschungszentrum Geesthacht, GmbH, Geesthacht, Germany.

Slovic, P., Fischhoff, B. and Lichtenstein, S., 1981. *The Assessment and Perception of Risk.* The Royal Society, London.

Smith, R.A., 1977. On the short crack limitations of fracture mechanics. *Int. J. Fract.* **13**: 717-720.

Sobczyk, K., 1986. Modelling of random fatigue crack growth. *Eng. Fract. Mech.* **24**: 609–623.

Sobczyk, K. and Spencer, B.F., 1992. *Random Fatigue: From Data to Theory.* Academic Press, Boston, MA.

Tang, J. and Spencer, B.F., 1989. Reliability solutions for the stochastic fatigue crack growth problem. *Eng. Fract. Mech.* **34**(2): 419–433.

Tanaka, K., Nakai, Y. and Yamashita, M., 1981. Fatigue growth threshold of small cracks. *Int. J. Fract.* **17**: 519–533.

Taylor, D., 1989. *Fatigue Thresholds.* Butterworths, UK.

Tober, G. and Klement W.B., 2000. NDI reliability rules used by transport aircraft – European view point. *15th World Conference of Nondestructive Testing – WCNDT*, Rome.

Tsurui, A. and Ichikawa, H., 1986. Application of Fokker–Plank equation to a stochastic fatigue crack growth. *Struct. Saf.* **4**:15–29.

UASF, 1984. *UASF Damage Tolerant Design Handbook: Guidelines for the Analysis and Design of Damage Tolerant Aircraft Structures. Revision B.* Authors: Gallagher, J.P., Giessler, F.J. and Berens. Dayton University Research Institute, Dayton, OH.

Varanasi, S.R. and Whittaker, I.C., 1976. Structural reliability prediction method considering crack growth and residual strength. In: *Fatigue Crack Growth under Spectrum Loads*, ASTM STP, **595**: 292–305. American Society for Testing and Materials, Philadelphia, PA.

Walker, E.K., 1970. The effect of stress ratio during crack propagation and fatigue for 2024-T3 and 7075-T6 aluminum. In: *Effect of Environment and Complex Load History on Fatigue Life*, ASTM STP **462**: 1–14. American Society for Testing and Materials, Philadelphia, PA.

Wang, D.Y., 1980. An investigation of initial fatigue quality. In: Design of Fatigue and Fracture Resistant Structures. ASTM STP **761**: 191–211. American Society for Testing and Materials, Philadelphia, PA.

Welch, K.M., Hajowski, L.L., Edghill, M.S. and Jeske, T.C., 2006. USAF F16A/B fuel shelf joint structural risk assessment. In: *USAF ASIP Conference*, 28–30 Nov. 2006.

Wu, W.F. and Ni, C.C., 2003. A study of stochastic fatigue crack growth modeling through experimental data. *Probab. Eng. Mech.* **18**: 107–118.

Yang, N.J. and Manning, S.D., 1980. Distribution of equivalent initial flaw size. In: *Proceeding of the 1980 Annual Reliability and Maintainability Symposium.* 112–120. IEEE, San Francisco, CA.

Yang, N.J., Salivar, G.C. and Annis, Jr. G.G., 1983. Statistical modeling of fatigue cracks in a nickel-base superalloy. *Eng. Fract. Mech.*, **18**: 257–270.

Yang, N.J. and Manning, S.D., Rudd, J.L. and Bader, R.M., 1990. Investigation of mechanistic based equivalent initial flaw size approach. In: *Proceedings of 18th Symposium. International Committee on Aeronautical Fatigue.* Eds Grandage, J.M. and Jost, G.M.: 385–404. Engineering Materials Advisory Service Ltd., London.

Yu, M.T., DuQuesnay, D.L. and Topper, T.H. 1988. Notch fatigue behavior of 1045 steel. *Int. J. Fatigue.* **10**:109–116.

Zener, C., 1948. *Fracturing of Metals.* ASM, Cleveland, OH.

ASSESSMENT OF LOCALISED CORROSION DAMAGING OF WELDED DISSIMILAR PIPES

IHOR DMYTRAKH*, VOLODYMYR PANASYUK
Karpenko Physico-Mechanical Institute of National Academy of Sciences of Ukraine, 5 Naukova Street, 79601, Lviv, Ukraine

Abstract The procedure of localised corrosion damaging assessment on internal surface of welded dissimilar pipes is developed. Method is based on numerical-analytic model of dissimilar welded joint as three-electrode electrochemical system with using of the fundamental electrochemical parameters of composing materials, which were received by standard potentiometric methods. The data on general characterisation and maximal depth of localised damaging on internal surface of pipes are presented and analysed with dependence of term exploitation and composition of operating environment.

Keywords: Welded joints of dissimilar pipes, local corrosion defects, potentiodynamic polarisation curves, corrosion current density distribution, depth of corrosion damaging

1. Introduction

The structural integrity of critical equipment operating under the conditions of combined action of mechanical loading and corrosion continues to provide a challenge for researches and practical engineers. For example, in heat-and-power engineering, various corrosion and corrosion-mechanical defects of the pipes of elements of water–steam circuit strongly affect the period of their safe operation.[1,2] At present, up to 70% of failures and, hence, idle periods of heat power generating units are explained by the initiation and development of unpredicted local corrosion-mechanical and

---

* E-mail: dmtr@ipm.lviv.ua

corrosion defects.[1] The problem of corrosion-induced defects of welded joints, in particular, of circular welded joints of the dissimilar pipes at the exit of steam super heaters is especially important.[3] Therefore, the development of the methods aimed at the analysis of the kinetics of the growth of corrosion-induced defects in objects of this type proves to be an actual problem.

In what follows, we propose a procedure for the numerical analysis and forecasting assessments of the maximum depth of corrosion-induced defects in the components of the analysed welded joints. It is based on the use of fundamental electrochemical parameters of the materials of welded joints obtained by using standard potentiometric methods and a numerical–analytic procedure of determination of the density of corrosion current on the inner surface of the pipe according to which the studied combined welded joint is regarded as a three-electrode electrochemical system.

2. Object of study and experimental procedure

The combined welded joints of the pipes of 12Kh1MF low-alloyed pearlite steel and Kh18N10T stainless austenitic steel were the object of study (Figure 1a). To do this, we tested joints of the following two typical sizes (Figure 1b): d = 38 mm, t = 4 mm and d = 42 mm, t = 4 mm in the as-received state after operation during 185,000 h at the hear power plant.

The tested specimens (with surfaces 10 × 10 mm in size) were cut out from the studied objects (12Kh1MF and Kh18N10T steels and the welding zone). The specimens cut out from the welding zone contained the material of the weld and the adjacent zones of melting and thermal influence (heat affected zones).

We studied the inner surface of the pipe operating in contact with working environment. The components of the welded joints were electrochemically studied according to the standard potentiometric methods with the help of a PI-50-1 Potentiostat equipped with a PR-8 programmer and connected with a personal computer.[4] In the course of the tests, the potentiodynamic polarization curves of the materials of welded joints were recorded under the action of working media of various compositions (Table 1). By using the potentiodynamic polarisation curves, we found the following basic (fundamental) electrochemical parameters (Table 2): corrosion potentials φ_k (as the points of intersection of the Tafel straight lines) and the cathodic R_c and anodic R_a polarisation resistances.

These data were used for the numerical–analytic determination of the density of corrosion current on the inner surface of the pipe in the vicinity of the welded joint.

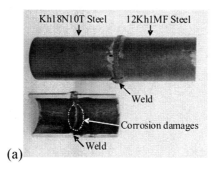

(a)

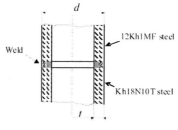

(b)

Figure 1. Object of study (a) and its schematic view (b).

TABLE 1. Parameters of operating environment.

Concentration NaCl, wt %	Conductivity σ, Sm/m	pH
0 (distilled water)	0.0009	5.9
0.03	0.0592	6.5
0.3	0.5570	6.8

TABLE 2. Basic electrochemical parameters of materials of welded joint.

NaCl Wt %	Corrosion potential φ_k, V	Polarisation resistance R_c, ohm m²	Polarisation resistance R_a, ohm m²	NaCl Wt %	Corrosion potential φ_k, V	Polarisation resistance R_c, ohm m²	Polarisation resistance R_a, ohm m²
	12Kh1MF steel (as received)				12Kh1MF steel (exploited)		
0	−0.375	3.937	4.132	0	−0.193	3.205	4.878
0.03	−0.427	0.460	0.340	0.03	−0.390	0.716	0.921
0.3	−0.433	0.339	0.297	0.3	−0.402	1.340	0.695
	Welding zone (as received)				Welding zone (exploited)		
0	−0.169	4.717	4.831	0	−0.217	3.571	4.975
0.03	−0.235	0.800	0.563	0.03	−0.255	0.821	0.526
0.3	−0.360	0.484	0.133	0.3	−0.314	0.181	0.160
	Kh18N10T steel (as received)				Kh18N10T steel (exploited)		
0	−0.021	8.772	11.278	0	−0.333	4.292	5.714
0.03	−0.312	3.279	4.566	0.03	−0.376	2.110	1.763
0.3	−0.369	1.976	2.232	0.3	−0.423	1.125	0.856

3. Numerical-analytic model for evaluation of corrosion current density in vicinity of welded joint

Under the conditions of electrochemical corrosion, the studied object is considered as a three-electrode system (Figure 2) characterised by the corrosion potentials φ_n, specific polarization resistances R_n, specific conductivity of the environment σ and dimensionless half width of the welding zone l, which related to the radius of the pipe r_0. Here and in what follows, the subscripts $n = 1,2,3$ mark the quantities in the regions: $-\infty \leq Z \leq -l$, $-l \leq Z \leq l$ and $l \leq Z \leq \infty$, respectively.

The indicated model scheme of combined welded joints is used as basic for the development of the numerical–analytic procedure of determination of the density of corrosion current on the inner surface of the pipe. This scheme is realised according to the method of equalised polarisation.[5] To do this, it is necessary first to establish some specific features of the exact solution of the posed problem. In the rigorous statement of the problem, the current density is analytically determined via the electric potentials of the medium $\psi_n(r, Z)$ satisfying the Laplace equation[5]:

$$\left(\frac{\partial^2}{\partial r^2} + \frac{1}{r} \frac{\partial}{\partial r} + \frac{\partial^2}{\partial Z^2} \right) \psi_n(r, Z) = 0, \quad n = 1, 2, 3. \quad (1)$$

In the period of establishment of the current density $j_n(1, Z)$ from the surface $r = 1$, $-\infty \leq Z \leq \infty$, the current directed from the metal into the environment is regarded as positive. Its density can be found according to the potential $\psi_n(r, Z)$ in two ways.

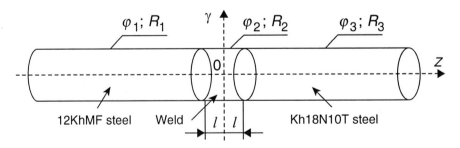

Figure 2. The scheme of object of study and the system of cylindrical coordinates.

Thus, if we take into account the fact that the conductivity of the metals is 10^6–10^7 times higher than the conductivity of the environment, then it is possible to assume that the electric potential of the metal surface:

$$\psi_m \approx const = 0. \qquad (2)$$

Then, according to work,[6] we have:

$$j_n(1, Z) = \frac{-[\psi_n(1,Z) + \varphi_n]}{R_n}, \ n = 1, 2, 3. \qquad (3)$$

In what follows, it is shown that $j_1(1,Z)$ and $j_3(1,Z)$ is almost exponentially vanishing as Z increases. Therefore, the region of effective action of corrosion currents is bounded and, hence, condition (2) and relation (3) can be considered as acceptable.

In the second case, according to the Ohm law in the differential form, by using the dimensionless coordinate r, we find

$$j_n(1, Z) = \frac{\sigma}{r_0} \left. \frac{\partial \psi_n(r, Z)}{\partial r} \right|_{r=1}. \qquad (4)$$

Equating the right-hand sides of equalities (3) and (4), we arrive at the following condition on the surface $r = 1$:

$$\left[\psi_n(r, Z) + K_n \frac{\partial \psi_n(r, Z)}{\partial r} \right]_{r=1} = -\varphi_n, \ n = 1, 2, 3, \text{ where } K_n = \frac{\sigma R_n}{r_0}. \qquad (5)$$

The potentials $\psi_n(r,Z)$ satisfy condition (5), the conditions of continuity of $\psi_n(r,Z)$ and $\partial\psi_n(r,Z)/\partial Z$, in the sections $0 \leq r < 1, Z = \pm l$, and the following condition at infinity:

$$\psi_1(r, Z = -\infty) = -\varphi_1, \ \psi_3(r, Z = \infty) = -\varphi_3. \qquad (6)$$

Conditions (6) follow from equality (3) and the obvious conditions $j_1(1,Z = -\infty) = 0$ and $j_3(1,Z = +\infty) = 0$. To find $\psi_n(r,Z)$, we use the method of equalised polarisation[5] according to which condition (5) with different K_n is replaced with a condition imposed for constant K. The role of this constant K is played by K_2. As a result, condition (5) takes the form:

$$\left[\psi(r,z) + K_2 \frac{\partial \psi(r,z)}{\partial r} \right]_{r=1} = \begin{cases} -(\varphi_1 + \tilde{\varphi}_1), & -\infty < z < -l, \\ -\varphi_2, & -l < z < l, \\ -(\varphi_3 + \tilde{\varphi}_3), & l < z < +\infty. \end{cases} \qquad (7)$$

Here, $\tilde{\varphi}_1$ and $\tilde{\varphi}_3$ are the corrections to the corrosion potentials K_1 and K_3 equivalent to the change $K_1 \to K_2$ and $K_3 \to K_2$, namely,

$$\tilde{\varphi}_1 = \frac{K_1 - K_2}{l_1} \int_{-\infty}^{-l} \left.\frac{\partial \psi_1(r, Z)}{\partial r}\right|_{r=1} dZ, \quad \tilde{\varphi}_3 = \frac{K_3 - K_2}{l_3} \int_{l}^{\infty} \left.\frac{\partial \psi_3(r, Z)}{\partial r}\right|_{r=1} dZ. \quad (8)$$

To find $\tilde{\varphi}_1$ and $\tilde{\varphi}_3$, we consider the surfaces of effective corrosion chosen in the form of the parts of the surfaces of the pipes l_1 and l_3 in length adjacent to the weld and concentrating more than 90% of corrosion current for each semi-infinite pipe. The lengths l_1 and l_3 are determined by using the analytic expressions for the current densities on the surfaces of both semi-infinite pipes obtained from relations (3) or (4). The Laplace equation for $\psi(r, Z)$ is solved under conditions (6) and (7) and the conditions of continuity of $\psi(r, Z)$ and $\partial \psi(r, Z)/\partial Z$, for $Z \pm l$ by the method of separation of variables. As a result, we obtain:

$$\left.\begin{array}{l}\psi_1(r, Z) = \sum_{m=1}^{\infty} [a_m \exp(\lambda_m Z) - \varphi_1 - \tilde{\varphi}_1] \\ \psi_2(r, Z) = \sum_{m=1}^{\infty} [b_m \exp(\lambda_m Z) + c_m \exp(-\lambda_m Z) - \varphi_2] \\ \psi_3(r, Z) = \sum_{m=1}^{\infty} [d_m \exp(-\lambda_m Z) - \varphi_3 - \tilde{\varphi}_3]\end{array}\right\} \frac{2 K_2^2 \lambda_m J_1(\lambda_m) J_0(\lambda_m r)}{(K_2^2 \lambda_m^2 + 1) J_0^2(\lambda_m)}. \quad (9)$$

$$\begin{array}{l}a_m = [(\varphi_1 + \tilde{\varphi}_1 - \varphi_2)\exp(\lambda_m l) + (\varphi_2 - \varphi_3 - \tilde{\varphi}_3)\exp(-\lambda_m l)]/2; \\ b_m = [(\varphi_2 - \varphi_3 - \tilde{\varphi}_3)\exp(-\lambda_m l)]/2; \\ c_m = [(\varphi_2 - \varphi_1 - \tilde{\varphi}_1)\exp(-\lambda_m l)]/2; \\ d_m = [(\varphi_3 + \tilde{\varphi}_3 - \varphi_2)\exp(\lambda_m l) + (\varphi_2 - \varphi_1 - \tilde{\varphi}_1)\exp(-\lambda_m l)]/2.\end{array} \quad (10)$$

We now determine the corrections $\tilde{\varphi}_1$ and $\tilde{\varphi}_3$. To this end, we substitute relations (9) and (10) in Eq. (8). This yields the system of equations:

$$\left(\frac{l_1}{K_2 - K_1} - A\right)\tilde{\varphi}_1 + B\tilde{\varphi}_3 = (\varphi_1 - \varphi_2)A + (\varphi_2 - \varphi_3)B,$$

$$\left(\frac{l_3}{K_3 - K_1} - A\right)\tilde{\varphi}_3 + B\tilde{\varphi}_1 = (\varphi_3 - \varphi_2)A + (\varphi_2 - \varphi_1)B.$$

Here

$$A = \sum_{m=1}^{\infty} \frac{1}{\lambda_m(K_2^2 \lambda_m^2 + 1)}, \quad B = \sum_{m=1}^{\infty} \frac{\exp(-2\lambda_m l)}{\lambda_m(K_2^2 \lambda_m^2 + 1)}.$$

ASSESSMENT OF LOCALISED CORROSION DAMAGING

Substituting $\psi(r,Z)$ in relations (3) or (4), we obtain the following expressions for the current densities:

$$j_1(1,Z) = \frac{\sigma}{r_0} \sum_{m=1}^{\infty} \frac{\left[(\phi_2 - \phi_1 - \tilde{\phi}_1)\exp(\lambda_m l) + (\phi_3 + \tilde{\phi}_3 - \phi_2)\exp(-\lambda_m l)\right]\exp(\lambda_m Z)}{K_2^2 \lambda_m^2 + 1}$$

(11)

$$j_2(1,Z) = \frac{\sigma}{r_0} \sum_{m=1}^{\infty} \frac{\left[(\phi_3 + \tilde{\phi}_3 - \phi_2)\exp(\lambda_m Z) + (\phi_1 + \tilde{\phi}_1 - \phi_2)\exp(-\lambda_m Z)\right]\exp(-\lambda_m l)}{K_2^2 \lambda_m^2 + 1}$$

(12)

$$j_3(1,Z) = \frac{\sigma}{r_0} \sum_{m=1}^{\infty} \frac{\left[(\phi_2 - \phi_3 - \tilde{\phi}_3)\exp(\lambda_m l) + (\phi_1 + \tilde{\phi}_1 - \phi_2)\exp(-\lambda_m l)\right]\exp(-\lambda_m Z)}{K_2^2 \lambda_m^2 + 1}$$

(13)

where λ_m are the roots of the equation $j_0(\lambda) - K_2 \cdot \lambda \cdot j_1(\lambda) = 0$ and $j_p(\lambda)$ are the Bessel functions of p-order.

It follows from relations (11)–(13) that the dependence of $j_1(1,Z)$ on Z is specified by the roots λ_m. The analysis of these roots computed for the ranges of conductivities and specific polarisation resistances presented in Tables 1 and 2 show that the quantity λ_m significantly increases beginning with $m = 2$. Note that, in view of the fact that the current density is an exponential function of Z, the principal contribution to the corrosion current in relations (11) and (13) is made by the first term. Hence, the length of the regions of effective corrosion in the pipe is also determined by the first term, i.e., by the root λ_1.

In finding the rigorous solution of the problem, i.e., by using the coefficients K_n specified for each region along the coordinate Z, the exponential dependences of $j_n(1,Z)$ ($n = 1,2,3$) on Z are preserved but the coefficients of the exponents are given by a fairly cumbersome expressions. In each region along the coordinate Z, these dependences are determined by the roots $\lambda_m^{(n)}$ of the equation $j_0(\lambda^{(n)}) - K_n \cdot \lambda^{(n)} \cdot j_1(\lambda^{(n)}) = 0$.

The analysis of the roots $\lambda_m^{(n)}$, $n = 1,2,3$, $m = \overline{1,\infty}$, shows that the behaviour of the quantities $\lambda_m^{(1)}$ and $\lambda_m^{(3)}$ within the ranges of the parameters specified in Tables 1 and 2 is such that the main contributions to the exact expressions for $j_n(1,Z)$ ($n = 1,2,3$) are made by the first terms. The lengths of the regions of effective corrosion (over 90% of the total current) on the surface of each semi-infinite pipe are given by the formulas $l_n = ln[10/\lambda_1^{(n)}]$, $n = 1,3$. On the basis of relations (11)–(13), we construct the dependences of the current densities $j(1,Z)$ for the new materials of welded joints and materials after long-term operation both in NaCl solutions and in distilled water (Figure 3).

It is shown (Figure 3a) that, for almost all analysed cases, the pipe made of 12Kh1MF steel plays the role of anode and the pipe made of Kh18N10T steel and the welding zone are cathodes. The dependences of the current density on Z for pipes made of both types of steel are almost exponentially decreasing. The maximum current density is attained on the boundaries of the welding zone. For distances from the welding zone greater than a certain value, the current density increases with the concentration of NaCl.

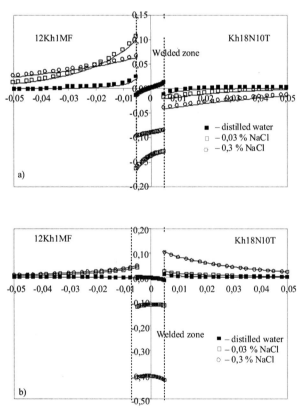

Figure 3. Current density $j(1,Z)$ for new (a) and exploited (b) dissimilar welded joint in the solutions of different composition.

For exploited materials, the welding zone plays the role of cathode (Figure 3b). It is shown that the pipes made of both types of steel in distilled water and, hence, in 0.03% and 0.3% NaCl solutions are anodes. The current density in the welding zone is much higher than for new materials. The dependences of $j(1,Z)$ for both pipes, as in the case of new materials, are described by exponentially decreasing functions.

4. Analysis of development of localised corrosion-induced defects

It is known, the process of anodic dissolution of alloys is characterised by the absence of any significant selective dissolution of individual components.[7] Thus, to predict the development of corrosion-induced defects of welded joints, we get the following dependence of the corrosion rate (m/s) on the parameters of the alloy:

$$v(Z) = \frac{1}{F \cdot D} \cdot \frac{j(Z)}{\sum_{p=1}^{P} \frac{d_p \cdot \Theta_p}{A_p}} \qquad (14)$$

where p is the number of a component, d_p is its relative mass content, A_p is the mass of an atom of the component in atomic units, Θ is the valence of the metal, P is the number of components, D is the density of the alloy, and F is the Faraday constant.

In practice, the corrosion rate is measured in mm/year. Therefore, it is necessary to transform m/s into mm/year. As a result, by using the value of the Faraday constant F, we rewrite equality (14) in the form:

$$v(Z) = \frac{326{,}8}{D} \cdot \frac{j(Z)}{\sum_{p=1}^{P} \frac{d_p \cdot Z_p}{A_p}} \qquad (15)$$

In the first approximation, we describe the time dependences of the electrochemical parameters R and φ by a linear law. Thus, by using relations (11)–(13) and (15), we find the depth of corrosion damage h of the components of the studied welded joint for a given period of operation T.

A typical example of this type calculations for a period of operation of the joint of 200,000 h in environment of different compositions (Figure 4) shows that the maximum depths of corrosion-induced defects is attained on the "metal-welding zone" boundaries and sufficiently rapidly decreases with the distance from the welding zone.

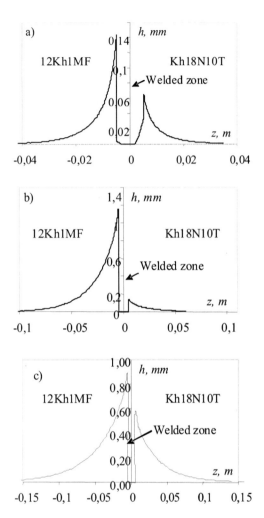

Figure 4. Profile of corrosion damaging h at the dissimilar welded joint on the base of 200,000 h in the solutions of different composition: (a) distilled water; (b) 0.03% NaCl; (c) 0.3% NaCl.

Therefore, the degree of corrosion damaging to combined welded joints is characterised by the maximum depth of the defects h_{max}^i max in its components (Figure 5), namely, h_{max}^1 for 12Kh1MF steel, h_{max}^2 for the welding zone neighbouring with this steel, h_{max}^3 for the welding zone neigh-

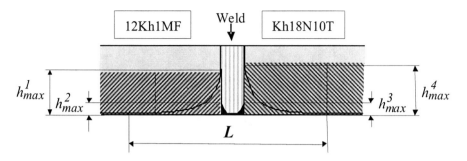

Figure 5. Schematic diagram of corrosion damaging in the walls of the dissimilar pipes welded by circumferential weld.

bouring with Kh18N10T steel, and h_{max}^4 for Kh18N10T steel. In addition, we introduce the notion of the zone of corrosion activity of a welded joint L (Figure 5) specified by the condition $h \geq 1\,\mu m$.[8]

The numerical analysis[9] of the parameter h_{max}^i carried out according to the proposed scheme on the time base $T = 400{,}000$ h enable us to establish the following prognostic estimates of corrosion damage to the components of welded joints of the pipes made of different materials in media of different chemical compositions (Figure 6).

The maximum depth of the defects in the base materials increases with the concentration of chlorides in the environment. Note that the welding zone is not damaged in NaCl solutions. At the same time, in distilled water the maximum depth of the defects is observed on the boundaries of the welding zone with the other components of the joint. On the other hand, all environments are characterised by the presence of two common tendencies. First, the material of the pipe produced with 12Kh1MF steel is intensely damaged in the initial stages of operation with subsequent stabilisation of this process at a certain level depending on the composition of the medium. The material of the pipe made of Kh18N10T steel remains practically intact for $T < 100{,}000$ h but then the character of its damage becomes similar to that exhibited by 12Kh1MF steel. Second, curves 1 and 4 intersect for all studied cases. This means that, beginning with a certain period of operation, the rate of corrosion damaging of stainless steel becomes higher than the rate of corrosion damage of low-alloy steel.

It should be emphasized that the length of the zone of corrosion activity of the welded joint L increases both with the period of operation and with the concentration of chlorides in the working medium (Figure 7 and Table 3).

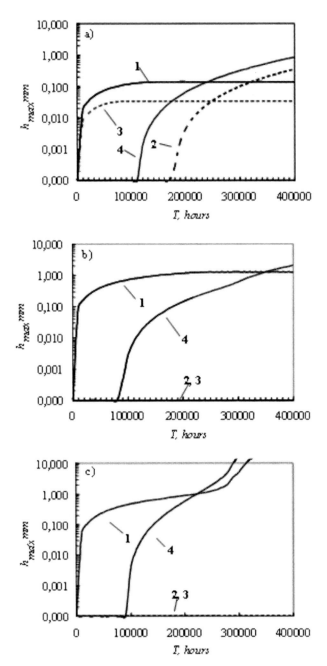

Figure 6. Maximum depths of corrosion damaging h_{max} in the components of a welded joint of dissimilar pipes in distilled water (a), 0.03% NaCl (b), 0.3% NaCl (c) as functions of the planned period of operation: 1 – 12Kh1MF steel, 2 – welding zone neighbouring with 12Kh1MF steel, 3 – welding zone neighbouring with Kh18N10T steel, 4 – Kh18N10T steel.

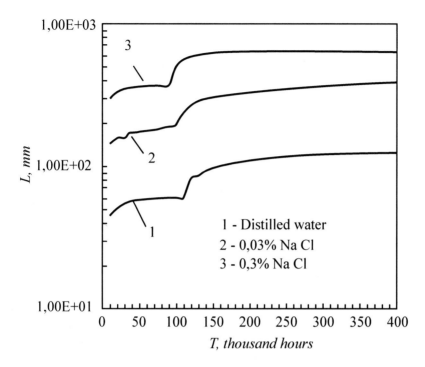

Figure 7. Length of zone of corrosion activity at welded joint of dissimilar pipes versus time of exploitation in the solutions of different composition.

TABLE 3. Length of zone of corrosion activity at welded zone on planned period of operation.

T, h	L, mm		
	Distilled water	0.03 % NaCl	0.3 % NaCl
50,000	58.15	175.02	363.52
100,000	60.16	196.72	496.69
150,000	98.04	307.24	623.84
200,000	112.31	334.87	639.35
250,000	117.89	350.65	639.35
300,000	121.61	364.47	639.35
350,000	124.09	380.26	639.35
400,000	125.95	388.15	639.35

5. Conclusions

Considering of welded joints of dissimilar pipes (low alloyed steel – stainless steel) as a three-electrode electrochemical system, an analytic procedure for the evaluation of corrosion current density distribution at the welded zone is proposed.

On this basis the experimental–numerical method for prediction of profile and maximum depth of local corrosion defects in the components of welded joints is developed.

The data on general characterisation and maximal depth of localised damaging on internal surface of pipes were received and analysed with dependence of term exploitation and composition of operating environment. In particular, the possibility of changing the polarity of local galvanic couples in the vicinity of the welding zone depending on the period of operation has been shown that cannot be predicted within the framework of the existed approaches.

ACKNOWLEGEMENT This study has been conducted within the Targeted Research and Engineering Programme "RESURS" of National Academy of Sciences of Ukraine (2007–2009).

References

1. I. M. Dmytrakh, A. B. Vainman, M. H. Stashchuk and L. Toth, *Reliability and Durability of Structural Elements for Heat-and-Power Engineering Equipment.* Reference manual, Edited by I. M. Dmytrakh (Publishing House of National Academy of Sciences of Ukraine "Academperiodyka", Kiev, 2005) (in Ukrainian).
2. I. M. Dmytrakh, A. M. Syrotyuk, B. P. Rusyn and Yu. V. Lysak, in: *Problems of Service Life and Safe Exploitation of Structures, Facilities and Machines*, Edited by B. Ye. Paton (National Academy of Sciences of Ukraine, Kyiv, 2006), pp. 221–225) (in Ukrainian).
3. I. M. Dmytrakh and V. V. Panasyuk Problems of lifetime assessment of water-steam circuit elements of power units, Bay Zoltan Institute for Logistics and Production Systems Miskolc, Hungary (11–12 April 2006); http:part.bzlogi.hu.
4. I. M. Dmytrakh and V. V. Panasyuk, *Influence of Corrosive Media on the Local Fracture of Metals Near Stress Concentrations*, (Karpenko Physico-Mechanical Institute, Ukrainian Academy of Sciences, Lviv, 1999) (in Ukrainian).
5. Yu. Ya. Iossel and G. Ye. Klenov, *Mathematical Methods for the Assessment of Electrochemical Corrosion and Protection of Metals. A Handbook*, (Metallurgiya, Moscow, 1984) (in Russian).

6. B. I. Kolodii, I. M. Dmytrakh, and O. L. Bilyi, Method of equivalent electrode for the evaluation of electrochemical currents in corrosion pits. *Physicochemical Mechanics of Materials*, 38 (5), 27–31 (2002). (in Ukrainian).
7. Ya. M. Kolotyrkin, *Metals and Corrosion*, (Metallurgiya, Moscow, 1985) (in Russian).
8. I. Dmytrakh, A. Syrotyuk and R. Leshchak, Assessment of surface corrosion fatigue damaging of pipeline steels. *Physicochemical Mechanics of Materials*, Special Issue No 4, 67–72 (2004) (in Ukrainian).
9. I. Dmytrakh, B. Kolodiy, O. Bilyy and R. Leshchak, Kinetics assessment of localised corrosion damages in pipe welds of water-steam circuit elements. *Physicochemical Mechanics of Materials*, Special Issue No 5, 17–25 (2006) (in Ukrainian).

PROCEDURES FOR STRUCTURE INTEGRITY ASSESSMENT

NENAD GUBELJAK[*], JOŽEF PREDAN
*University of Maribor, Faculty of Mechanical Engineering,
Smetanova 17, SI-2000 Maribor, Slovenia*

Abstract Quite few analytical flow assessments methods as specific standard and guidelines there have been developed in recent years. Today, as one of a most comprehensive assessment procedure is SINTAP – Structural INTegrity Assessment Procedure. The SINTAP introduced the basic principles of R6 (rev. 4) and ETM. SINTAP procedure is possible to performed assessment for inhomogeneous configurations such as strength mis-matched weldments and an effect of residual stresses. Nevertheless, the SINTAP procedure take into account temperature transition region from ductile-to-brittle behavior of material. In the paper the SINTAP procedure was applied to the failure analysis of cracked component. By results of used example was shown that SINTAP procedure gives reliable conservative results where the safe factor decreasing by increasing the quality of input data.

Keywords: Integrity of structure, crack driving force, stress intensity factor, pipelines, material properties, structural steel, high strength low alloy steel

1. Introduction

The SINTAP procedure is taken overcome of interdisciplinary Brite–Euram project aimed at examining and unifying the fracture mechanics based flaw assessment approached available in Europe and propose a procedure which should form the basis of future European standard.[1] Among many other publications a special issue of the journal (Engineering Fracture Mechanics 67, 2000, 479–668) contains a number of papers, which describe the main features of the SINTAP procedure. In the SINTAP procedure the implicit background assumption is that the component is defect-free. In this case when a real of assumed crack or crack like flow affects the load carrying capacity, the fracture mechanics principles have to be applied. Then the comparison between the applied and the material side has to be carried out

---

[*] E-mail: nenad.gubeljak@uni-mb.si

on the basis of crack tip parameters such as the linear elastic stress intensity factor, K, the J integral or the crack tip opening displacement ($CTOD$). As a result, the fracture behavior of the component can be predicted in terms of a critical applied load or a critical crack size.

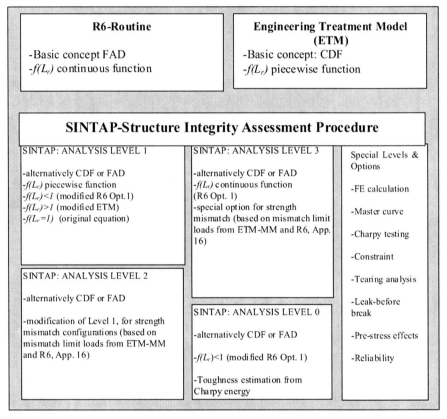

Figure 1. Overall structure of the SINTAP procedure.[9]

Standard solutions for the crack tip parameters are available for the specimens which are used for measuring the material's resistance to fracture. As long as the deformation behavior of the structural component is linear elastic, then the relevant applied parameter in the component is K, available in comprehensive compendia of K factor solutions.[2–5] If the component behaves in an elastic-plastic manner the situation is much more complex because the crack tip loading is additionally influenced by the deformation pattern of the material as given by its stress–strain curve. This makes the generation of handbook solutions an expensive tasks. To a limited extent this tasks has been realized for a few component configurations, for example, in the Electronic Power Research Institute-EPRI handbook.[6] SINTAP procedure also consist solutions for cracked plates, bars and pipes with different configurations of loading and crack positions. Overall structure of SINTAP is shown in Figure 1. For more detailed discussion see Refs.[7,8] The aim of this paper is give a basic principles of SINTAP procedure application by using broken forklift as working example. The fork of a forklift is example for using "Default" level, level "1" and level "3".

2. SINTAP procedure

The SINTAP procedure based on fracture mechanics principles, as shown in Figure 2. If two of the input parameters are known the third can be determined theoretically. This principle allows for different tasks of a fracture mechanics analysis:

1. A crack is detected in a component under service. The question to be answered is whether this law will lead to component failure or not. In certain circumstances the critical state can be suspended by reducing the load.
2. In the design stage a component can be set-off with respect of a hypothetical crack the dimensions of which have to be chosen such that the crack will be detected by NDT whether at final quality control or in-service.
3. Vice versa critical crack dimensions for subsequent NDT testing can be determined.

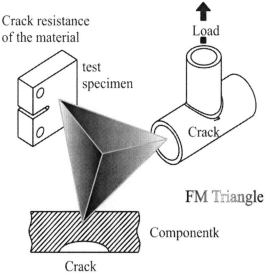

Figure 2. Fracture mechanics principles in design.

The determination of the critical crack size is illustrated in Figure 3. In the following the application of SINTAP to the forklift as well as some basic items of the procedure will be addressed following this flow chart. Note, that some of the features and analysis steps shown in the flowchart will not be part of the present analysis. Nevertheless they are shown for completeness and will be addressed briefly because they provide important information for many other cases of failure analysis.

In order to determine a critical crack size the following input information is required:

- Geometry and dimensions of the component
- Applied loading including secondary load components such as residual stresses
- Information on crack type and orientation
- The stress–strain curve and fracture toughness of the material

(A) GEOMETRY AND DIMENSIONS OF THE COMPONENT

The geometry and dimensions of the component normally is different by case to the case, but for analytical-handbook approach it is necessary to determine geometrical features of component. For example, the geometry of the fork as given in Figure 4 is essentially a thick plate. The dimensions of the relevant cross section where fracture occurred is shown in Figure 5.

(B) APPLIED LOADING INCLUDING SECONDARY LOAD COMPONENTS

In the SINTAP procedure the applied load can be introduced as a single load such as a tensile force, a bending moment or an internal pressure. The consideration as a stress profile, which is, i.e., determined by a finite element analysis is, however, more common, Figure 6. Note, that such a stress profile refers to the component without crack. In the present case the loading type was predominant bending which would have allowed for the application of a simple analytical model for determining the bending stress. However, in order to consider the membrane stress component too, a finite element analysis was carried out, which yielded the stress profile shown in Figure 7 which was characterized by the stress values σ_1 and σ_2 at the front and back surfaces of the plate. Based on these information a membrane stress component σ_b and a membrane stress component σ_m of

$$\sigma_b = 209 \text{ MPa}; \quad \sigma_m = 2 \text{ MPa}$$

were determined. These values refer to one half of the nominal applied load of 35 kN, the fork lift was designed for.

PROCEDURES FOR STRUCTURE INTEGRITY ASSESSMENT 95

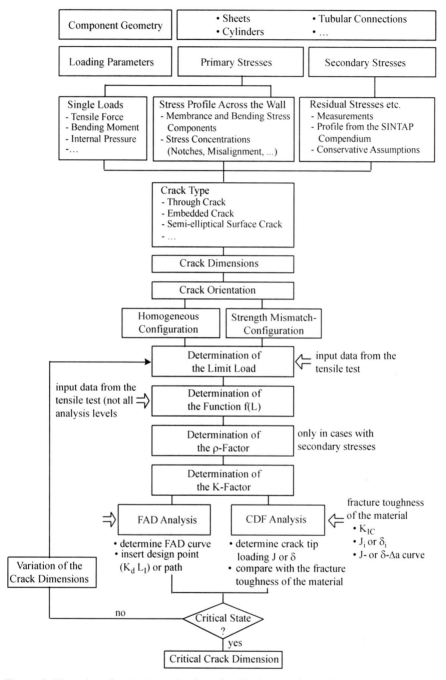

Figure 3. Flaw chart for the determination of critical crack sizes using the European flow assessment procedure SINTAP.

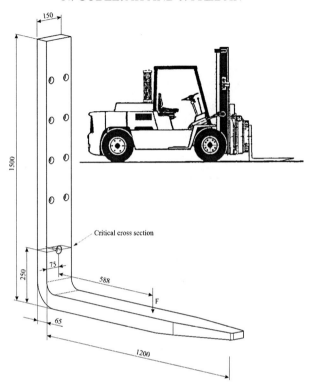

Figure 4. Geometry and dimensions (all measures in mm) of the broken fork lift.

In the present case only primary stresses had to be considered. In practice there are, however, many cases – for particular weldments – where these have to be complimented by secondary stresses. In general primary stresses arise from mechanical applied loads including the weight of the structure whereas secondary stresses are due to suspended stresses. Typical examples of secondary stresses are welding residual stresses. Secondary stresses are of minor significance for common strength analyses because they are self equilibrating across the section. This is, however, no longer true when the same cross section contains a crack. In such a case secondary stresses can be a major loading component, which has to be considered in any analysis. Secondary stresses are taken into account in determining the K factor but not in determining the limit load, F_Y, or the degree of ligament plasticity, L_r.

1. Linear-elastic deformation behavior

For linear-elastic deformation behavior the crack tip loading can simply be determined by superposition the K factor due to primary and the K factor due to secondary stress, provided the crack opening mode are identical:

$$K_I = K_I^P + K_I^S \qquad (1)$$

2. Elastic-plastic deformation behavior

In the general case the assessment is more sophisticated because interaction effects between the primary and secondary stresses must be accounted for. Secondary stresses do not yield plastic collapse, however they may well contribute to plastic deformation. If they reach yield strength magnitude the resultant crack tip loading is larger than $K_I^P + K_I^S$. On the other hand, the secondary stresses may be partly relived due to relaxation effects combined with ligament yielding. The interaction effect is modeled by a correction term ρ, which in the FAD approach is defined as

$$K_r = \frac{K_I^P + K_I^S}{K_{mat}} + \rho \qquad (2)$$

and in the CDF route as,

$$J = \frac{1}{E'} \times \left[\frac{K_I^P + K_I^S}{f(L_r) - \rho}\right]^2 \text{ or } \delta = \frac{1}{E' \cdot \sigma_Y} \times \left[\frac{K_I^P + K_I^S}{f(L_r) - \rho}\right]^2 \qquad (3,4)$$

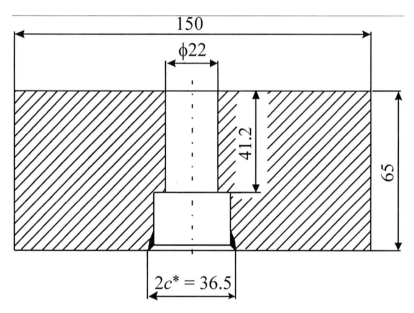

Figure 5. Geometry and dimensions (all measures in millimetre) of the cross section where fracture occured.

The quantity ρ characterizes the difference between the actual crack tip loading and the crack tip loading which would result from simple superposition of K_I^P and K_I^S.

By using

$$L_r = \frac{\sigma_{ref}^p}{\sigma_Y} \tag{5}$$

it is dependent on the ligament of plasticity, L_r (which is a function on primary loading, on the magnitude of the secondary stresses, and on the equation applied for $f(L_r)$.

Therefore, the correction term ρ is possible to determine from plot in Figure 8.

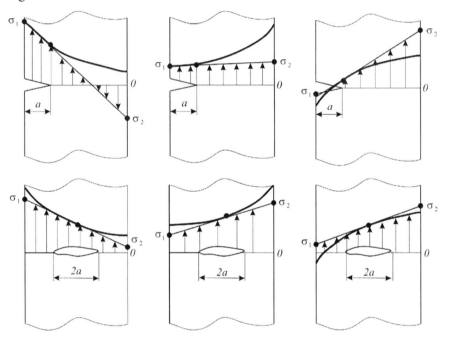

Membrane stress component:

$$\sigma_m = \frac{1}{2}(\sigma_1 + \sigma_2)$$

Bending stress component:

$$\sigma_b = \frac{1}{2}(\sigma_1 - \sigma_2)$$

Figure 6. Membrane and stress profile through the thickness.

Because secondary stresses are of no relevance in the context of this paper they shall only be mentioned without going in details. Note, that the SINTAP procedure gives elaborate guidance on the treatment of secondary stresses.

(C) CRACK TYPE AND ORIENTATION

In a fracture mechanics analysis it is distinguished between through cracks, embedded cracks and surface cracks. Real crack shapes are idealized by substitute geometries such as rectangles ellipses and semi-ellipses. The idealization has to been done such that the crack tip loading will be overestimated. Sometimes a crack or conglomerations of cracks have to be re-characterized if they interfere one with each other or with a free surface. Real, irregular cracks are modeled by "ideal" straight, elliptical or semi-elliptical cracks the dimensions of which are defined by their containment rectangles, as demonstrated in Figure 9. Most important is that the idealized flaws yield conservative results of the FE analyses as compared to the original crack. Conglomerations of multiple flaws may interact. If multiple cracks are located adjacent on the same cross section they will be more severe than the single cracks would be. This is taken into account by interaction criteria. If the spacing between single cracks is below a certain value these cracks have to be replaced by a larger crack such as if they already have coincided, as demonstrated in Figure 10. In a similar way interaction effect between cracks and free surfaces are treated.

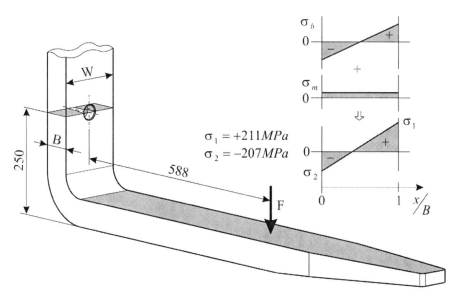

Figure 7. Stress profile across the fork section containing the crack (all measures in millimetre).

In the present case of fork the two edge cracks have been substituted by one through crack the dimensions of which include the hole diameter as demonstrated in Figure 11. For simplicity the crack was assumed to be of constant length 2c over the wall thickness.

Usually the crack plane is assumed to be perpendicular to the larger of the two principle stresses. There are, however cases, where a real crack will not growth within these plane because of mechanical reasons, i.e, both principal stresses are of a magnitude in the same order, or because of heterogeneity of the material. In such cases a more complicated mixed-mode analysis has to be carried out. In the present case the situation is quite simple because the maximum principle stress direction is identical to the axis of the fork.

In the flow chart in Figure 3 the crack dimensions are introduced as input information. Actually this refers to a default crack size, which is then varied iteratively. At each iteration step the question has to be answered whether the actual crack size is already critical or not.

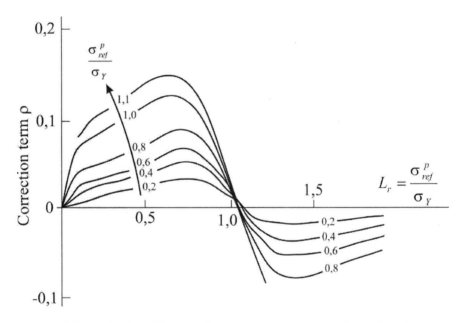

Figure 8. Determination of the correction term ρ on the treatment of secondary stresses.

(D) HOMOGENOUS OR STRENGTH MISMATCHED CONFIGURATION

Strength mismatch means that, e.g. in a weldment, the base plate and the weld material are of different strength with the consequence of local strain concentrations within the softer area. If the yield strength of weld metal differing by more than 10% from that of the base material; if this difference is less than 10%, the use of SINTAP procedure for homogeneous material based on the base plate properties is recommended. In any another case the strength mismatched configuration should be take in account. The weld metal is commonly produced with yield strength σ_{YW}, greater than that of the base plate σ_{YB}; (see Figure 12b) this case is designated as OVERMATCHING (OM) with the mis-match factor M.

$$M = \frac{\sigma_{YW}}{\sigma_{YB}} > 1 \qquad (6)$$

UNDERMATCHING (UM) (see Figure 12a) is defined by

$$M < 1 \qquad (7)$$

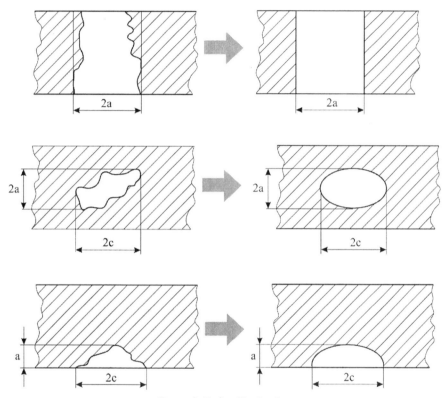

Figure 9. Defect idealisation.

The mechanical consequences of mismatch are obvious from Figure 13.

Overmatching reduces the strain in the weld metal as compared to the base plate, thus leading to a shielding of a defect in the weld metal. Undermatching gives rise to a strain concentration in the weld metal.

There are, however, many cases where strength mismatch is of paramount interest. The mismatch play important role:

- To fracture toughness of material (weld metal and base metal), (K_{mat}, J_{mat}, $CTOD_{mat}$)

- To stress intensity factor solution (K_I) in linear-elastic and elastic-plastic deformation behavior
- To limit load solution given by appropriate terms (F_Y, p_Y, σ_{ref}, etc.)

Therefore, the SINTAP procedure offers separate assessment options for the analysis of such cases.

For the present example mismatch does not play any role.

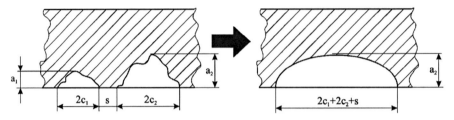

Interaction citerion ($c_1 < c_2$): $s \leq 2c_1$ for a_1/c_1 or $a_2/c_2 > 1$

                                  $s = 0$ for a_1/c_1 and $a_2/c_2 \leq 1$

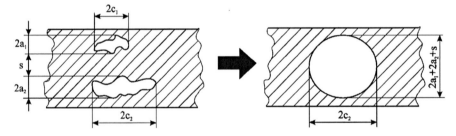

Interaction criterion: $s \leq (a_1 + a_2)$

Figure 10. Defect idalisation of multiple flaws.

(E) PLASTIC LIMIT LOAD F_Y

The plastic limit load of the component with crack is one of the key parameters of the SINTAP analysis. Here some remarks are due. In solid mechanics the limit load is usually determined for ideally plastic materials. When the limit load is reached the deformation becomes unbounded over the cross section. Real materials, however, work harden with the consequence that the applied force can increase beyond the value given by the non-hardening limit load. Therefore, in the frame of a fracture mechanics analysis it has to be distinguished between a plastic collapse load which is identical to the maximum load which the structure with crack can sustain and a net section yield load which refers roughly to that load at which the still unbroken ligament ahead of the crack is first fully plastic and the local load-deformation curve becomes non-linearly. This parameter is what is designated above as the plastic limit load F_Y. In practice it is usually determined under the assumption of an ideally plastic material inserting the yield strength as the maximum sustainable stress. This is supposed to represent the attainment of net section yielding, i.e. each point in the ligament ahead of the crack is supposed to have just reached the yield condition. This is correct for ideally plastic material, however, for hardening materials there are some points, which are still under elastic deformation condition. Therefore, the thus determined value of F_Y represents a lower bound to the real yield load of component's materials.

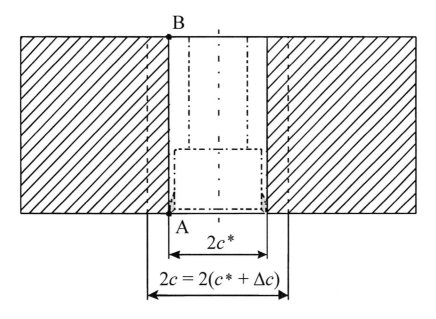

Figure 11. Definition of the idealized crack dimension 2c.

Within the SINTAP procedure a compendium of limit loads is provided. Other compilations are available in the literature.[11] For cases, which are not covered by this compilation conservative estimates are possible based on substitute geometries. In such cases the stress profiles in the components without crack are taken as input information.

Within the SINTAP procedure a loading parameter L_r is used which is defined as the ratio of the applied load F and the limit load F_Y or respectively as the ratio of an applied net section stress σ_{ref} and the yield strength of the material, σ_Y:

$$L_r = \frac{F}{F_Y} = \frac{\sigma_{ref}}{\sigma_Y} \qquad (8)$$

the latter being given as $\sigma_Y = R_{eL}$ for materials with and $\sigma_Y = R_{p0.2}$ for materials without a Lüders plateau.

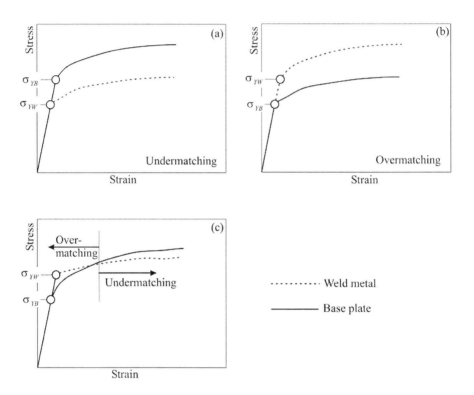

Figure 12. Definition of strength mismatched configuration.

The reference stress of the plate geometry considered within this paper can easily be determined as

$$\sigma_{ref} = \frac{1}{1-(2c/W)} \left\{ \frac{\sigma_b}{3} + \sqrt{\frac{\sigma_b^2}{9} + \sigma_m^2} \right\} \qquad (9)$$

(F) STRESS INTENSITY FACTOR (K FACTOR)

As in the case of the limit load numerous solutions for K factors are available in compendia.[2,3] The SINTAP procedure provides an own compilation of such solutions. Stress intensity factors can be determined for single loads such as forces, bending moments, internal pressures etc. as well as for stress profiles. The latter alternative allows to handle geometrically complex components by using substitute structures, i.e., the stress profile is determined for the real structure without crack whereas the determination of the K-factor is based on a simpler geometry like a plate, a cylinder etc. The K factor for the fork in the present paper was determined by

$$K_I(c,F) = \sqrt{\pi c}(\sigma_m \cdot f_m + \sigma_b \cdot f_b) \qquad (10)$$

with f_m and f_b are shape functions being defined for a plate with a through crack. They are $f_m^A = 1$ and $f_b^A = 1$ for point A and $f_m^B = 1$ and $f_b^B = -1$ for point B (Figure 11).

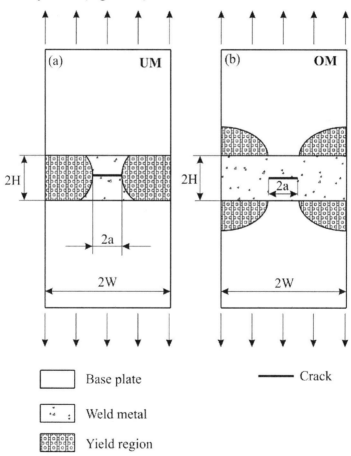

a) **Undermatching** (UM) gives rise to a strain concentration in the weld metal
b) **Overmatching** (OM) reduces the strain in teh weld metal as compared to the base plate

Figure 13. Strength mismatch caused different spreadof the plastic region.

(G) CORRECTION FUNCTION $F(L_R)$

Under conditions of small scale yielding (roughly up to 0.6 times the limit load) a fracture mechanics analysis can be based on the linear-elastic K factor. This is, however, not possible for contained and net section yielding where the plastic zone is no more limited to a small region ahead of the crack tip. Under this condition any application of the K concept would lead to a significant underestimation of the real crack tip loading in terms of the J-integral or CTOD. Irrespective of this general statement the application of a formal K concept becomes possible when the linear-elastic K factor is corrected with respect of the yield effect. This is the essential of the correction function $f(L_r)$. With respect of $f(L_r)$ the SINTAP procedure is structured in a hierarchic manner consisting of various analysis levels constituted by the quality and completeness of the required input information. Higher levels are more advanced than lower levels: they need more complex input information but the user is "rewarded" by less conservative results. An unacceptable result provides a motivation for repeating the analysis at the next higher level rather than claiming the component to be unsafe. The SINTAP standard analysis levels are:

Basic Level	Only the toughness and the yield strength and the ultimate tensile strength of the material need to be known. Different sets of equations are offered for materials with and without Lüders plateau.
Mismatch Level	This is a modification of the Basic Level for inhomogeneous configurations such as strength-mismatched weldments.
Advanced or Stress–Strain Level	This requires toughness data and the complete stress–strain curve of the material. Both, homogenous and strength mismatched components can be treated.

There are additional levels:

Default Level	Only the yield strength of the material is required. The fracture resistance of the material can be conservatively estimated from Charpy data.
Constraint Level	Within this level, the effect of loss of constraint in thin sections or predominately tensile loading on fracture resistance is considered.
J-Integral Analysis Level	This level includes a complete numerical analysis of the defect structure.

Within the present paper the Default Level, the Basic Level and the Advanced Level are applied. The according equations for $f(L_r)$ for ferritic steels without Lüders plateau are:

- **Default Level**

$$f(L_r) = \left[1 + \frac{1}{2} L_r^2\right]^{-1/2} \cdot \left[0.3 + 0.7 \cdot \exp(-0.6 \cdot L_r^6)\right] \quad \text{for } 0 \leq L_r \leq L_r^{max} \quad (11)$$

$$L_r^{max} = 1 + \left[\frac{150}{\sigma_Y}\right]^{2.5} \qquad \sigma_Y = R_{p0,2} \quad in \quad [MPa] \quad (12)$$

The fracture toughness is estimated in a conservative manner from Charpy data by

$$K_{mat} = \left[(12 \cdot \sqrt{KV} - 20) \cdot \left(\frac{25}{B}\right)^{1/4}\right] + 20 \quad in \quad [MPa \cdot \sqrt{m}] \quad (13)$$

on the lower shelf and by

$$K_{mat} = K_{J0,2} = \sqrt{\frac{E(0.53 \cdot KV^{1.28}) \cdot 0.2^{(0.133 \cdot KV^{0.256})}}{1000 \cdot (1 - \nu^2)}} \quad in \quad [MPa \cdot \sqrt{m}] \quad (14)$$

on the upper shelf.

K_{mat} – relevant fracture toughness of material .... [MPa√m]
B – specimen thickness .... [mm]
KV – Charpy energy .... [J]

In addition SINTAP offers a correlation for the ductile-to-brittle transition based on the Charpy transition temperature for 28 J.

In the present analysis Eq. (13) was applied for estimating fracture toughness from the Charpy energy.

- **Basic Level**

$$f(L_r) = \left[1 + \frac{1}{2} L_r^2\right]^{-1/2} \cdot \left[0.3 + 0.7 \cdot \exp(-\mu \cdot L_r^6)\right] \quad \text{for } 0 \leq L_r \leq L_r^{max} \quad (15)$$

with

$$\mu = \min \begin{cases} 0.001 \cdot E / \sigma_Y \\ 0.6 \end{cases} \quad (16)$$

and

$$f(L_r) = f(L_r = 1) \cdot L_r^{(N-1)/(2N)} \quad \text{for } 1 \leq L_r \leq L_r^{\max} \tag{17}$$

$$N = 0.3\left[1 - \frac{\sigma_Y}{R_m}\right] \quad \sigma_Y = R_{p0,2} \quad \text{in } [MPa] \tag{18}$$

$$L_r^{\max} = \frac{1}{2}\left[\frac{\sigma_Y + R_m}{\sigma_Y}\right] \tag{19}$$

with $L_r^{\max}$ being the limit against plastic collapse.

- **Advanced Level**

$$f(L_r) = \left[\frac{E \cdot \varepsilon_{ref}}{\sigma_{ref}} + \frac{1}{2}\frac{L_r^2}{\left(\frac{E \cdot \varepsilon_{ref}}{\sigma_{ref}}\right)}\right]^{-1/2} \quad \text{for } 0 \leq L_r \leq L_r^{\max} \tag{20}$$

$$L_r^{\max} = \frac{1}{2}\left[\frac{\sigma_Y + R_m}{\sigma_Y}\right]. \tag{21}$$

Different to the Levels above f(L$_r$) is a continuous function, which follows point-wise the true stress–strain curve. Each value of σ_{ref} is assigned to an L$_r$ value by

$$\sigma_{ref} = L_r \sigma_Y \tag{22}$$

The corresponding reference strain ε_{ref} is obtained from the true stress–strain curve as illustrated in Figure 14. No distinction is necessary between materials with and without a Lüders plateau. On the other hand $\sigma_{ref}/\varepsilon_{ref}$ values have to be available at L_r = 0.7/0.9/0.98/1/1.02/1.1 and other values of L_r.

(H) THE TRUE STRESS–STRAIN CURVE OF THE MATERIAL

The engineering stress–strain curve of the material is shown in Figure 15. Five tests were carried out but only the lowest curve was used for the SINTAP analysis. The mechanical properties derived from these curve are summarized in Table 1. With σ and ε designating the engineering stress and strain, the true stress and strain values, σ_t and ε_t are determined by

$$\varepsilon_t = \ln(1+\varepsilon) \quad \text{and} \quad \sigma_t = \sigma(1+\varepsilon) \tag{23}$$

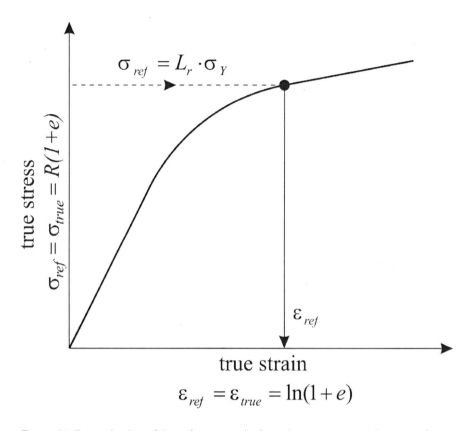

Figure 14. Determination of the reference strain from the true stress–strain curve of ε_{true} for the assessment of the normalized load.

TABLE 1. Mechanical properties obtained by tensile test and Charpy impact toughness values Cv.

E (GPa)	ν (−)	$R_{p0,2}$ (MPa)	R_m (MPa)	A_g (%)	A_t (%)	Z (%)	N (−)
2.1	0.3	446	720	6.89	14.95	53.60	0.176
		448	735	8.89	18.19	53.77	0.192
		578	754	7.52	20.95	59.05	0.125
		474	764	9.72	19.48	55.81	0.187
		440	716	6.74	14.45	52.59	0.195

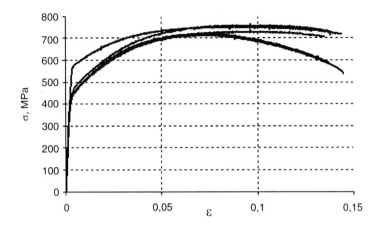

Figure 15. Engineering stress–strain curves of the fork material.

(I) CDF VERSUS FAD ANALYSES

An important feature of the procedure is that the analyses can alternatively be based on a Failure Assessment Diagram (FAD) or on a Crack Driving Force (CDF) philosophy. Applying the FAD philosophy a failure line is constructed by normalising the crack tip loading by the material's fracture resistance. The assessment of the component is then based on the relative location of a geometry dependent assessment point with respect to this failure line. In the simplest application the component is regarded as safe as long as the assessment point lies within the area enclosed by the failure line. It is potentially unsafe if it is located on or above the failure line. In contrast to this in the CDF route the crack tip loading in the component is determined in a separate step. It is then compared with the fracture resistance of the material. If the crack tip loading is less than the fracture resistance the component is safe, otherwise it is potentially unsafe.

The basic equations of the FAD and CDF routes are set out in Sections 2.1 and 2.2.

2.1. FAD ROUTE

In the FAD route (Figure 16) a failure assessment curve (FAC), K_r versus L_r, is described by the equation

$$K_r = f(L_r) \tag{24}$$

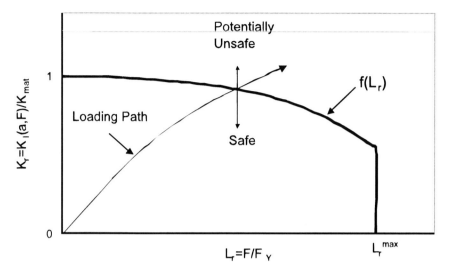

Figure 16. Failure assessment based on a FAD philosophy.

To assess for crack initiation and growth, two parameters need to be calculated. The first one K_r, is defined by

$$K_r = \frac{K_I(a,F)}{K_{mat}} \qquad (25)$$

where $K_I(a,F)$ is the linear-elastic stress intensity factor of the defective component and K_{mat} is fracture toughness.

The second parameter L_r is defined by:

$$L_r = \frac{F}{F_Y} \qquad (26)$$

where F_Y is the yield load of the cracked configuration.

2.2. CDF ROUTE

In the CDF route (Figure 17), an applied parameter such as the J-integral or crack tip opening displacement (CTOD = δ) is determined, which characterises the stresses and strains ahead of the crack tip in a specimen or component:

$$J = J_e [f(L_r)]^{-2} \text{ or } \delta = \delta_e [f(L_r)]^{-2} \qquad (27, 28)$$

where J_e and δ_e are the elastic values of the crack tip parameters which can be deduced from the stress intensity factor $K_I(a,F)$ as

$$J_e = \frac{[K_I(a,F)]^2}{E'} \quad \text{or} \quad \delta_e = \frac{[K_I(a,F)]^2}{E' \cdot \sigma_Y} \quad (29, 30)$$

with E' being Young's modulus E in plane stress and $E/(1-v^2)$ in plane strain. The quantity v is Poisson's ratio.

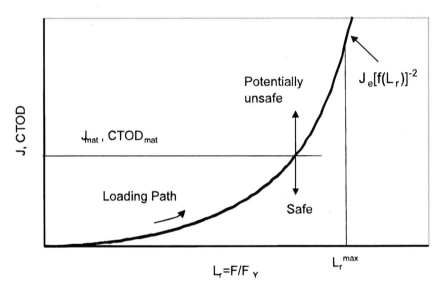

Figure 17. Failure assessment based on the CDF philosophy (Note: The function $f(L_r)$ is identical for the FAD and CDF routes).

In Eq. (2) the fracture resistance of the material is used in terms of the K factor, K_{mat}.

This quantity is obtained formally from the J-integral or CTOD by

$$K_{mat} = \sqrt{\frac{J_{mat} \cdot E}{(1-v^2)}} = \sqrt{\frac{\sigma_y \cdot \delta_{mat} \cdot E}{(1-v^2)}} \quad (31)$$

The SINTAP procedure includes different analysis levels. The main differences between these levels is the function $f(L_r)$. It is defined such that the lower levels can be applied even with relatively poor input information.

Due to this the output is more conservative as compared with the more advanced levels which require more detailed input information but "reward" the user with more realistic results. The Default Level is the lowest level of the SINTAP procedure. Its use is recommended only if no other data than the yield strength of the material and Charpy data are available.

The $f(L_r)$ function, which is the same for both the FAD and CDF routes, is given by:

- In cases where the material exhibits a Lüders strain

$$f(L_r) = \left(1 + \frac{1}{2}L_r^2\right)^{-1/2} \quad \text{for} \quad 0 \le L_r \le L_r^{max} \tag{32}$$

The cut-off L_r^{max} is defined in a conservative manner as

$$L_r^{max} = 1 \tag{33}$$

- If the material *does not exhibit a Lüders strain* the failure assessment curve $f(L_r)$ is described by the equation

$$f(L_r) = \left(1 + \frac{1}{2}L_r^2\right)^{-1/2} \times \left[0.3 + 0.7\exp(-0.6 \cdot L_r^6)\right] \quad \text{for} \quad 0 \le L_r \le L_r^{max} \tag{34}$$

The cut-off L_r^{max} is defined slightly less conservatively as

$$L_r^{max} = 1 + \left(\frac{150}{R_{p0.2}}\right)^{2.5} \tag{35}$$

(J) FRACTURE TOUGHNESS

The fracture toughness was determined in terms of the crack tip opening displacement CTOD (δ) according to the British standard BS 7448, Part 1.[10] Four tests were carried out using three-point bend specimens at room temperature. The test setup is shown in Figure 18; a typical test record in Figure 19. It shows typical pop-in behaviour. Pop-ins are cleavage fracture events disrupting the ductile tearing process. The crack is arrested subsequent to each pop-in. Note, however, that the specimen is subjected to displacement control in the test machine whereas in reality load control might occur. Usually the crack would not be arrested in such a case but cause failure. Therefore, no benefit can be taken from the crack arrest following a pop-in which was specified as such by the test standard.

For the SINTAP analyses the lowest of the pop-in fracture toughness values was chosen. This was δ_c = 0.02 mm or corresponding K_{mat} = 49.7 MPa$\sqrt{m}$.

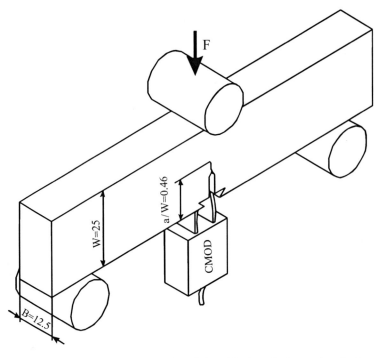

Figure 18. Test setup for the determination of the critical crack tip opening displacement.

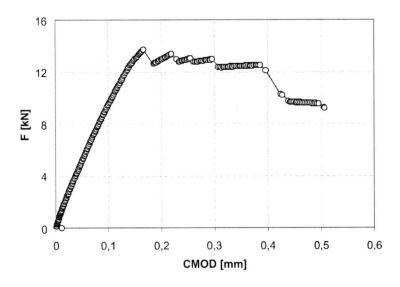

Figure 19. Typical test record of a CTOD test.

(K) CHARPY ENERGY

Information on the Charpy energy is necessary for the SINTAP Default Level assessment. Nine tests have been carried out at three temperatures. The result is summarised in Table 2. The SINTAP analysis was based on the minimum room temperature Charpy energy of 6 J.

TABLE 2. Charpy energy of the fork material at different temperatures.

Charpy impact toughness J/80 mm^2		
+10°C	+20°C	+50°C
6, 6, 6,	7, 6, 7	9, 8, 9

3. Failure analysis of the component, results and discussion

In Figure 20 and 21 the CDF and FAD analyses are demonstrated for a crack size of 2c = 45.5 mm. That crack length corresponds to real crack length measured at the fracture surface of the broken fork, as shown in Figure 22. It is shown that the higher analysis levels yield less conservative results. The Default Level, which uses fracture toughness values estimated from Charpy energy give much lower critical loads than the higher levels.

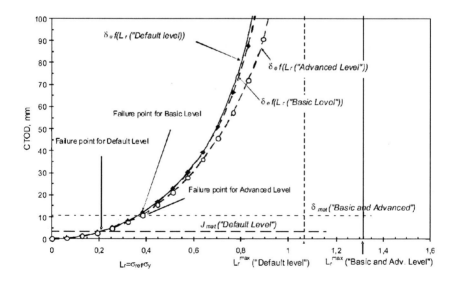

Figure 20. CDF analysis of the fork assuming a crack width of 2c = 45.5 mm. Failure is predicted for an applied load of 15 kN (Default Level), 27.38 kN (Basic Level) and 29.9 kN (Advanced Level).

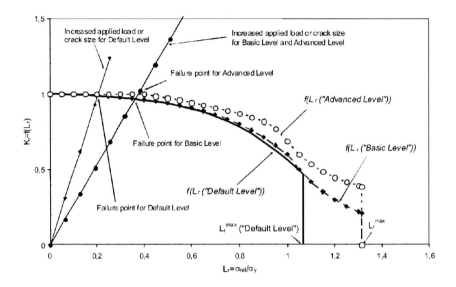

Figure 21. FAD analysis of the fork assuming a crack width of 2c = 45.5 mm. The predicted failure loads are identical to those obtained by the CDF analysis in Figure 20.

Figure 22. Fracture surface of the broken fork. Failure originated at two edge cracks left and right from the hole at the top side.

According to Figure 3 the analysis was repeated for stepwise increased crack sizes 2c. The critical crack size was than determined as the value of 2c that caused failure at half the nominal applied load the forklift was designed for. The bisection was necessary because the forklift contained two forks.

As the final result the critical crack size was determined to be

- 2c = 10.35 mm (Default Level analysis)
- 2c = 33.2 mm (Basic Level analysis)
- 2c = 35.6 mm (Advanced Level analysis)

Compared to the real overall surface dimension of the edge cracks at failure of 45.5 mm (Figure 2) the predictions were conservative by

- 77.28% (Default Level analysis)
- 27.03% (Basic Level analysis)
- 21.75% (Advanced Level analysis),

which is not so much at the highest level because critical crack sizes are used to be very sensitive with respect to the input information. At the highest level the conservatism was mostly due to the simplified crack model used as the substitute geometry (Figure 11).

As the consequence it can be concluded that the failure occurred as the consequence of inadequate design and not of inadmissible handling such as overloading. The failure could have been avoided by applying fracture mechanics in the design stage. The SINTAP algorithm was shown to be an easy but suitable tool for this purpose.

4. Conclusions

The basic principle of the recently developed European SINTAP procedure has been reviewed. The SINTAP procedure has been developed as a contribution towards the development of a European Committee for Standardization (CEN) fitness for service standard. In the paper the SINTAP procedure was applied to the failure analysis of cracked component. The procedure was applied to predict the critical size of cracks on both sides of a bore hole in a fork of a forklift. Assuming loading by the design load an overall critical crack size (including both cracks and the hole) of 35.6 mm was predicted at the highest analysis level whereas the real fork broke in service after the overall crack size had reached a length of 45.5 mm. By this result it was shown that inadequate design could give a sufficient account of the failure without any need to imply further reasons such as inadmissible handling of the forklift in service. By results of used example was shown that SINTAP procedure gives reliable conservative results where the conservatism (e.g. safe factor) decreasing by increasing the quality of input data.

References

1. SINTAP: Structural Integrity Assessment Procedure. Final Report. EU-Project BE 95-1462. Brite Euram Programme, Brussels, 1999.
2. H. Tada, P.C. Paris and G.R. Irwin, The Stress Analysis of Cracks Handbook. ASME Press, New York, 1998.
3. Murakami Y., Hasebe N., Itoh Y., Kishimoto K., Miyata H., Miyazaki N., Noda N., Sakae C., Shindo H., Tohgo K.: Stress Intensity Factor Handbook. Vols 1 and 2, 1987, Vol. 3, 1992, Pergamon Press, Oxford.
4. Schwalbe K-H, Kim Y-J, Hao S, Cornec A, Koçak M., EFAM, ETM-MM96: the ETM method for assessing significance of crack-like defects in joints with mechanical heterogeneity (strength mismatch), GKSS Report 97/E/9, 1997.
5. Schwalbe K-H, Zerbst U, Kim Y-J, Brocks W, Cornec A, Heerens J, Amstutz H.: EFAM ETM 97: the ETM method for assessing crack-like defects in engineering structures. GKSS Report GKSS 98/E/6, 1988.
6. Kumar V, German MD, Shih CF. An engineering approach for elastic-plastic fracture analysis. EPRI-Report NP-1931, EPRI, Palo Alto, CA, 1981.
7. SINTAP Special Issue of Eng. Fract. Mech. 67 (6), 476–668, Dec. 2000.
8. Ainsworth, R.A, Bannister, A.C, Zerbst, U. (2001): An overview of the European flaw assessment procedure SINTAP and its validation. Int. J. Pres. Ves. Piping 77, 869–876.
9. Zerbst U, Hamann R, Wohlschlegel A. (2000): Application of the European flaw assessment procedure SINTAP to pipes. Int. J. Pres. Ves. Piping 77, 697–702.
10. BS 7448: Part 2: 1997: Fracture mechanics toughness tests, Part 2. Method for determination of K_{Ic}, critical CTOD and critical J values of welds in metallic materials, British standards institution, London, 1997.
11. A, G. Miller, Review of limit loads of structures containing defects. Int. J. Pres. Ves. Piping, 1988, 32, 197–327.

EXPERIENCE IN NON DESTRUCTIVE TESTING OF PROCESS EQUIPMENT

JANO KURAI
Petrohemija, Spoljnostarčevačka 82, 23000 Pančevo

MIODRAG KIRIĆ*
The Innovation Center of the Faculty of Mechanical Engineering, Kraljice Marije 16, 11120 Belgrade

Abstract The paper gives an overview of common non-destructive methods used in-service and during shut-down of the plant. Their choice is dependent on the aim of the testing, component of process equipment and material. Ultrasonic testing is used as the method for materials evaluation, too.

Keywords: Ultrasonic, attenuation, storage tank, LPG, pipeline, crack, lamination

1. Introduction

Various non-destructive methods can be used to characterize the condition of material in operating equipment. They are especially useful for quantifying flaws. The information on flaw type, size, and location then can be used to evaluate equipment integrity and remaining life.

Managing equipment integrity is essential to the safe, reliable operation of process plant equipment. Using the initial data on the plant equipment, the current condition of equipment and remaining life must be evaluated to perform the risk assessment of the plant. By determining e.g. stress state of the material, and/or other methods, one evaluates structural components to determine if they are fit to continue operation for some desired future period. This approach, known as the fitness-for-service assessment, may involve fracture mechanics if the component of interest contains flaws or other damages, which are non-allowable for quality control standards.

---

* E-mail: mkiric@mas.bg.ac.yu

In this paper examples are given of the methodology used by the author organizations and techniques developed over the past years to achieve the above goals. Examples are given of non-destructive testing (NDT) methods and techniques used in the examination of storage tanks, pipelines, tubes in heat transfer or processing equipment, techniques needed for implementation, and discussed some problems in performing NDT.

2. The Directive 97/23 EC and the classification of pressure vessels

Directive 97/23 EC (PED)[1] applies to the design, manufacture and conformity assessment of pressure equipment with a maximum allowable pressure greater than 0.5 bar. Large number of examined pressure vessels can be classified by category in accordance with Annex II of PED on the basis of their maximum allowable pressure, volume V or their nominal size DN and the group of fluids for which they are intended. Among others, most frequent fluids in process industry are liquified petrol gas (LPG), hydrogen and nitrogen.

According to the Article 9 of the PED, pressure vessels are classified according to ascending level of hazard. For the purposes of such classification fluids are divided into two groups. The Group 1 comprises dangerous fluids, explosive, extremely flammable, highly flammable, flammable, very toxic, toxic and oxidizing. The importance of NDT of pressure vessels for the Group 1 fluids is obvious. According to PED non-destructive tests of permanent joints is carried out by suitable qualified personnel. Final assessment of pressure equipment must include a test for the pressure containment aspect, which will normally take the form of a hydrostatic pressure test (HT). Where the HT is harmful or impractical, PED allows other tests of a recognized value may be carried out. For tests other than HT, additional measures, such as NDT or other methods of equivalent validity, must be applied before those tests are carried out.

European law distinguishes clearly between the law of dangerous goods and the law of hazardous materials. The first refers primarily to the transport of the respective goods including the interim storage, if caused by the transport. The latter describes the requirements of storage (including warehousing) and usage of hazardous materials.

LPG, being flammable, belongs to the Group 1 of dangerous fluids. Large amounts of LPG can be stored in bulk tanks. LPG burns cleanly with no soot and very few sulfur emissions. It has a typical specific calorific value of 46.1 MJ/kg compared to 42.5 MJ/kg for diesel and 43.5 MJ/kg for premium grade petrol (gasoline). Large, spherical LPG containers may have up to a 15 cm steel wall thickness. The relief valve on the top is designed to vent off excess pressure in order to prevent the rupture of the tank itself.

Due to the increased threat of terrorism in the early twenty-first century, funding for greater hazardous materials-handling capabilities was increased throughout the world, in recognition of the fact that flammable, poisonous, explosive, or radioactive substances in particular could make attractive weapons for terrorist attacks.

3. An overview of our NDT experiences

Many in-service and also pre-service inspection findings on pressure vessels indicate a basis for consideration of continuing reliability and environment safety, especially when an extending usage is required. It is well known that a crack can lead to a critical failure with a tremendous destroying capacity for a large vessel, and heavy consequences to humans, property and biological environment. Many examples of catastrophic failures over the world may be quoted.

An optimistic approach to high strength steels (HSS) application for pressure vessels since 1960s led to some degree of disappointment later. Poor weldability has caused fabrication problems, mostly in form of cold cracking, and even more problems has appeared during service, when welded joints have been exposed to corrosive environment, causing stress corrosion cracking.

In Serbia a large experience on HSS cracking was gained in last four decades, especially on sphere reservoirs. The majority of vessels were made of Nb, V micro-alloyed steel of domestic and foreign production. In many spherical tanks in-service inspection revealed cracks. Number of revealed cracks was greater than ten per vessel. The most frequent cracks were surface cracks on pressure retaining surface and internal cracks inside the material. Removal of cracks has been done by grinding and only when cracks were deep, arc-air gouging and grinding was applied. The repairs of deeper cracks by welding in prepared grooves were followed by re-inspection, hydrostatic testing and new inspection. In some cases cracks were not detected by NDT during fabrication because magnetic particle testing or flux leakage (MT) was not performed. The originally used radiography testing (RT) and ultrasonic testing (UT) could not detect surface cracks which were less than 3 mm deep. The next problem was in fact that NDT after proof hydro test (HT) was not prescribed, although indications of new cracks, found after HT, were obtained.[2]

After the first experience with the leakage failures in 1982, the requirements of in-service inspection were more severe.[3] All HSS vessels were submitted to 100% UT and MT. This level of NDT has been used after weld repair and HT, too. Magnetic particle testing with the alternating current (AC) is considered as the most sensitive, but the opening of a crack is not a measure of its depth in wall thickness direction. Surface cracks deeper than

3 mm were also detected by ultrasonics. The wet fluorescent MT, using fluorescent wet particles, AC and black light, gave the best results for detection of surface cracks. Dye penetrant was used rarely. Radiography was used only for detection of volume discontinuities (defects) revealed by ultrasonics, as an additional method for their characterization. At the middle of seventies, a large number of such discontinuities was revealed in the Trans-Alaska oil pipeline girth welds, and the only alternative for the assessment of their depth was to apply comparative radiographic density measurements. The necessary assumption that these defects are to be considered as sharp cracks of the same depth, led to conservative analysis and to significant lowering the savings of applying the fitness-for-purpose approach.[4]

Acoustic emission (AE) monitoring during HT was used to detect possible crack propagation. If a crack was located by AE, this location was further examined by UT and MT. Most measurements by using strain gauges during HT found acceptable values. Replica prints, taken from the pressure retaining surface after polishing were useful for the analysis of microstructure and detection of new micro cracks, not detectable by MT. The experiences shown that manual UT is the best method for discontinuity characterization because of its versatility and possibility of quantification and storage of results enabled by digital ultrasonic flaw detectors.[5–7]

A number of laboratories for NDT in Serbia, included in process plant maintenance, is not large; those are for instance The Institute for welding, Belgrade, HIP Fertilizer factory, Pancevo, The Institute for materials, Belgrade, Zavarivac, Vranje and others. The attention is payed to the education and sertification of NDT personel in accordance with PED and EN 473. Corresponding bodies are CertLab, Pancevo, School for IBR of Vinca, Belgrade, The Institute for welding, Institute Kirilo Savic, Belgrade, and so on.

The terms defect, discontinuity and flaw are used in EN 1330-4.[8] while in EN 1714 the term imperfections (as well discontinuities) is used in its Clause 12.3, too.[9] Standards EN 1712–1714 also use the term indication, and ISO 5577:2000 in Clause 015 defines flaw and defect as 'discontinuity which is deemed to be recorded'.[10] The standard ASTM E164 – 08,[11] uses the term discontinuity. All of these terms are used here.

4. NDT methods and techniques

According to EN 12062 and EN 1330-4, mettalic materials and welded joints are examined for imperfections by following NDT methods: Visual testing (VT), UT, Liquid penetrant testing (PT), MT, AE and RT.

Each testing method has its features, thus the scope and range of products and materials to be tested. Not all of them are suitable to evaluate cracks, but many of them are very sensitive for crack detection, Figures 1–3.

VT, which usually precedes the other testing methods, is particularly important. For remote VT various boroscopes and endoscopes are used, Figure 1. Boroscopes (or borescopes) can be with a rigid tube or with flexible tube when they are called fiberscopes. Tools also used are magnifying glasses and mirrors.

Rigid borescopes are the ideal choice when straight-line access to the area of interest is available. These instruments use an optical lens system to transmit an image from the inspection area back to the eye and a non-coherent fibre bundle to illuminate the object. They are available in a range of diameters from 0.9 to 16 mm and working lengths up to 1.5 m. They are ease for use and are used for visual checks in hard-to-access places such as air conditioning ducts, ventilation systems, machines, motors and petrochemical plants.

Fiberscopes use flexible fiber bundle to transmit an image back to the inspector's eye. A separate fiber bundle is used for transferring the illumination from the light source to the inspection area. Fiberscopes can also be attached to closed-circuit television (CCTV) and digital imaging equipment so that movie or still images can be recorded for future reference, used in associated documentation or send by Internet to the user or examination laboratory for expertize.

PT has some similar features with MT, but it is applied also to non-ferromagnetic and non-metallic materials. It has less sensitivity in detecting fine cracks and it is restricted to discontinuities open to the surface such as cracks, seams, cold shuts, laminations, through leaks, or lack of fusion and is applicable to in-process, final and maintenance testing, as given in Figure 2.

Figure 1. A versatile scope for general inspection with rigid borescope, remote control and fiberscope (from left to right).

MT is used with dry or wet magnetic particles for detecting cracks and other discontinuities at or near the surface in ferromagnetic materials. MT is applied to raw material, half-finished material (billets, blooms, casting and forgings), finished material and welds, regardless of heat treatment or lack thereof. It is useful for preventive maintenance testing. After indications have been produced, they must be interpreted or classified and then evaluated. For this purpose a separate codes, specifications or specific agreements define the type, size, location, degree of alignment and spacing, area concentration and orientation of indications that are unacceptable in a specific part versus those which need not be removed before part acceptance. The following MT techniques are in use: dry magnetic powder, wet magnetic particle, magnetic slurry/paint magnetic particle and polymer magnetic particle.

Figure 2. Liquid penetrant testing of different pieces.

Typical magnetic particle indications are: surface discontinuities, which usually produce sharp, distinct patterns and near-surface discontinuities which produce less distinct indications than those open to the surface. The patterns are broad, rather than sharp and the particles are less tightly held. Fluorescent wet magnetic particles used are smaller than dry magnetic particles, thus wet method techniques are generally used to locate smaller cracks than by dry method. The illustration is given in Figure 3. It is used the region of green visible light because of greatest sensitivity of human eye to corresponding wavelengths.[12]

Figure 3. MT of products with different geometries.

AE is used for post-welding control of pressure vessels, or monitoring during continuous welding. The technique is developing and there are some obstacles in its routinely application on production welding. The technique is applicable to the detection and location of AE sources in weldments and in their heat-affecting zone during fabrication (like crack growth), particularly in those cases where the time duration of welding is such that fusion and solidification take place while welding is still in progress. The effectiveness of AE to detect discontinuities in the weldment and the heat-affected zone is dependent on the design of the AE system, the calibration procedure, the weld process and the material type. Materials that have been monitored include low-carbon steels, especially HSLA steels, low-alloy steels, stainless steels, some aluminum alloys and composites. The system performance was verified for each application by demonstrating that the discontinuities of concern can be detected with the desired reliability.

Eddy current testing (ET), UT or MT techniques are used to inspect welded, extruded, and seamless tubing. The choice depends on the non-conforming conditions one wishes to detect and the size, including wall thickness, and characteristics of the material. Carbon steels, stainless alloys, copper, aluminum, titanium and all other non ferrous metals can be inspected using these methods. For detecting typical defects such as small, short incomplete welds, inclusions, voids or cavities and some subsurface cracks in carbon steel or non ferrous tube, a standard eddy current instruments can be used with segment or encircling test coils. The choice between segment or encircling coils is usually determined by the wall thickness.

While an UT is generally the first choice for internal surface (ID) defects, some applications, particularly thin wall tubing, can be very successfully handled using eddy current instruments. In these cases, careful setup using phase and amplitude thresholds, and other selective circuits can give accurate separation between signals for ID and signals for outside surface (OD) defects. For detecting long, continuous surface defects such as seams and laps in tube, rotary probe systems are the most appropriate. By rotating multiple test probes around the tubing, even relatively short flaws can be reliably detected in many applications, without sacrificing throughput speed.

Inspecting for defects in heavy wall pipe and tube is best done using MT or ultrasonic techniques. For internal longitudinal and transverse flaws and ID defects, UT techniques are often the best choice. The testing is usually automated with rotary transducer assembly, with its unique rotating sealless water coupling system permits ultrasonic inspection at high throughput speeds. For those applications that require the detection of both short and long continuous surface defects, the ET tester with several test channels, allows simultaneous detection of both types of defects in magnetic or nonmagnetic material. To detect defects such as corrosive pitting, holes, erosion, fatigue cracks, and OD wear at the tube supports in non magnetic heat exchanger tubing, an eddy current tester is appropriate. With optional tube mapping software, immediate tube maps can be printed to show test results. For applications where more than one of the conditions described above must be met, combined systems using several testers or technologies can be assembled. In these cases, each tester or technology is used to find the types of defects or conditions that it is best suited to detect. The result is often a more accurate test and fewer rejects to allow more of the product to be shipped to the customer.

A comparison of some NDT methods regard internal cracks detectability is given in Table 1 for optimal conditions for their detection. For UT it is the orientation relative to the direction of wave propagation and the threshold of detection is proportional to the wave length, thus transversal waves are favorable. Crack widths given in Table 1 mean crack thickness; for UT they are minimum due to the strong dependence of ultrasonic reflection coefficient on difference of acoustical resistance.

TABLE 1. Size of internal cracks which can be detected by three NDT methods.[13]

Dimension, mm	4.1.1. NDT method		
	4.1.2. MT	4.1.3. UT	4.1.4. RT
Width	0.3	0.001	0.01
Length	5	3	5
Height	0.5	2	2

UT is capable of detecting very small internal discontinuities and tight cracks. Subsurface planar discontinuities perpendicular to the test surface are difficult to detect with single angle probe techniques. For such imperfections specific testing techniques are applied, particularly for welds in thicker materials. The suitable technique, known as tandem technique, is a scanning technique involving the use of two or more angle probes, usually having the same angle of incidence, facing in the same direction and having their ultrasonic beam axes in the same plane perpendicular to the surface of the object under testing, where one probe is used for transmission and the other for reception of ultrasonic energy.[8,14]

5. Cracks in spherical storage tanks

Most examined spherical storage tanks (SST) have been produced of fine grained microalloyed steels. The analysis of performed tests of SST for LPG revealed the following facts. By in-service testing cracks in welded joints of these SST have been detected. Cracks had different directions respect to welded joints, lengths and depths. The cracks were found in these SST, but also in tanks made of carbon steels after long term service, as well in tanks for ammonia made of fine grained structural steels.[2]

Possible causes of crack initiation are:

- Cold cracks induced by welding, due to high hydrogen content in consumables.
- Deviation from prescribed welding technology, or inadequate welding technology.
- Stress corrosion and damages in exploitation.
- Detrimental effect of hydro test (cold water pressure test).
- In the case of one SST and one road trailer tank for hydrogen transport it is definitely confirmed that the cracks initiated during exploitation of the vessel, since in the previous periodic inspections non-allowable defects had not been detected. These data also confirm the reference data on the existence of incubation period for stress corrosion cracks nucleation.

The cracks are typically located on the ID as longitudinal cracks in weld metal or in the heat affected zone (HAZ). Among all tested SST only on one vessel cracks were detected in the joint between the mantle of the SST and the lid of the SST on the OD, and on one vessel in the joint between the support reinforcement and mantle. Only on one sphere the defects are detected in supporting columns.

In three cases the cracks were detected in parent metal whose origin is probably the tack weld of temporary holders during sphere manufacturing. In some cases during grinding operation it had been confirmed that the crack ends in the pore or inclusion, invoved during manufacturing. In all tested SST cracks occured most frequently in radial welded joints and typically in their middle (upper) part, at the border between liquid and gaseous phase.

The testing of the tanks before and after the HT has clearly shown that HT in service can cause new cracks in the positions of "old" (but not repaired) welded joints. For that reason the national authorized Boiler Inspection Office has been adviced several times to reduce the HT pressure, especially for tests in service.

The experience has shown that the pressure vessel repaired by welding should not be subjected to HT, but only to test periodically by UT the typical repaired positions from the OD, e.g.:

- Immediately after repair (so called initial state).
- After operating parameters are reached; if in these tests no crack has been detected, the test should be repeated every 6–12 months, until the term of regular HT.
- Based on testing of several car cylindrical tanks and tank wagons for LPG with residual stress (RS) relieved after manufacturing, it had been found that HT has not caused the occurence of new cracks in these vessels.

Testing of SST for vinilchloridemonomer (VCM) supported the conclusion that the cause of crack initiation is local overstressing induced by the effect of HT. Therefore, the revised repair technology for this tank included hammering of welded joints as well, in order to reduce the RS.

6. Experiences with UT of crude oil pipelines

Material degradation of pipelines for gas or liquid transportation, such as corrosion or cracking, can lead to premature failure with potentially catastrophic impact on man and environment. NDT for the detection and sizing of material damage are required, if the integrity of pipelines is to be reliably assessed. This can be achieved by means of so-called intelligent pigs which allow to inspect up to several hundred pipeline kilometers in one run with respect to special damage types such as e.g. corrosion damage.

The integrity of older pipelines, in particular, has to be proved by the operating companies at periodic intervals. In the past, integrity testing was usually performed by means of HT, which reveals cracks that could cause failure under normal operating conditions. However, since no information on sub-critical cracks is obtained, the estimation of the safe future service life becomes rather uncertain. Moreover, HT, as discussed, can cause crack growth of near-critical cracks, thus reducing the expected safety margin. Additionally, HTs are expensive and time consuming as the line has to be taken out of service.

More than one decade ago the German regulations for pressurized vessels also allow the application of NDT instead of HT provided that at least equivalent results are obtained. This means that HT can be replaced by NDT if defects which would lead to rupture during a HT can be detected with sufficient reliability. The most critical defects associated with pipeline ruptures are axial crack-like defects (fatigue cracks, stress corrosion cracking, SCC, and crack-like weld defects). In order to replace HT retesting by intelligent inline inspection, it was therefore decided to develop an inline inspection tool for the detection of this type of defects.[15]

The following defect specification was defined using conservative fracture mechanics calculations for the crack detection tool:

- Defect type: axial crack-like defects (±15° deviation from axial direction)
- Minimum defect length: 30 mm
- Minimum defect depth: 2 mm (for wall thickness ≥8 mm)

The crack detection tool UltraScan CD, a digital ultrasonic apparatus, was developed between 1991 and 1994 and adapted for greater wall diameters in 1996. The inspection is performed using the angle beam 45°-shear waves transducers. Because axial cracks are to be detected, the ultrasonic pulses are transmitted in circumferential direction.

The probe of the crack detection tool is designed such that the complete pipe circumference is uniformly scanned in both clockwise and counter-clockwise directions for pipe diameters 600–1,420 mm for crack detection. This arrangement provides multiple wall coverage which ensures that relevant reflectors are detected by several probes. Additionally, two additional probes serve to continually measure the actual wall thickness and to detect girth welds, the latter information being used both for pipe marking and to precisely locate defects with respect to the nearest girth weld.

After having completed an extended verification program, the TÜV experts formally stated the equivalence of the inline UT program and approved the integrity of the pipelines considered. The verification program included HT of a 22 km/660 mm section. In accordance with the findings of the crack inspection, no rupture occurred.

A very important step in the process of data evaluation is therefore discrimination of various types of indications, specifically discriminating between injurious and non-injurious reflectors. In most cases, the type of indications detected by the tool can be classified in the following way:

- Crack-like (cracks, other surface-breaking defects like laps or shells)
- Notch-like (scratches, grooves, undercuts etc.)
- Inclusion-like (inclusions, laminations)
- Geometry-related (in particular indications caused by geometry of longitudinal weld)

Based on the results of the data interpretation, 30 verification digs were carried out by TAL in order to verify the overall performance of the crack detection tool. The results were satisfactory.

The analysis considered the significance of various defects, too.

Inclusion-like indications are non-injurious defects with longitudinal orientation such as e.g. elongated, surface-parallel inclusions or laminations. In more detail, however, the B-scan patterns look different and the overall information given by the tool is such that discrimination can positively be made. C-scan can distinguish between pipe joint which is relatively inclusion free from a joint with a high inclusion/lamination density.[16] Such pipe joints with high content of imperfections result from improper manufacturing. Since for pipelines also the wall thickness data are available (which enable an easy identification of surface-parallel reflectors), such features can be verified indirectly by using the information from the wall thickness inspection data. A surface-breaking lamination, however, which has to be treated as a crack-like defect, can be identified as such on the basis of the associated corner reflection.

Notch-like indications – as opposed to crack-like indications, notch-like indications are identified in the B-scans by a more homogeneous shape as well as amplitude pattern. Such indications are in particular caused by scratches, gouges and undercuts etc., usually resulting from transportation or manufacturing of pipe joints. Typically, the depths of these defects are below 1 mm. They are however a preferential site for crack initiation due to their shape concentrating stress. A notch-like indication was caused by a 1.5 mm deep scratch as revealed by the verification dig. Even such relatively small defects are picked up by typically four to eight probes.

Crack-like indications can be classified as follows (based on the verifications):

- Shells, laps, surface-breaking laminations
- Cracks close to the longitudinal weld (within ± 10 cm, but usually not in the HAZ)
- Cold cracks in the weld seam

The origin of these defects is in all cases attributed to the manufacturing process. In particular, neither fatigue cracks or nor SCC were found. An interesting example was given of a cold crack detected in a 1,024 mm crude oil pipeline located in the middle of the weld seam. Length was 175 mm, depth was 4–5 mm. The crack consists of a sequence of inter-crystalline smaller cracks with mainly axial orientation. The crack was very difficult to find by manual inspection and amplitude based depth sizing would have been totally wrong. The example illustrates the advantages of the automated inspection system providing multiple defect detection, high signal dynamic by logarithmic data compression and context-related data visualization, compared with manual UT. The results obtained from inspection of almost 2,000 km of operational pipelines (crude oil and gas) confirm that the reported inspection tool can detect axial crack-like defects with lengths >30 mm and depths >1 mm in the base material as well as in the weld area with high reliability.

7. Ultrasonic attenuation measurement

Ultrasonic measurements enable to evaluate the steel structure e.g. to detect hydrogen attack (HA) or to distinguish between steels with different structures.[17] For the first aim various techniques are applied; some of them are based on changes of velocities of both longitudinal (v_L) and shear (v_S) waves, but this method is time-consuming to take normally incident S-wave transit time measurements, and the method can be applied only on sections that produce a well-resolved back-surface reflection whose transit-time measurement can be taken.[17] Well-resolved back-surface reflections are not produced on boiler tubes, because hydrogen damage is usually associated with inner surface roughened by corrosion. Another method of HA detection is based on attenuation measurement (AM). The reliability of AM also deteriorates on rough surfaces that scatter ultrasonic.[18] The Standard Practice of the ASTM E 664-78 (1989) describes a procedure for measuring the apparent attenuation of ultrasonic in materials or components with flat, parallel surfaces using conventional pulse-echo ultrasonic flaw detection equipment. The Standard Practice is concerned with the attenuation associated with L-waves introduced into the specimen by the immersion method.[19] It can be used for the determination of relative attenuation between materials.

Hydrogen attack is a process produced in plain carbon and low alloy steels exposed to a high-pressure hydrogen environment at high temperatures. A chemical reaction between hydrogen and carbides in the steel produces methane gas bubbles in the grain boundaries. The reaction may occur at the surface, resulting in decarburization with an attendant loss in strength. As the bubbles grow, they interlink to form intergranular fissures or microcracks, which result in a loss in both strength and ductility.

Hydrogen blistering occurs when atomic hydrogen deposits as molecular hydrogen at a defect, such as a lamination or band of nonmetallic inclusions. High pressures of molecular hydrogen can build up at that site as atomic hydrogen continues to enter the steel, ultimately forming a blister. Blistering generally occurs in more ductile steels under conditions where hydrogen embrittlement does not occur.

The apparent attenuation in decibels per unit length as defined by the units of thickness is given by the relationship

$$\text{Attenuation } V = \frac{dB}{2(n-m)t}(dB/mm) \quad (1)$$

where $dB = 20 \log_{10}(A_m/A_n)$ the apparent attenuation in dB

- m and n are ordinal numbers of back reflections ($n > m$).
- A_m and A_n are corresponding (mth and nth respectively) echo-amplitudes.
- t is specimen thickness (mm).

The values of the attenuation at 5 MHz for unattacked steel are less than 0.16 dB/mm, for 94% of unattacked steel specimen 0.16 dB/mm and the average attenuation level for attacked specimens is 0.35 dB/mm.[20]

The pressure vessel A for hydrogen transport (road trailer tank) was made in the 1981 year from a Mn-V steel Č.8380 (C 0.28–0.34%, Mn 1.1–1.4%, V 0.16–0.22%, S, P max 0.035%, Si max 0.4%) in the tubular form with inside diameter $2R_i = 300$ mm and wall thickness $t = 8.8$ mm. In the late 1980s the thickness was increased to 10 mm (pressure vessels B, C and D). The minimum measured wall thickness for the pressure vessel A was

7.5 mm, thus it was rejected and used for liquid penetrant and fluorescent magnetic particle testing, as well for mechanical testing, hardness and AM. Chemical analysis found 0.24% Cu in vessel A material and 0.19% Cu in vessels B and C materials. The steel (Č.8380) is normalized with minimum yield strength 560 MPa, tensile strength 700 MPa and minimum elongation of 14%.

The cylindrical part of pressure vessels is hot rolled and their ends are formed by forging. Working pressure is 15 MPa and working temperature is the ambient temperature. The International Gaseous Committee recommends Cr-Mo steels for the same application.[21]

Samples cut off from the pressure vessel A were prepared from outer and inner vessel side as follows:

- Surface ground by abrasive paper No 400 (the preparation 1)
- Surface cleaned by sand blasting (the preparation 2)
- Surface fine ground by diamond paste (the preparation 3)
- Surface fine ground by abrasive paper No 600 (the preparation 4)

Series (sets) of n AM are taken from outside with a probe of 6.3 mm diameter and the frequency 5 MHz, and given in Table 2.

Mean AV, α, for each series of n measurements is given with standard deviation σ_{n-1} as well the confidence interval for probability 0.95 in Table 1. Finer both surfaces polish decreased σ_{n-1}, the interval and mean value, but rougher back (inner) surface increased α.

TABLE 2. Mean attenuation values α for 5 MHz/6.3 mm probe and vessel A (in dB/mm).[22]

Series No.	8. Preparation No Outer/inner	n	α dB/mm	σ_{n-1} dB/mm	Interval for measured values expected with probability 0.95
1	1/2	33	0.245	0.058	0.129–0.361
2	3/2 or 4/2	6	0.392	0.044	0.304–0.480
3	3/4 or 4/4	10	0.177	0.035	0.107–0.248

The measurement with the 2.25 MHz/6.3 mm ultrasonic probe has shown similar influences of surface preparation, except that AV scatter are less, indicating lower sensitivity of smaller frequency probe to surface condition. Mean AV are greater than 0.160 dB/mm regardless the surface preparation.

Metallographic examination from the outer side of the vessel A has shown inhomogeneous ferrite–pearlite microstructure with micro segregations of impurities, while the examination of longitudinal section revealed ferrite–pearlite structure of striped appearance with micro segregations of impurities building nests.

The microstructure of the cross section on cylindrical part of the vessel A contains also microcracks, Figure 4. The wall material at rounded end of the vessel contains a few cracks. The echogram taken on cylindrical part of the pressure vessel A with 5 MHz probe and preparation 3 of outer surface (inner surface is sand blasted) is given in Figure 5.

The echograms obtained on cylindrical part for the same surfaces preparation for very fine grained structure of pressure vessel C material, at the location of minimum attenuation, is shown in Figure 6 and for the pressure vessel D, made from homogeneous and fine grained material, too, in Figure 7. The content of phosphorus in vessel C material is only 0.002%.[21]

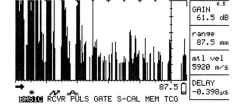

Figure 4. Microstructure of the cross section on cylindrical part of vessel A, etched (100×).[17]

Figure 5. Echogram obtained on vessel A with 5 MHz/6.3 mm probe, $\alpha = 0.453$ dB/mm.[17]

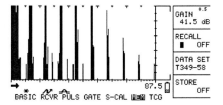
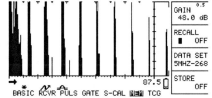

Figure 6. Echogram for the vessel C 5 MHz/6.3 mm probe, $\alpha = 0.164$ dB/mm.

Figure 7. Echogram for the vessel D 5 MHz/6.3 mm probe, $\alpha = 0.174$ dB/mm.

Tensile testing gave yield strength between 496.8 and 547.6 MPa, less than required minimum value for Č.8380, satisfactory tensile strength and elongation only 13.4%, less than required minimum value. Impact testing at −20°C showed that the energy for crack propagation is negligible compared to the energy for crack initiation and that total impact energy is less than or equal to 32.6 J.[22]

The results show that increased AVs arise from inhomogeneous structure of pressure vessel A produced by an inadequate manufacturing process.

The application of AM requires larger number of measurement probe positions and statistical data analysis. Thus the mean quadratic error of the mean AV will be decreased and repeatability improved. If small flaw echoes are present, well distinguished fifth successive back-surface echo can be used for more precisely AV calculation.

Mean AVs for the vessel A steel are consistent with results of metallographic analysis and mechanical testing. The use of lower frequency probes (2.25 MHz) is recommendable if surfaces are corroded, because AV scatter less than at higher frequencies (5 MHz). Probe diameter is less significant parameter than frequency if imperfections are much less than the diameter. AV for pressure vessel outer diameter 318 mm are higher than 0.160 dB/mm even for pressure vessels C and D, but AV for the fittings material are less than 0.160 dB/mm where blisters are found. This is not in contradiction with results for the calibration block 1: similar probe 2 MHz/10 mm gave AV between 0.138 and 0.208 dB/mm for calibration block 1, made from a fine grained steel.[23]

AM is frequent in Oil Refinery Pančevo. Working fluid in examined fittings of coolers is processed gasoline with 65% mol H_2 with organic chlorides added. During the regeneration process at 300°C the fluid contains N_2, CO, CO_2 and moisture. Calculated temperature and pressure for the examined fittings N7 and N2 are: 454°C and 38 bar for N7 and 371°C and 38 bar for N2. AM were performed by using two straight beam probes for longitudinal waves with measurement points in three parallel planes at different heights along the vertical fitting axis. The fittings N7 and N2 are made of A-182F1 steel with wall thickness of 45 and 35 mm respectively, and outside diameter of about 550 mm. Blisters close to the inner surface of the fitting N7 are detected at the middle plane with the 5 MHz/12.5 mm probe.

It is concluded that the damaged layer thickness does not exceed 6 mm and that the inner and outer surfaces are roughened by corrosion. Mean AV are less than 0.160 dB/mm and increase with frequency, even at increasing probe diameter, i.e. at a smaller beam spread.

8. Discontinuity evaluation using UT

The assessment of discontinuities for acceptance purposes according to EN 1714,[9] by agreement between the contracting parties, is performed by either of the following methods:

- Evaluation based primarily on length and echo amplitude, known as the quality control concept.
- Evaluation based on characterization and sizing of the discontinuity by probe movement methods, known as the fitness-for-purpose concept.

By the principle of 'fitness for purpose' a weld in a particular fabrication is considered to be adequate for its purpose provided the conditions to cause failure are not reached, after allowing for some measure of abuse in service. It is well known that some aggressive working media and stress induce cracks to grow, thus their size is likely to be the condition to cause failure after a time.

Quality control levels are considered to be both arbitrary and usually conservative. If flaws more severe than the quality control levels are revealed, rejection is not necessarily automatic, but decisions on whether rejection and/or repairs are justified may be based on fitness for purpose, for instance using engineering critical assessment, such as PD 6493.[24]

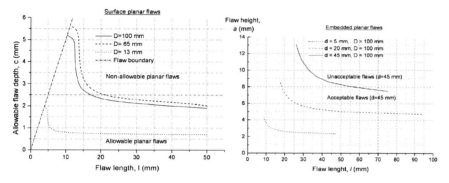

Figure 8. The dependence of allowable height a on length ℓ for surface planar flaws (left) and for embedded flaws (right) according to ASME Section XI.

Evaluation of discontinuities may include discrimination between planar and non-planar discontinuities as the primary discrimination of an acceptable or rejectable indication. In this case *all* discontinuities above the evaluation level shall be characterized and if characterized as planar, are rejected.[9]

ASME Boiler and Pressure Vessel Code, Section XI[25] uses the stress intensity factor to take into account the stress and crack size. The acceptance criteria are given for different components such as: welds in reactor and other vessels, vessel nozzle welds, dissimilar and similar metal welds in piping, steam generator tubing etc. and are applied to the components following the pre-service testing and each service testing. The component containing a crack that is not allowable, may be evaluated by analytical procedure given as a non-mandatory appendix to the Section XI. The Section makes a difference between surface, as more dangerous, and embedded (internal) planar flaws and gives their allowed dimensions.[25] The dependence of acceptable planar surface flaw height on the flaw length (ℓ) is shown in Figure 8: falling curves for three values of wall thickness D as a parameter in diagram on the left hand side and falling curves for three values of flaw depth d in the diagram on right. The curves become almost flat for higher flaw lengths. In other words, the length of surface flaw is less critical for the acceptance criteria, than its height. There is a resemblance between curves in Figure 8, given according to ASME Section XI, and acceptance levels of EN 1712 represented by diagrams, as in Figure 9.[26-28]

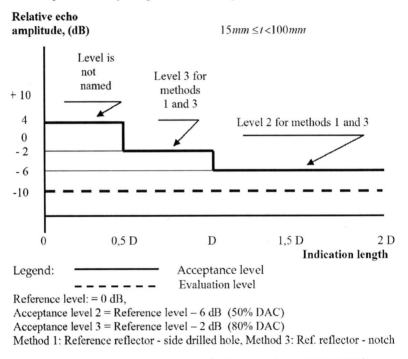

Figure 9. Acceptance levels 2 and 3 for the methods 1 and 3 (EN 1712).

9. Conclusions and discussions

The technology in use today continuously monitors exchangers, valves, rotating equipment, pipelines, offshore platforms and reactors. The test and monitoring techniques described here are often used because of the need to inspect for particular damage mechanisms in order to reduce risk of in-service failure.

Financial demands require that plant down-time is minimized, so shut-downs should involve advance planning and in-service testing to take as much testing and inspection work as possible out of the shut-down period.

One of the things different in the NDT-industry to the established practice in the welding industry, is validation. When welds are made, validation is something normal, when a fabricator of vessels wants to introduce a new welding process, than the procedure is validated by determination of the sample's strength, hardness and toughness of the material. If these obtained values are within the specification limits, then the welding procedure and therefore the new process will be accepted. In NDT world this is generally not possible. One exception is the nuclear code ASME XI and particularly its Appendix 8.

The advantages of techniques such as radiography and manual UT are well-know and generally accepted. These techniques have however limited reliability and accuracy. Today more reliable and more accurate techniques are available that could fulfil the roll of a verification system much better such as mechanized UT, UT using C-scan and Time of Flight Diffraction (TOFD). The other useful technical improvements are industrial endoscopes for TV monitor inspection. Less then two decades ago these systems could be called expensive, cumbersome and inflexible but in the present time, these prejudices can no longer be held up.

References

1. Directive 97/23 EC – Non-Simple Pressure Vessels Directive, 1999.
2. J. Kurai and B. Aleksić, Proof pressure test as a cause of crack occurrence in pressurized equipment in service, (in English and Serbo), Integritet i vek konstrukcija, 3(2), 65–71 (2003)
3. Z. Lukačević, Quality assurance requirements for high strength steel weldments, in: The application of fracture mechanics to life estimation of power plant components (IFMASS 5), editor S. Sedmak, EMAS, U.K., 1990, pp. 191–210.
4. R.P. Reed, H.I. McHenry, M.B. Kasen, A fracture mechanics evaluation of flaws in pipeline girth welds, Welding Research Council Bulletin 245, Jan. 1979.
5. M. Kirić, *The Application of Fitness-for Purpose Concept to Welded Joints of Pressure Vessels Made of Microalloyed Steel*, Ph.D. Thesis, (in Serbo), The Faculty of Mechanical Engineering, Belgrade 2000, pp. 1–251.
6. M. Kirić, IIW-type calibration blocks and an analysis of their use in the ultrasonic examination of welds (in English and Serbo), Zavarivanje i zavarene konstrukcije, 41(4), 329–340 (1996).
7. A. Fertilio, V. Aleksić, M. Kirić, "Experiences in contemporary documenting of ultrasonic testing reports", *The 4th International Conference on Accomplishments of Electrical and Mechanical Industries "DEMI 2001"* (in English), 25–26. April 2001, Banja Luka, Republika Srpska, Proceedings, pp. 149–154.
8. EN 1330-4:2000 Non destructive testing – Terminology – Part 4: Terms used in ultrasonic testing, CEN.
9. EN 1714:1997 Non destructive examination of welds – Ultrasonic examination of welded joints, CEN.
10. ISO 5577:2000 Non destructive testing – Ultrasonic inspection – Vocabulary, ISO.
11. ASTM E164-08 Standard Practice for Contact Ultrasonic Testing of Weldments, Annual Book of ASTM Standards Vol. 03.03., American Society for Materials, West Conshohocken, PA, 1990.
12. M. Kirić, *UV zračenje i njegova primena u ispitivanju bez razaranja*, Conference with international participation IBR 2002 "European trends – the application in Yugoslavia", Tara 25–29 Nov. 2002. god. (CD)/

13. S. Sedmak, Primena parametara prsline u oceni integriteta konstrukcija (in Serbo), Zavarivanje i zavarene konstrukcije, 42(3), 189–196 (1997).
14. EN 583-4: Non destructive testing–Ultrasonic examination – Part 4: Examination for imperfections perpendicular to the surface, CEN.
15. H. Willems, O.A. Barbian, N. Vatter, Operational Experience with Inline Ultrasonic Crack Inspection of German Crude Oil Pipelines, 7th European Conference on Non-destructive Testing, Copenhagen 26–29 May 1998.
16. M. Kirić, Ultrasonic testing with C-scan, The application to pipelines and welded joints, (in English and Serbian), Integritet i vek konstrukcija, 6(1–2), 195–206 (2006).
17. M. Kirić, D. Đukanović, A. Fertilio, Experiences in steel structure evaluation by means of ultrasonic attenuation measurement, The 4th International conference on accomplishments of electrical and mechanical industries "DEMI 2001", Banja Luka, 25–26 April 2001, Proceedings, pp. 127–132.
18. A.S. Birring, D.G. Alcazar, J.J. Hanley, S. Gehl, "Ultrasonic Detection of Hydrogen Damage," *Materials Evaluation*, 47(3), pp. 345–350 (1989).
19. ASTM E 664-78 (Reap. 1989), Standard Practice for the Measurement of the Apparent Attenuation of Longitudinal Ultrasonic Waves by Immersion Method, Annual Book of ASTM Standards Vol. 03.03., 1990, pp. 263–264.
20. API Publication 941/83 'Development and application of a nondestructive ultrasonic test of detecting high-temperature hydrogen attack of steels' Order No. 820-00004.
21. International Gaseous Committee (IGC) Doc. T.N. 26/81.
22. M. Kirić, Z. Burzić, J. Kurai, Ultrasonic attenuation measurement as the control method for steel evaluation, the paper TMT07-430, 11th International Research/Expert Conference "Trends in the Development of Machinery and Associated Technology", TMT 2007, Hammamet, Tunisia, 5–9 Sept 2007, Proceedings, pp. 267–270.
23. D. Markučić et al., The analysis of ultrasonic attenuation measurements in steel, in: MATEST '97, Rovinj (Croatia) 1997, Proceedings, pp. 25–30.
24. PD 6493:1991, Guidance on methods for assessing the acceptability of flaws in fusion welded structures, BSI Standards.
25. ASME Boiler and Pressure Vessel Code, Section XI Division 1, Rules for Inservice Inspection of Nuclear Power Plant Components, Boiler and Pressure Vessel Code, SI Edition 1983 and later editions.
26. EN 1712:1997 Non destructive examination of welds – Ultrasonic examination of welded joints – Acceptable levels, CEN, Brussels.
27. M. Kirić, Structural Integrity Assessment Applying Ultrasonic Testing, Paper kir-722, in: ECF 16, Alexandroupolis, Greece, 3–7 July 2006, CD, and: *Proceedings of Nano and Engineering Materials and Structures*, editor E. Gdoutos, Springer-Verlag, pp. 1063–1065.
28. M. Kirić, J. Kurai, Testing for detection and evaluation of cracks, in: *The challenge of materials and weldment-Structural integrity and life assessment* (IFMASS 9), editors S. Sedmak, Z. Radaković, J. Lozanović, Belgrade 2008, Proceedings, pp. 151–168.

CRACK–INTERFACE INTERACTION IN COMPOSITE MATERIALS

LIVIU MARSAVINA[*]
POLITEHNICA University of Timisoara, Blvd. M. Viteazu, Nr. 1, Timisoara 300222, Romania

TOMASZ SADOWSKI
Lublin University of Technology, ul. Nadbystrzycka 40, 20-618 Lublin, Poland

Abstract The presence of cracks has a major impact on the reliability of advanced materials, like fiber or particle reinforced composites or laminated composites. This paper presents different aspects of the interaction between crack and interface: stress field and fracture parameters for a crack approaching the interface and the crack deflection versus penetration for a crack with the tip on the interface. Different material combinations and mixed mode loads were considered by using a bi-axial bi-material specimen. The stress filed and fracture parameters were obtained numerically using the finite element method.

Keywords: Bi-material interface, fracture parameters, crack path

1. Introduction

The presence of cracks has a major impact on the reliability of advanced materials, like fiber or particle reinforced ceramic composites, ceramic interfaces, laminated ceramics. The understanding of the failure mechanisms is very important, as much as the estimation of fracture parameters at a tip of the crack approaching an interface and crack propagation path. Different researchers have investigated the interaction between an interface and a perpendicular or inclined crack. Zak and Williams (1963) showed that the

---

[*] E-mail: msvina@mec.upt.ro

stress field singularity at the tip of a crack terminating perpendicular to an interface is of order $r^{-\lambda}$, where λ is the real part of the eigenvalue and depends on the elastic properties of the bi-material. Cook and Erdogan (1972) used the Mellin transform method to derive the governing equation of a finite crack perpendicular to the interface and obtained the stress intensity factors. Erdogan and Biricikoglu (1973) solved the problem of two bounded half planes with a crack going through the interface. Bogy (1971) investigated the stress singularity of an infinite crack terminated at the interface with an arbitrary angle. Wang and Chen (1993) used photoelasticity to determine the stress distribution and the stress intensity factors of a crack perpendicular to the interface. Lin and Mar (1976), Ahmad (1991) and Tan and Meguid (1996) used finite element to analyze cracks perpendicular to bi-material in finite elastic body. Chen (1994) used the body force method to determine the stress intensity factors for a normal crack terminated at a bi-material interface. Chen et al. (2003) used the dislocation simulation approach in order to investigate the crack tip parameters for a crack perpendicular to an interface of a finite solid. He and Hutchinson (1993) also considered cracks approaching the interface at oblique angles. Chang and Xu (2007) presented the singular stress field and the stress intensity factors solution for an inclined crack terminating at a bi-material interface. A theoretically description of the stress singularity of an inclined crack terminating at an anisotropic bi-material interface was proposed by Lin and Sung (1997). Wang and Stahle (1998) using a dislocation approach presented the complete solution of the stress field ahead of a crack approaching a bimaterial interface. They calculate the stress intensity factor solutions and the T-stress. Liu et al. (2004) determined the mixed mode stress intensity factors for a bi-material interface crack in the infinite strip configuration and in the case where both phases are fully anisotropic. Kaddouri et al. (2006) and Madani et al. (2007) used the finite element analysis to investigate the interaction between a crack and an interface in a ceramic/metal bi-material. They investigated the effects of the elastic properties of the two bounded materials and the crack deflection at the interface using the energy release rate.

Simha et al. (2003) and Predan et al. (2007) highlight the phenomena of shielding and anti-shielding produced by the bimaterial interfaces. They also highlight the strong influence of the material inhomogeneity on the crack-driving force and introduced the inhomogeneity parameter C_{inh}.

However, all the above studies are limited to the solution of the stress singularity and crack tip parameters. For engineering application is very important to estimate the crack path, and the influence of the interface on the crack trajectory. In this way an interesting study was presented by Gunnars et al. (1997), they considered an initial crack perpendicular to an interface which will grow unstable and will not reach the interface.

2. Simulating crack propagation near bi-material interfaces

The presence of material interfaces can have a significant role on crack propagation. The bonding between two materials is never perfect. In most cases, when the loading is normal to the interface, the interface is less tough than the adjacent materials. If this type of loading is present, once a crack propagates into the interface, it is energetically favorable for the crack to continue propagation in the interface. However, empirical observations indicate that the toughness of the interface is a strong function of the relative amounts of shear and normal loading across the interface. The interface becomes much tougher as the proportion of shear loading increases. In general, as a crack propagates along an interface, the relative amounts of shear and normal loading will change. At some point, the energy required to continue to propagate in the interface will be greater than that required to propagate into one of the surrounding materials. At this point, the crack will kink out of the interface and continue to propagate in the other material.

An additional complexity present in many interface problems is the existence of significant residual stresses in the interface region. Bonding between the materials is often attained at elevated temperatures. Differential rates of shrink of the materials as they cool can "lock" in residual stresses, Wawrzynek and Ingraffea (1991).

2.1. CLASESS OF INTERFACE PROPAGATION CONFIGURATIONS

We consider the general case of a crack approaching the interface between two materials, Figure 1a. As the crack propagates could be push back by interface (Figure 1b) or can reach the interface (Figure 1c).

The prediction of the initial direction of crack propagation can be done using standard criteria. The presence of the interface will affect the computed stress field and fracture parameters, Marsavina and Sadowski (2008).

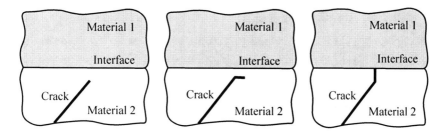

Figure 1. Crack position relative to interface. (a) Approaching interface. (b) Push-back by interface. (c) Into interface.

If the crack tip is initially located on the interface, the predicting of crack evolution becomes more complicated. There are now a number of possibilities:

- The crack tip could stay in the interface (Figure 2a).
- Crack is deflected by the interface (Figure 2b).
- The crack tip could penetrate in the adjacent material (Fig. 2c).

Criteria to assess if crack penetrates or deflects will be presented later.

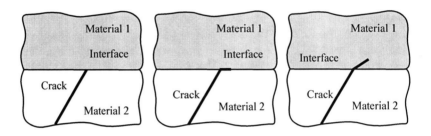

Figure 2. Crack propagation from interface. (a) Crack tip at interface. (b) Deflected crack, (c) Penetrated crack.

2.2. STRESS FIELD AROUND A CRACK APPROACING AN INTERFACE

Jin and Noda (1994) showed that the singular stress fields in non-homo-geneous materials are the same like in homogeneous case. This applies ahead of a crack approaching an interface (Figure 1a) and the stress and displacements could be expressed by standard formulation:

$$\sigma_{ij} = \frac{K_\alpha}{\sqrt{2\pi r}} f_{ij}^\alpha(\theta) + \text{other terms, with } i,j = x,y, \quad (1)$$

$$u_i = \frac{K_\alpha(2 + 2\nu_2)}{E_2} \sqrt{\frac{8r}{\pi}} g_i^\alpha(\theta) \quad (2)$$

with K_α (α = I, II) the stress intensity factors, (r, θ) polar coordinates from crack tip, E_2 and ν_2 are the Young modulus and Poisson's ratio of the material at crack tip, $f_{ij}^\alpha(\theta)$ and $g_i^\alpha(\theta)$ are angular functions.

The energy release rate corresponding to a unit advance of the crack is:

$$G = \frac{1}{\overline{E_2}}\left(K_I^2 + K_{II}^2\right) \quad (3)$$

with $\overline{E_2} = E_2$ for plane stress or $\overline{E_2} = E_2/(1-\nu_2^2)$ for plane strain.

However, the presence of interface influences the stress field and the fracture parameters K and G. An in-homogeneity parameter C_{inh} was introduced by Simha et al. (2003), and was introduced trough a configurational forces approach.

2.3. STRESS FIELD FOR AN INTERFACE CRACK

Dundurs (1967) shows that the stress and displacement fields for an interface crack can be characterized by the following bi-material parameters:

$$\alpha = \frac{\mu_1(\kappa_2+1)-\mu_2(\kappa_1+1)}{\mu_1(\kappa_2+1)+\mu_2(\kappa_1+1)}, \quad \beta = \frac{\mu_1(\kappa_2-1)-\mu_2(\kappa_1-1)}{\mu_1(\kappa_2+1)+\mu_2(\kappa_1+1)}, \quad (4)$$

$$\varepsilon = \frac{1}{2\pi}\ln\left(\frac{\mu_1+\kappa_1\mu_2}{\kappa_2\mu_1+\mu_2}\right),$$

where $\kappa_j = 3-4v_j$ for plane strain, $\kappa_j = (3-v_j)/(1+v_j)$ for plane stress, and μ_j and v_j, $(j=1,2)$ are the shear modulus and Poisson's ratios of the constituent materials.

Explicit expressions for state of stress within the singular zone at the crack tip may be expressed in the form:

$$\sigma_{mn} = \frac{1}{\sqrt{2\pi r}}\left[\mathrm{Re}(K_{int}r^{i\varepsilon})f^1_{mn}(\theta,\varepsilon)+\mathrm{Im}(K_{int}r^{i\varepsilon})f^2_{mn}(\theta,\varepsilon)\right] \quad (5)$$

where $(m, n) = (x, y)$ for Cartesian coordinates or $(m, n) = (r, \theta)$ for polar coordinates, $K_{int} = K_1 + iK_2$ is the complex stress intensity factor at the tip of the interface crack, ε is the oscillating index, $i = \sqrt{-1}$. The functions f^1_{mn} and f^2_{mn} are given by Shih and Asaro (1988) in Cartesian coordinates or by Rice et al. (1990) in polar coordinates.

The energy release rate for an interface crack is:

$$G_{int} = \frac{1}{2}(1-\beta^2)\left(\frac{1-v_1^2}{E_1}+\frac{1-v_2^2}{E_2}\right)(K_1^2+K_2^2), \quad (6)$$

with E_j $(j = 1,2)$ the Young modulus of the two materials.

2.4. INTERFACE CRACK PROPAGATION CRITERION

The interface resistance to fracture is a function of loading on the interface, Suo and Hutchinson (1990). We consider $\phi = \arctan(v/u)$ the phase angle, which describes the relative amount of opening (v) and sliding (u) at the crack tip. A related quantity is the load angle $\psi = \arctan(K_{II}/K_I)$. If the materials on both sides of the interface have the same elastic constants, these two quantities will be equal. If the elastic constants of the two adjacent materials are different, a K_{II} will be induced, even if the crack is loaded in pure mode I. In this case $\phi \neq \psi$.

The relationship between the load angle ψ and the critical energy release rate G_C is a property of the type of material on either side of the interface and the nature of the bonding. This is determined experimentally, or some analytical expressions could be used. Usually in the literature a relation of the following form is used, Banks-Sills and Ashkenazi (2000):

$$G_C(\psi) = G_{C0}(1 + \tan^2 \psi) \tag{7}$$

here G_{C0} represents the toughness when the shear loading is zero.

An extended review of fracture criteria for interface crack was presented by Banks-Sills and Ashkenazi (2000).

If we consider that the crack meets an interface (Figure 2a) a criterion to assess if the crack deflects along the interface (Figure 2b) or penetrates the adjacent material (Figure 2c) was expressed by He and Hutchinson (1993):

$$\frac{G_d}{G_p} > \frac{G_C(\psi)}{G_{jC}} \tag{8}$$

where $G_C(\psi)$ represents the fracture toughness of the interface, ψ is the load angle, G_{jC} the fracture toughness of the penetrated material j (with $j = 1, 2$), $G_d = G_{int}$ represents the energy release rate at the tip of the deflected crack and could be calculated with Eq. (5), and $G_p = G$ is the energy release rate for the crack penetrated, Eq. (3).

3. Interaction between crack and interface

3.1. BI-MATERIAL MODEL

The biaxial specimen with an inclined crack at 45° has been used successfully to growth mixed mode cracks and to investigate the singular stress field in mixed mode conditions, Bold et al. (1991). In contrast with the specimens containing inclined cracks in monoaxial tension, this type of specimen has the advantage of creating different type of mixed modes at the crack tip on the same geometry, only by changing the applied loads on the two axes. An initial study for determining the fracture parameters for this homogeneous specimen was carried out. Then a biaxial specimen with an interface was numerically investigated with FRANC2DL code, Iesulauro (2002). Eight node isoparametric elements were used to model a quarter of the biaxial specimen, Figure 3. Eight singular elements were placed around the crack tip as a common technique to model the stress singularity. The model was loaded with different combinations of stresses σ_x and σ_y in order to produce mixed modes. The symmetric boundary conditions were imposed. The initial crack a_0 length was considered 10 mm. Then the same problem was analyzed by considering the two parts of the model from different material combinations.

For propagating the crack the maximum circumferential stress theory $\sigma_{\theta\theta max}$ was used, Erdogan and Sih (1963), and the stress intensity factors at the crack tip were estimated using the modified crack closure integral, both implemented in FRANC2DL.

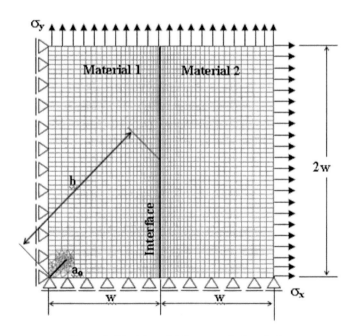

Figure 3. Biaxial model for numerical analysis.

The considered material combinations for the bi-axial model are shown in Table 1.

TABLE 1. Material properties combinations.

Material Combination	Material 1		Material 2	
	E_1	v_1	E_1	v_2
$E_1/E_2 = 10$	200,000	0.25	20,000	0.25
$E_1/E_2 = 2$	400,000	0.22	200,000	0.25
$E_1/E_2 = 1$	200,000	0.25	200,000	0.25
$E_1/E_2 = 0.5$	200,000	0.25	400,000	0.22
$E_1/E_2 = 0.1$	20,000	0.22	200,000	0.25

3.2. FRACTURE PARAMETERS FOR A CRACK APPROACHING THE INTERFACE

Finite Element Method (FEM) was used to estimate the fracture parameters at the tip of the cracks. The Modified Crack Closure Integral (MCCI) was used to compute the Energy Release Rate (ERR), and to extract the stress intensity factors (SIF) results from the numerical results. The concept of closure integral was first used by Irwin (1957) to relate the global ERRs to the crack tip SIFs. Rybicki and Kanninen (1977) modified the procedure so that only one analysis was necessary. They observed that the stress field in front of the crack-tip is similar to the stress field that would exist over a closed portion of the crack. They proposed determining ERRs from the integrals:

$$G_I = \lim_{\Delta a \to 0} \frac{1}{\Delta a} \int_0^{\Delta a} \sigma_{yy}(x,0) \cdot u_y(\Delta a - x, 0) dx$$
$$G_{II} = \lim_{\Delta a \to 0} \frac{1}{\Delta a} \int_0^{\Delta a} \sigma_{xy}(x,0) \cdot u_x(\Delta a - x, 0) dx$$
(9)

where G_I and G_{II} are the ERR rates, u_x and u_y are the opening, respectively sliding displacements of two points from the opposite crack flanks situated at a small distance behind the crack tip Δa. With the quarter-point elements having symmetric nodal positions the crack closure integral can be performed independently for the opening (Mode I) and sliding (Mode II) displacements. This yields decoupled values for G_I and G_{II}, which are used to compute K_I and K_{II}. The determination of the SIF based on MCCI method implemented in FRANC2DL code use the formulation of Ramamurthy et al. (1986)

for quarter point elements, Figure 4. They expressed the crack-tip displacement and the stress fields in terms of second order polynomials that were consistent with the quarter point behavior. The ERR could be express:

$$G_I = \frac{1}{\Delta a}\left[\left(C_{11}F_y^A + C_{12}F_y^F + C_{13}F_y^G\right)\left(u_y^B - u_y^E\right) + \left(C_{21}F_y^A + C_{22}F_y^F + C_{23}F_y^G\right)\left(u_y^C - u_y^D\right)\right]$$

$$G_{II} = \frac{1}{\Delta a}\left[\left(C_{11}F_x^A + C_{12}F_x^F + C_{13}F_x^G\right)\left(u_x^B - u_x^E\right) + \left(C_{21}F_x^A + C_{22}F_x^F + C_{23}F_x^G\right)\left(u_x^C - u_x^D\right)\right]$$

(10)

where F_x^n, F_y^n and u_x^n, u_y^n represent the nodal forces and displacements on node "n" and C_{ij} are coefficients:

$$C_{11} = \frac{33\pi}{2} - 52, \ C_{12} = 17 - \frac{21\pi}{4}, \ C_{13} = \frac{21\pi}{2} - 32,$$
$$C_{21} = 14 - \frac{33\pi}{8}, \ C_{22} = \frac{7\pi}{2} - \frac{7}{2}, \ C_{23} = 8 - \frac{21\pi}{8}.$$

(11)

The SIFs can be obtained:

$$K_I = \sqrt{\overline{E}G_I} \ , \ K_{II} = \sqrt{\overline{E}G_{II}}$$

(12)

with $\overline{E} = E$ for plane stress or $\overline{E} = E/(1-v_2^2)$ for plane strain.

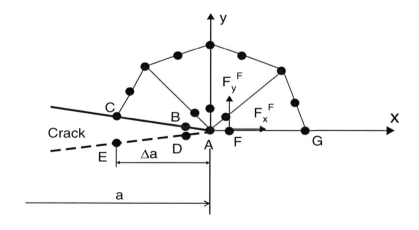

Figure 4. Crack tip mesh with singular elements.

The MCCI method was used successfully for determination of SIF's and ERR in bi-material models with cracks approaching the interface Masavina and Sadowski (2008), and for kinked cracks from interface Masavina and Sadowski (2009).

The numerical model was loaded with different combinations of stresses σ_x and σ_y in order to produce mixed modes from pure Mode I ($\sigma_x/\sigma_y = 1$) to pure Mode II ($\sigma_x/\sigma_y = -1$). The symmetric boundary conditions were imposed. The crack was extended at 45° in 13 increments of 5 mm starting from 5 to 65 mm.

In order to investigate the influence of the interface on the stress field Figure 5 presents the stress distributions σ_x, σ_y and τ_{xy} for three material combination ($E_1/E_2 = 0.5$, 1 and 2). The stress distributions are shown for example presented for a crack length of 50 mm and for mixed mode loading $\sigma_x/\sigma_y = 1/3$. Different stress distributions were obtained for the homogeneous and bi-material cases. It was observed that the influence of the interface become more important when the crack is close to the interface.

In order to quantify the influence of interface on the fracture parameters normalized stress intensity factor (SIF) were defined as the ratio between the stress intensity factors for the bi-material case ($K_{I,bi}$, $K_{II,bi}$) and homogeneous case ($K_{I,hom}$, $K_{II,hom}$), Marsavina and Sadowski (2008):

$$C_I = \frac{K_{I,bi}}{K_{I,hom}}, \quad C_{II} = \frac{K_{II,bi}}{K_{II,hom}}. \tag{13}$$

Figure 6 shows the variation of normalized SIF's versus normalized crack length (a/b) for the material combination $E_1/E_2 = 2$. It can be observed for all considered mixed modes that the Mode I SIF in the presence of interface has values above the homogeneous case. For predominantly mode I the normalized SIF K_I increases with the crack length. In opposite, for predominantly mode II loads the K_I deceases with increasing the crack length. For the Mode II SIF the presence of the interface produces an increase in the K_{II} only when the crack became close to the interface (for $a/b > 0.5$). This phenomenon is known as anti-shielding or amplification effect ($C_\alpha > 1$, $\alpha = $ I, II).

The normalized stress intensity factors for the material combination with $E_1/E_2 = 0.5$ are presented in Fig. 7. It can be observed that the SIF's results for the bi-material case are below those of the homogeneous case. This is a typical example of shielding effect ($C_\alpha < 1$, $\alpha = $ I, II). Different behaviour could be observed for the K_I normalized SIF. When predominantly mode I load are applied C_I decrease with crack length, and when predominantly mode II loads are applied, the C_I increase with crack length. The mode II normalized SIF is close to 1 for short cracks and the decrease with increasing crack length. This variation is more relevant for mode I loads. The maximum decrease is obtained for pure Mode I load, and the minimum one for pure Mode II load.

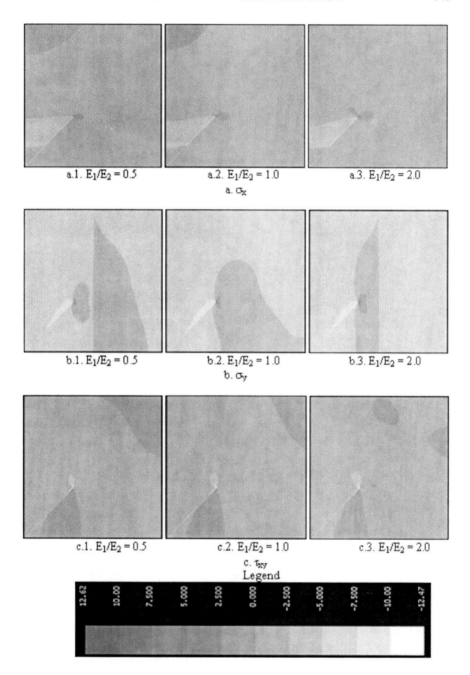

Figure 5. Stress distributions around a 50 mm crack, near to an interface loaded in mixed mode.

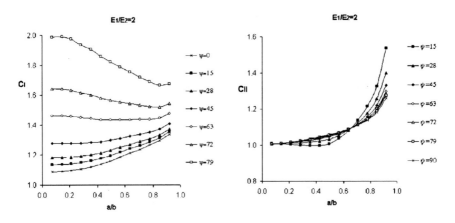

Figure 6. The normalized stress intensity factor C_i versus a/b for $E_1/E_2 = 2$. (a) Mode I. (b) Mode II.

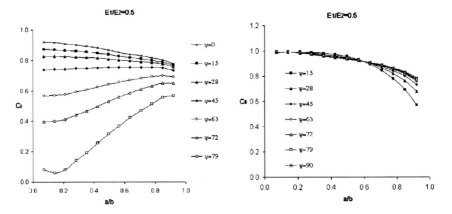

Figure 7. The normalized stress intensity factor C_i versus a/b for $E_1/E_2 = 0.5$. (a) Mode I. (b) Mode II.

3.3. INFLUENCE OF MATERIAL PROPERTIES AND LOADING MODE ON CRACK PROPAGATION

Study of crack propagation was performed starting from an initial crack of 10 mm. The crack was extended based on delete and fill algorithm for remeshing, implemented in FRANC2DL.

In Figure 8 are shown the crack paths for different material combinations, according with Table 1. It can be observed that the presence of interface influences the crack path, and this influence increases with decreasing the ratio between modulus of elasticity of the two materials, Figure 8. Figure 9 show the stress intensity factors values versus crack length during the propagation process.

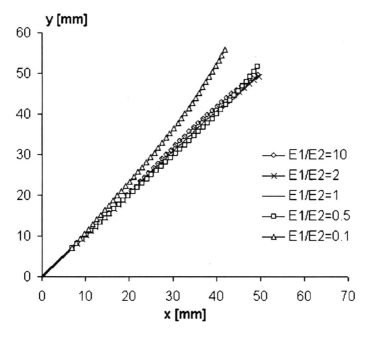

Figure 8. Crack propagation paths for a Mode I loading and different material combinations.

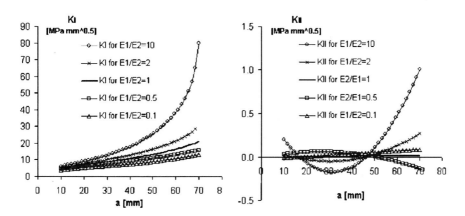

Figure 9. Stress intensity factors for propagating cracks. (a) Mode I. (b) Mode II.

Figure 10 presents the numerical results of the crack propagation paths for different applied loads (ratio $\psi = 0°$, 18.4° and 26.5° corresponding to applied load ratio $\sigma_x/\sigma_y = 1$, 2 and 3) and material combinations $E_1/E_2 = 2$ and 0.5. Different crack paths were obtained for different applied load and material combinations. It can be observed that the crack path is curvilinear, with the exception of the homogeneous case ($E_1/E_2 = 1$), loaded in mode I ($\sigma_y/\sigma_x = 1$).

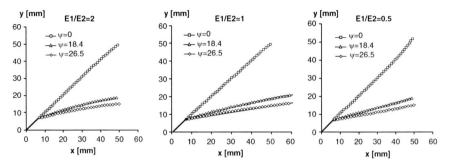

Figure 10. Crack propagation paths for different applied loads and material combinations. (a) $E_1/E_2=2$. (b) $E_1/E_2 = 1$. (c) $E_1/E_2 = 0.5$.

3.4. CRACK DEFLECTION VERSUS PENETRATION

After the 45° inclined crack reaches the interface two situations were modeled crack was extended with an increment of 5 mm along the interface, respectively was penetrated in the adjacent material, using Maximum circumferential stress criteria of Erdogan–Sih (1963). The energy release rate for deflected and penetrated cracks was calculated according with relations (5) and (3) using the values of the stress intensity factors obtained numerically from FRANC2DL. The Modified Crack Closure Integral was used for penetrated crack, and the Crack Flank Displacement method for the deflected crack (Marsavina and Sadowski 2009). A convergence study was carried on in order to determine the size of the singular elements for the case of penetrated crack and the distance at which the crack flank displacements were collected for the case of the deflected crack.

Figure 11 presents the ratio G_d/G_p versus applied mixed mode ψ for the two material combinations (I: 1 – Al_2O_3 and 2 – ZrO_2, Figure 11a, respectively 1 – ZrO_2 and 2 – Al_2O_3, Figure 12b). In order to apply the He–Hutchinson criteria, Eq. (7) the fracture toughness of Al_2O_3 (4.5 MPa m$^{1/2}$) and ZrO_2 (11 MPa m$^{1/2}$) were considered. We assume that the toughness of the interface is half of the lower fracture toughness of the constituents (for our case Al_2O_3) corresponding to weak interfaces. It can be observed that the ratio G_d/G_p has higher values if the crack wants to penetrate the stiffer material ($E_1/E_2 = 0.5$). The minimum values for the ratio G_d/G_p were obtained to an applied load angle between 35° and 45°. For the case I when the initial crack is in the stiffer material (Al_2O_3) the crack will be deflected by the interface. If the initial crack is ZrO_2 the crack will deflect if the applied load angle ψ is lower than 20° and higher than 60°. For 20°< ψ <60° the crack penetrates the adjacent material (Al_2O_3 – for this case).

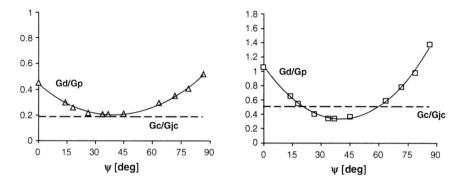

Figure 11. Variation of G_d/G_p versus applied mixed mode and comparison with fracture toughness ratio $G_c(\psi)/G_{jc}$. (a) $E_1/E_2 = 2$. (b) $E_1/E_2 = 0.5$.

4. Conclusions

For investigating the influence of the inclined interface on fracture parameters a biaxial model with an inclined crack at 45° was considered. Different combinations of applied load were applied in order to produce mixed modes from pure Mode I to pure Mode II. The results for the normalized stress intensity factors are presented for two different material combinations.

The influence of the interface on the fracture parameters at the tip of the crack approaching the interface is highlighted. An inhomogeneous parameter is introduced:

- $(C_I, C_{II}) > 1$ for $E_1/E_2 = 2$ (anti-shielding effect)
- $(C_I, C_{II}) < 1$ for $E_1/E_2 = 0.5$ (shielding effect)

It was observed that when a crack in a stiffer material approaches a bi-material interface with a compliant material the stress intensity factor increases, and vice versa. The effects of shielding and anti-shielding is higher when the crack tip is close to the interface $a/b > 0.6$.

Simulation of the crack propagation around an interface is presented with emphasizing the possibilities of crack path. The fracture criteria are presented for non-homogeneous/interface case.

A 45° inclined crack approaching an interface in a bi-axial loaded specimen is modeled using Finite Element Method. Different materials combinations are considered for the two materials and the influence of the ratio between the two Young's modules on the crack paths are numerically investigated. A study of applied bi-axial load on the crack path is also presented.

The variations of the stress intensity factors at the tip of the propagating crack are plotted. Different behaviors were observed for the Mode I SIF with ratios E_1/E_2.

The variation of the energy release rate G_d/G_p was plotted against the applied mixed mode for the two material combinations. Higher values for the ratio G_d/G_p were obtained when the initial crack is in ZrO_2 and deflects on the interface or penetrates in Al_2O_3. Energy release rate ratio G_d/G_p was compared with the ratio between fracture toughness of interface versus fracture toughness of adjacent material $G_C(\psi)/G_{jC}$.

ACKNOWLEDGMENTS This research was supported by the Marie Curie Transfer of Knowledge project MTKD-CT-2004-014058 and by Research for Excellence Grant 202/2006 awarded by the Romanian National Authority for Scientific Research.

References

Ahmad J., 1991, A micromechanics analysis of cracks in unidirectional fibre composite, *J. Appl. Mech.* **58**:964–972.
Banks-Sills L., Ashkenazi D., 2000, A note on fracture criteria for interface fracture, *Int. J. Fract.* **103**:177–188.
Bogy D.B., 1971, On the plane elastic problem of a loaded crack terminating a material interface, *J. Int. Fract.* **38**:911–918.
Bold P.E, Brown M.W., Allen R.J., 1991, Shear Mode crack growth and rolling contact fatigue, *Wear.* **144**:307–317.
Chang J., Xu J.-Q., 2007, The singular stress field and stress intensity factors of a crack terminating at a bimaterial interface, *Int. J. Mech. Sci.* **49**:888–897.
Chen D.H., 1994, A crack normal to and terminating at a bimaterial interface, *Eng. Fract. Mech.* **19**:517–532.
Chen S.H., Wang T.C., Kao – Walter S., 2003, A crack perpendicular to the bi-material interface in finite solid, *Int. J. Solids Struct.* **40**:2731–2755.
Cook T.S., Erdogan F., 1972, Stress in bonded materials with a crack perpendicular to the interface, *Int. J. Eng. Sci.* **10**:677–697.
Dundurs J., 1967, Effect of elastic constants on stress in a composite under plane deformation. *J. Compos. Mater.* **1**:310–322.
Erdogan F., Biricikoglu V., 1973, Two bonded half planes with a crack going through the interface, *Int. J. Eng. Sci.* **11**:745–766.
Erdogan F., Sih G.C., 1963, On the crack extension in plates under plane loading and transverse shear, *J. Basic Eng.* **85** (4):519–525.
Gunnars J., Stahle P., Wang T. C., 1997, On crack path stability in a layered material, *Comput Mech.* **19**:545–552.
He M.Y., Hutchinson J.W., 1993, Crack deflection at an interface between dissimilar elastic materials, *Int. J. Solids Struct.* **25**:1053–1067.
Iesulauro E., 2002, FRANC2D/L a Crack Propagation simulator for plane layered materials, Cornell University, Ithaca.
Irwin G.R., 1957, Analysis of stress and strain near the end of a crack transversing a plate. *ASME J. Appl. Mech.* **24**:361–364.

Jin Z.-H., Noda N., 1994, Crack-tip singular fields in nonhomogeneous materials, *J. Appl. Mech.* **61**:738–740.

Kaddouri K., Belhouari M., Bachir Bouiadjra B., Serier B., 2006, Finite element analysis of crack perpendicular to bi-material interface: case of couple ceramic-metal, *Comput. Mater. Sci.* **35**:53–60.

Lin K.Y., Mar J.W., 1976, Finite element analysis of stress intensity factors for crack at a bimaterial interface, *Int. J. Fract.* **12**:451–531.

Lin Y.Y., Sung J.C., 1997, Singularities of an inclined crack terminating at an anisotropic biomaterial interface, *Int. J. Solids Struct.* **38**:3727–3754.

Liu L., Kardomateas G.A., Holmes J.W., 2004, Mixed – mode stress intensity factors for a crack in an anisotropic bi-material strip, *Int. J. Solids Struct.* **41**:3095–3017.

Madani K., Belhouari M., Bachir Bouiadjra B., Serier B., Benguediab M., 2007, Crack deflection at an interface of alumina/metal joint: a numerical analysis, *Comput. Mater. Sci.* **35**:625–630.

Marsavina L., Sadowski T., 2008, Fracture parameters at bi-material ceramic interfaces under bi-axial state of stress, *Comput. Mater. Sci.* doi: 10.1016/j.commatsci. 2008.06.005.

Marsavina L., Sadowski T., 2009, Numerical determination of fracture parameters for a kinked crack at a bi-material ceramic, *Comput. Mater. Sci.* **44**(3):941–950.

Predan J., Gubeljak N., Kolednik O., 2007, On the local variation of the crack driving force in a double mismatched weld, *Eng. Fract. Mech.* **74**:1739–1757.

Ramamurthy T.S., Krishnamurthy T., Badari Narayana K., Vijayakumar K., Dattaguru B., 1986, Modified crack closure integral method with quarter point elements, *Mech. Res. Commun.* **13**:179–186.

Rice J.R., Suo Z., Wang J.S., 1990, Mechanics and Thermodynamics of Brittle Interface Failure in Bimaterial Systems, in: Metal–Ceramic Interfaces, M. Ruhle, A.G. Evans, M.F. Ashby, J.P. Hirth (Eds.), Pergamon Press, Oxford, pp. 269–294.

Rybicki E.R., Kanninen M., 1977, A finite element calculation of stress intensity factors by a modified crack closure integral. *Eng. Fract. Mech.* **9**:931–938.

Shih C.F., Asaro R.J., 1988, Elastic–plastic analysis of cracks on bimaterial interfaces: Part 1 – Small scale yielding, *J. Appl. Mech.* **55**:299–315.

Simha N.K., Fischer F.D., Kolednik O., Predan J., Shan G.X., 2003, Inhomogenity effects on the crack driving force in elastic and elastic-plastic materials, *J. Mech. Phy solids.* **51**:209–240.

Suo Z., Hutchinson J.W., 1990, Interface crack between two elastic layers, *Int. J. Fract.* **43**:1–18.

Tan M., Meguid S.A., 1996, Dynamic analysis of cracks perpendicular to bimaterial interfaces using new singular finite element, *Finite Elem Anal Des.* **22**:69–83.

Wang W.C., Chen J.T., 1993, Theoretical and experimental re-examination of a crack at a bimaterial interface, *J. Strain Anal.* **28**:53–61.

Wang T.C., Stahle P., 1998, Stress state in front of a crack perpendicular to bi-material interface, *Eng. Fract. Mech.* **4**:471–485.

Wawrzynek P.A., Ingraffea A.R., 1991, Discrete modeling of crack propagation: theoretical aspects and implementation issues in two and three dimensions, Ph.D. thesis, Cornell University, Ithaca.

Zak A.R., Williams M.L., Crack point stress singularities at a bi-material interface, 1963, *J. Appl. Mech.* **30**:142–143.

MEASUREMENT OF THE RESISTANCE TO FRACTURE EMANATING FROM SCRATCHES IN GAS PIPES USING NON-STANDARD CURVED SPECIMENS

J. CAPELLE[1], J. GILGERT[1], YU.G. MATVIENKO[2], G. PLUVINAGE[1]
[1] Laboratoire de Fiabilité Mécanique (LFM), Université Paul Verlaine – Metz – ENIM, 57045 Metz, France
[2] Mechanical Engineering Research Institute of the Russian Academy of Sciences, 4 M. Kharitonievsky Per., 101990 Moscow, Russia

Abstract The experimental procedure for measurement of the notch fracture toughness of the API 5L X52 steel using a non-standard curved specimen with a notch simulating the expected scratch damage of gas pipelines has been developed.

The concept of the notch stress intensity factor based on the volumetric method has been employed and the corresponding notch fracture toughness $K_{\rho,c}$ was obtained. In this case, the notch fracture toughness was calculated using the critical load determined by acoustic emission and the results of finite element analysis of the elastic-plastic stress distribution ahead of the notch tip. The notch fracture toughness $J_{\rho,c}$ in terms of the J-integral has been estimated by means of the load separation method which allows measuring the η-factor. In this case, the original representation of the J-integral as total energy release rate and the tests records have been used. The proposed procedure allows avoiding calculation of the stress intensity factor for non-standard specimens to determine the notch fracture toughness $J_{\rho,c}$.

Development of the present methodology encourages replacing the conservative crack-like defects approach. The results on the notch fracture toughness of the API 5L X52 steel can be used in the modified failure assessment diagram for structural integrity assessment of gas pipelines damaged by scratches.

Keywords: Notch fracture toughness, J-integral, non-standard curved specimen, volumetric method, load separation

1. Introduction

The causes of gas pipe failures have various natures. They can appear either by a fracture, plastic instability or by a leak. The majority of these failures are caused by corrosion pitting or stress corrosion cracking, but also related to welding defects or impact of foreign objects. Movements of ground (landslip, earthquake, etc.) can also lead to damage in the buried pipelines. The owners of pipelines have study these problems for a long time and have a good knowledge of the methods allowing managing them.

It can be seen that the failure caused by corrosion is most important, expressed in 56% cases (Figure 1). But, very important fraction of failures is also connected with mechanical damage due to soil digging or excavating by machines. Frequently, notch-like defects are formed in pipelines by machines used in ground removal during construction. Together with cracks caused by corrosion (general, pitting, stress corrosion) these notches are stress raisers, reducing the material resistance to fatigue and fracture. Stress

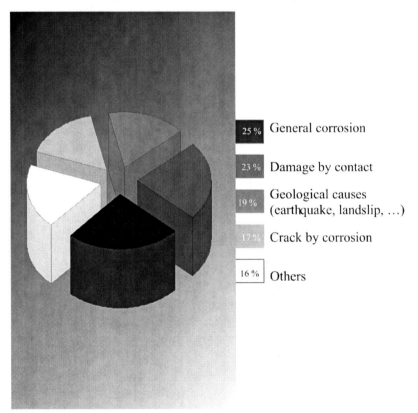

Figure 1. Causes of the fracture of pipelines in the course of exploitation recorded by the members of the ACPRÉ from 1985 to 1995.[1]

concentration is considered as the origin in more than 90% of the failures in service. Therefore, structural integrity of pipelines under various service conditions including the presence of notch-like defects should be evaluated.

The defect assessment methods pursue two different philosophies, which can be designated as a failure assessment diagram (FAD) and crack driving force (CDF).[2-4] In the CDF approach based on the J-integral concept, the crack driving force is plotted and compared directly with the material's fracture toughness and separate analysis carried out for the plastic limit. In the FAD approach, both the comparison of the crack tip driving force with the material's fracture toughness and with the plastic load limit analysis is performed at the same time. The basic failure curve of the FAD is written as $K_r = f(L_r)$, where $K_r = K / K_{1C}$ is the ratio of the applied stress intensity factor K to the material's fracture toughness K_{1C} and L_r is equal to the ratio of applied load P to plastic collapse load P_Y. If the assessment point (K_r, L_r) is situated within the non-critical region enclosed by the line of FAD, failure of the cracked structure does not occur. Overview on some methods for analytical defect assessment and their industrial realizations within guidelines and standards has been mentioned.[2] The SINTAP procedure[5] offers both a FAD and CDF routes which are complementary and give identical results.[3]

At the present time the SINTAP procedure has been modified using the concept of the notch intensity factor and a notch-based failure assessment diagram (NFAD) for a notch-like defect taking into account a finite notch tip radius.[6,7] In this case, the fracture toughness or so-called the notch fracture toughness, which is applied to the NFAD, should be measured for a structural component. The notch stress intensity factor is described by effective stress and effective distance and can be used for measurement of the notch fracture toughness by the volumetric method.[8] It should be noted that some expression for the notch fracture toughness have been proposed in Refs [9,10]. Moreover, the FITNET assessment of a structural component by the standard and advanced J-integral based options also requires notch fracture toughness data in terms of the J-integral derived from tests of notched specimens.[11]

It is usual to measure the fracture toughness of materials, standard specimens, like the compact tensile specimen, can be used. But, when it is necessary to obtain the fracture toughness for manufactured goods similar to a pipeline with a small thickness and a high curvature, difficulties arise. In this case, a non-standard specimen has to be employed.

The aim of the present paper is to develop the experimental procedure for measurement of the notch fracture toughness of the API 5L X52 steel using a non-standard curved specimen with a notch simulating the expected scratch damage of gas pipelines. The concept of the notch stress intensity factor based on the volumetric method has been employed and the corresponding notch fracture toughness $K_{\rho,c}$ was measured. Furthermore, the notch fracture toughness $J_{\rho,c}$ in terms of the J-integral has been determined by means of the load separation method described in Section 3.2.

2. Material, non-standard notched specimens and test procedure

The API 5L X52 steel is traditionally used for the manufacture of pipelines. Moreover, the API 5L X52 steel was the most common gas pipelines material for transmission of oil and gas during 1950–1960. To determine mechanical properties of API 5L X52 steel pipes in the transversal direction, tensile tests were performed using specimens extracted from a pipe with 219.1 mm outer diameter and 6.1 mm wall thickness W. Chemical composition and tensile properties of the present API 5L X52 steel are summarized in Tables 1 and 2, respectively. This steel follows the Ludwik law $\sigma = K\varepsilon_p^n$.

TABLE 1. Chemical composition of the present API 5L X52 steel (wt %).

C	Mn	Si	Cr	Ni	Mo	S	Cu	Ti	Nb	Al
0.22	1.22	0.24	0.16	0.14	0.06	0.036	0.19	0.04	<0.05	0.032

The notch fracture toughness of the API 5L X52 steel has been measured in radial direction at room temperature using non-standard curved notched specimens, namely, "Roman tile" specimens because the pipe dimensions do not permit to measure through thickness mechanical characteristics. These properties are needed in the case of pipe longitudinal defect assessment.

TABLE 2. Mechanical tensile properties at room temperature of the API 5L X52 steel used in the present work.

Young's modulus E (GPa)	Yield strength σ_Y (MPa)	Ultimate strength σ_U (MPa)	Elongation δ (%)	Strain hardening exponent	Constant K (MPa)
203	410	528	32	0.0446	587.3

The "Roman tile" specimen extracted from the pipe is presented in Figure 2. The specimen shape is a circle arc, corresponding to central angle of 160° of 60 mm length. The V-notch with the notch opening angle of 45° and root radius of 0.15 mm was machined to a depth of size a simulating the expected scratch damage. Test specimen sets include specimens with the initial notch aspect ratio $a/W = 0.1; 0.15; 0.25; 0.3; 0.5; 0.7$.

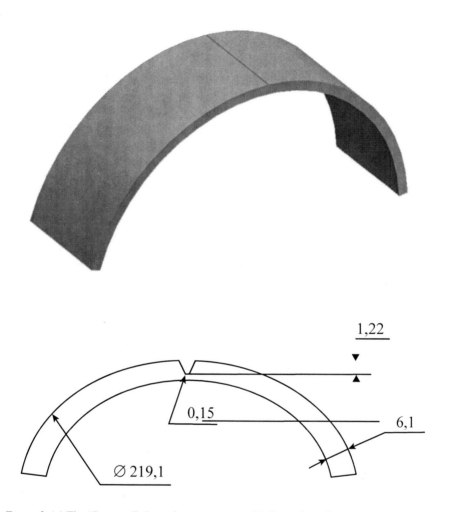

Figure 2. (a) The "Roman tile" specimen geometry, (b) dimensions (in mm) of the specimen and the notch.

The special device and specimens have been developed since it is not possible to get flat specimens from small diameter pipes. Test set-up of three-point bend test for "Roman tile" specimens and testing machine with the bend-test fixture are given in Figure 3. The bend-test fixture was positioned on the closed loop hydraulic testing machine with a load cell of capacity ± 10 kN (Figure 3b). The applied load, frequency and the signal type (sinusoidal, trapezoidal or rectangular) were monitored on the control panel.

The specimen was loaded by three-point bending through a support A (Figure 3a) and supporting rollers B and C. Support and rollers were produced from polyvinyl chloride (PVC) to reduce a friction. All the performed tests are static imposed displacement tests. Displacement rate was monitored for a constant value of at 0.01 mm/s. Test duration was of about 30 min.

To obtain the fracture toughness, apparatus was required for measurement of applied load and load-line displacement. Load versus load-line displacement was recorded digitally for processing by computer.

The test procedure also allowed measuring the critical load corresponding to crack initiation. In the present case, acoustic emission technique has been employed for this purpose. Comparison of dependences of the load versus time and duration of acoustic emission versus time indicates crack initiation

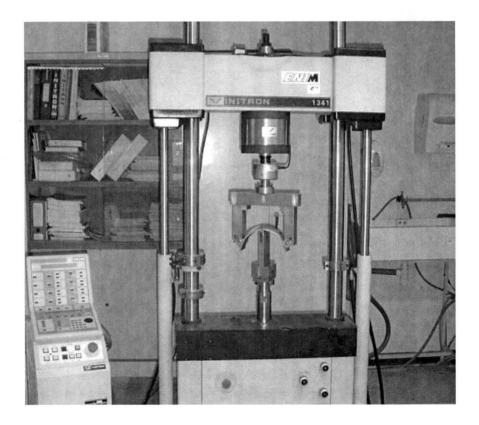

Figure 3. (a) "Roman tile" specimen's fixture assembly (1 – connection with load cell, 2 – transmitting component with rounded tip, 3 – connection of test assembly with the testing machine bottom, 4 – "Roman tile" specimen); (b) testing machine with the bend-test fixture.

which was closed to "pop-in" (Figure 4a and b). For this event, acoustic salves with the highest duration and the most important number of acoustic hits are easily detectable.

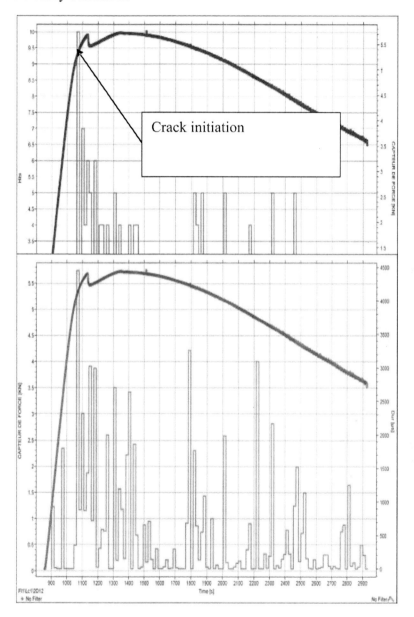

Figure 4. (a) Typical Dependence of hits versus time and the load versus time; (b) dependence of duration of the acoustic emission versus time and the load versus time.

3. The volumetric and the load separation methods in measurement of the notch fracture toughness

The notch fracture toughness of the API 5L X52 steel was measured in terms of the notch stress intensity factor and the J-integral. The first approach is based on the volumetric method. For measurement the notch fracture toughness in terms of the J-integral, the load separation method is considered as a very attractive in the case of non-standard specimens.[12]

3.1. THE VOLUMETRIC METHOD

The concept of the critical notch stress intensity factor and corresponding local fracture criterion assume that the fracture process requires a certain fracture process volume.[8] This volume is assumed as a cylinder with a diameter called the effective distance. Determination of the effective distance is based on the bi-logarithmic elastic-plastic stress distribution on the continuation of the notch because the fracture process zone is the highest stressed zone. This zone is characterized by an inflexion point in the stress distribution (1) at the limit of zones II and III in Figure 5.

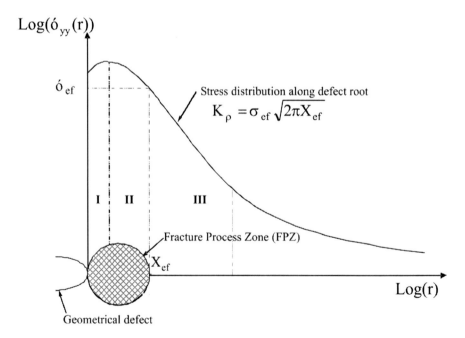

Figure 5. Schematic distribution of elastic-plastic stress ahead of the notch tip on the line of notch extension and the notch stress intensity virtual crack concept.

$$\sigma_{eff} = \frac{1}{X_{eff}} \int_0^{X_{eff}} \sigma_{yy}(r)\Phi(r)dr \qquad (1)$$

Here, σ_{eff}, X_{eff}, $\sigma_{yy}(r)$ and $\Phi(r)$ are effective stress, effective distance, maximum principal stress and weight function, respectively. This stress distribution (1) is corrected by a weight function in order to take into account the distance from notch tip of the acting point and the stress gradient at this point.

The effective distance corresponds is to the inflexion point with the minimum of the relative stress gradient χ which can be written as

$$\chi(r) = \frac{1}{\sigma_{yy}(r)} \frac{\partial \sigma_{yy}(r)}{\partial r} \qquad (2)$$

The effective stress is considered as the average value of the stress distribution within the fracture process zone.

The notch stress intensity factor is defined as a function of the effective distance and the effective stress[8]

$$K_\rho = \sigma_{eff} \sqrt{2\pi X_{eff}} \qquad (3)$$

and describes the stress distribution in zone III as given by the following equation

$$\sigma_{yy} = \frac{K_\rho}{(2\pi r)^\alpha} \qquad (4)$$

where K_ρ is the notch intensity factor, α is a constant. Failure occurs when the notch stress intensity factor K_ρ reaches the critical value, i.e. the notch fracture toughness $K_{\rho,c}$ which reflects the resistance to fracture initiation from the notch tip.

3.2. THE LOAD SEPARATION METHOD

3.2.1. *The J-integral representation*

The load separation method was adopted to measure the notch fracture toughness in terms of the J-integral. The J-integral can be calculated as the sum of elastic J_e and plastic J_p components[13]

$$J = J_e + J_p = \eta_{el} \frac{A_{el}}{B(W-a)} + \eta_{pl} \frac{A_{pl}}{B(W-a)} \qquad (5)$$

where $E' = \dfrac{E}{1-v^2}$, E is Young's modulus; v is Poisson ratio; η_{el} and η_{pl} are elastic and plastic correction factors, respectively; A_{el}, A_{pl} are elastic and plastic area under the load versus load-line displacement curve, respectively; W and B are thickness (direction of the notch continuation) and width of the "Roman tile" specimen, respectively; a is notch depth in thickness direction. More recognized equation is given by

$$J = J_e + J_p = \frac{K^2}{E'} + \eta_{pl}\frac{A_{pl}}{B(W-a)} \qquad (6)$$

where K is the notch stress intensity factor. Such dividing the J-integral could be useful to analyse a contribution of elastic and plastic J-components in deformation and fracture process.

The J-integral can be also written using its energy rate interpretation

$$J_{pl} = -\frac{1}{B}\frac{dA_{pl}}{da} \qquad (7)$$

The classical and numerical methods to determine η_{pl}-factors are based on the energy rate interpretation of the J-integral for a body with a crack. In this case, from Eqs. (6) and (7) the η_{pl}-factor is found to be

$$\eta_{pl} = -\frac{(W-a)}{A_{pl}}\frac{dA_{pl}}{da} = \frac{b}{A_{pl}}\frac{dA_{pl}}{da} \qquad (8)$$

where $W - a = b$ is remaining ligament length, $A_{pl} = \int_0^{v_{pl}} P dv$. In the Eq. (8) the value of P represents the load applied during the test and the value of v_{pl} is the plastic load-line displacement.

3.2.2. The η_{pl}-factor estimation

The procedure of calculating the η_{pl}-factor can be also based on the load separation method which was proposed.[13] The method assumes that the load can be represented as a product of two functions, namely, a crack geometry function G and a material deformation function H. To evaluate the plastic η-factor (η_{pl}) for planar specimens with cracks, Sharobeam and Landes[14] suggested an experimental procedure using the load separation concept. They assumed the load to be in the form

$$P = G\left(\frac{b}{W}\right) H\left(\frac{v_{pl}}{W}\right) \qquad (9)$$

Using this load separation form, it is possible from Eqs. (8) and (9) to represent the η_{pl}-factor as follows

$$\eta_{pl} = \frac{b}{\int G\left(\frac{b}{W}\right) H\left(\frac{v_{pl}}{W}\right) dv_{pl}} \frac{d\left(\int G\left(\frac{b}{W}\right) H\left(\frac{v_{pl}}{W}\right) dv_{pl}\right)}{db} = b \frac{G'\left(\frac{b}{W}\right)}{G\left(\frac{b}{W}\right)} \quad (10)$$

The load separation concept introduces a separation parameter S_{ij} as the ratio of loads $P(a, v_{pl})$ of same specimens but with two different crack lengths a_i and a_j over the whole domain of the plastic displacement, namely

$$S_{ij} = \frac{P(a_i, v_{pl})}{P(a_j, v_{pl})}\bigg|_{v_{pl}=const} \quad (11)$$

Substituting Eq. (9) into Eq. (11), we obtain another presentation of the separation parameter S_{ij}

$$S_{ij}\left(\frac{b_i}{W}\right)\bigg|_{v_{pl}} = \frac{G\left(\frac{b_i}{W}\right) H\left(\frac{v_{pl}}{W}\right)}{G\left(\frac{b_j}{W}\right) H\left(\frac{v_{pl}}{W}\right)}\bigg|_{v_{pl}} = \frac{G\left(\frac{b_i}{W}\right)}{G\left(\frac{b_j}{W}\right)}\bigg|_{v_{pl}} = const \quad (12)$$

It was suggested[14] that for a given material in the separation region, the $S_{ij}\left(\frac{b_i}{W}\right)$ parameter as a function of remaining ligament b_i determined at constant plastic load-line displacement can be fitted with a power law $G\left(\frac{b}{W}\right) = C\left(\frac{b}{W}\right)^{\eta_{pl}}$, i.e.

$$S_{ij}\left(\frac{b_i}{W}\right)\bigg|_{v_{pl}} = \frac{C\left(\frac{b_i}{W}\right)^{\eta_{pl}}}{C\left(\frac{b_j}{W}\right)^{\eta_{pl}}}\bigg|_{v_{pl}} = A_m\left(\frac{b_i}{W}\right)^{\eta_{pl}} \quad (13)$$

where C is a constant. Thus, the slope of the $S_{ij} - b/W$ curve is the η_{pl}-factor in Eq. (6).

3.2.3. The η_{pl}- and η-factors

Recently, the load separation method has been developed for the theoretical and experimental estimation of the η_{pl}-factor for standard and non-standard specimens with certain combinations of crack size, material and specimen configuration and recommended to obtain η_{pl}-values for new configurations.[12,15–17]

Unfortunately, there are not equations to calculate the stress intensity factor in Eq. (6) for the presented non-standard curved notched specimen and to estimate the J-integral and the notch fracture toughness $J_{\rho,c}$. It has been proposed to come back to the original representation of the J-integral as total energy release rate, namely,

$$J = -\frac{1}{B}\frac{dA}{da} \qquad (14)$$

and

$$J = \eta \frac{A}{B(W-a)} \qquad (15)$$

where A is total area under the load versus total load-line displacement curve, η is a total η-factor. It is attractive to estimate the difference between the elastic and plastic components of the η-factor. Equation (5) leads to the expression

$$\eta_{pl} = \eta + (\eta - \eta_{el})\frac{A_{el}}{A_{pl}} \qquad (16)$$

If the value of the plastic work done is dominant on the load versus load-line displacement curve, i.e. $A_{pl} >> A_{el}$, that it is a typical for ductile materials, than the following equation $\eta_{pl} \approx \eta$ is valid. In this case, the J-integral can be estimated as

$$J = \eta \frac{A}{B(W-a)} \approx \eta_{pl}\frac{A}{B(W-a)} \qquad (17)$$

Experimental results of the work done to fracture in the case of the steel XC 38 show that the difference between the η-factor and its plastic component is negligible, i.e. less than 4%.[18]

$$-0.014 \leq \eta - \eta_{pl} \leq 0.007 \qquad (18)$$

Moreover, same conclusions can be made from experimental results published.[15]

Thus, procedure of calculating the η-factor and the J-integral (Eq. (17)) can be based on the load separation method and allows experimental estimation of the notch fracture toughness $J_{\rho,c}$ for the non-standard notched specimen of the present API 5L X52 steel.

4. Results and discussion

The stress distribution ahead of the notch tip and along notch ligament is computed by the Finite Element Method taking into account the critical load defined by acoustic emission technique (Figure 6). The critical notch stress intensity factor $K_{\rho,c}$ has been calculated using the effective distance and the effective stress obtained from the relative stress gradient as described in Section 3.1. The results of calculations are summarized in Table 3 for 11 specimens with notch aspect ratio $a/W = 0.2$.

The mean value of the notch fracture toughness (the critical notch stress intensity factor) is 57.21 MPa √m with a standard deviation of 0.67 MPa √m and a coefficient of variation of 1.17%. The scatter of the notch fracture toughness is low and does not exceed 10%.

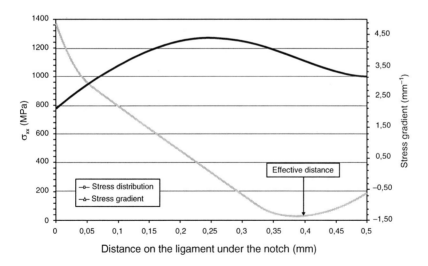

Figure 6. Maximum principal stress and relative stress gradient distributions ahead of the notch tip on the line of crack extension ($a/W = 0.2$).

TABLE 3. The notch fracture toughness of the API 5L X52 steel calculated by the volumetric method for the specimens with the notch aspect ratio $a/W = 0.2$.

Test no	$X_{ef\ initiation}$ (mm)	$K_{p,\ c}$ (MPa√m)
1	0.399	58.41
2	0.394	57.22
3	0.394	57.22
4	0.399	58.41
5	0.394	57.22
6	0.393	56.20
7	0.394	56.86
8	0.394	56.86
9	0.394	57.05
10	0.394	57.22
11	0.393	56.63
Mean	0.395	57.21
Standard deviation	0.002	0.672
Coefficient of variation	0.5%	1.17%

The load separation method has been adopted to measure η_{pl}-factor for the notched curved specimen. The η_{pl}-factor has been estimated by testing ten specimens with the notch aspect ratio $a/W = 0.15, 0.25, 0.3, 0.5$ and 0.7. Figure 7 shows the original test records for the notched "Roman tile" specimens. To determine the load versus plastic displacement v_{pl} record, the elastic displacement was subtracted from the total displacement, i.e. $v_{pl} - P \cdot C$. The compliance C was calculated from the elastic part of the load versus total displacement curves.

The separation parameter S_{ij} for each notched specimen test record was obtained by dividing the notched specimen load record by the reference notched specimen load record for the same plastic displacement. The notched curved specimen with the crack aspect ratio $a/W = 0.15$ was used as the reference specimen. Therefore, three values of S_{ij} have been obtained ($S_{15/15}$, $S_{25/15}$ and $S_{30/15}$). Figure 8 shows the separation parameters S_{ij} of the curved specimen with respect to plastic displacement. It can be seen that the S_{ij} plot had some unseparable region at the beginning of the record, and then the S_{ij} parameter becomes approximately independent of plastic displacement. Therefore, the separation constants S_{ij} versus the ligament b_i/W were estimated from the approximately constant separation parameter region in Figure 8.

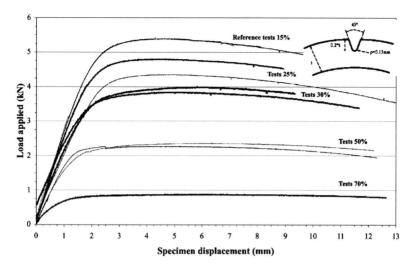

Figure 7. Original test records of the notched curved specimens of the API 5L X52 steel pipe.

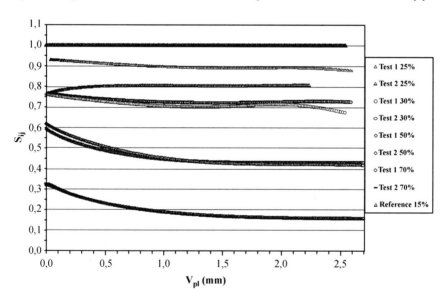

Figure 8. Variation of separation parameters S_{ij} of the notched curved specimens of the API 5L X52 steel pipe versus plastic displacement.

The straight line fits through these points and the slope of the $\log(S_{ij}) - \log(b_i/W)$ curve for the notched curved specimen (Figure 9) according to Eq. (13) is the η_{pl}-factor, which equals to 1.78 for the value of $a/W = 0.15...0.7$. The obtained η_{pl}-factor is closed to 2, i.e. the theoretical value which is found for deep sharp notch bending solution. The η_{pl}-factor was estimated for the specimen with root radius of 0.15 mm. But, it should be noted[18] that the η_{pl}-factor should be a function of root radius.

The η-factor is assumed to be constant for the present notch aspect ratios and equal to the η_{pl}-factor (see Section 3.2.3).

The notch fracture toughness (in terms of the J-integral) of the curved "Roman tile" specimen has been calculated using the load versus total load-line displacement records, the predicted η-factor and Eq. (15). The results

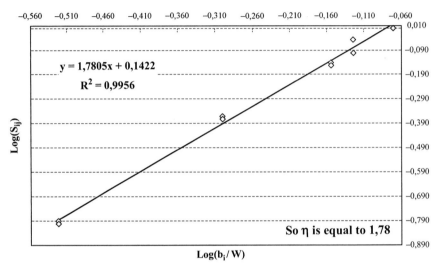

Figure 9. Separation constants of the notched curved specimens of the API 5L X52 steel pipe versus non-notched ligament.

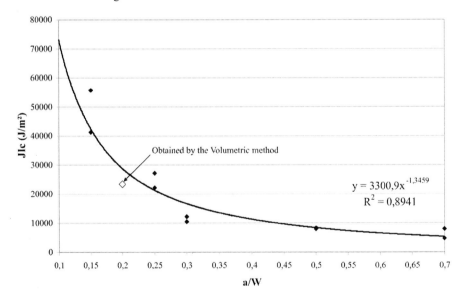

Figure 10. Dependence of the notch fracture toughness $J_{\rho,c}$ of the "Roman tile" specimens of the API 5L X52 steel pipe versus the notch aspect ratio. The value of $J_{\rho,c}$ for $a/W = 0.2$ was represented as a result of the tests of 11 specimens.

of the notch fracture toughness $J_{\rho,c}$ as a function of the notch aspect ratio are presented in Figure 10. It can be seen that the value of $J_{\rho,c}$ increases with the decrease of the notch aspect ratio.

Thus, the present results on the notch fracture toughness of the API 5L X52 steel can be used in the modified failure assessment diagram for structural integrity assessment of gas pipelines damaged by scratches.

5. Concluding remarks

A non-standard curved specimen, so called "Roman tile" specimen, and the bend-test fixture have been suggested as it is not possible to get flat specimens from small diameter pipes. These curved specimens have been used to measure the notch fracture toughness of the API 5L X52 steel in presence of a notch simulating a scratch which can be produced during service of pipelines.

The load versus load-line displacement was recorded digitally for processing by computer. The test procedure allowed measuring the critical load corresponding to crack initiation. In the present case, acoustic emission technique has been employed for this purpose. Comparison of dependences of the load versus time and duration of acoustic emission versus time indicates crack initiation which was closed to "pop-in".

The notch fracture toughness was calculated in terms of the notch stress intensity factor and the J-integral. The notch fracture toughness $K_{\rho,c}$ has been obtained by means of the volumetric method. To estimate the notch fracture toughness $J_{\rho,c}$, the test procedure based on the load separation method has been attracted to determine the η_{pl}-factor of the non-standard curved notched specimen. In this case, the original representation of the J-integral as total energy release rate (Eq. 15) and the tests records, namely, load versus total load-line displacement have been used taking into account an aquality between the η_{pl}- and η-factors. Thus, the proposed procedure allows avoiding calculation of the stress intensity factor for non-standard specimens.

Development of the present methodology encourages replacing the conservative crack-like defects approach. The results on the notch fracture toughness of the API 5L X52 steel can be used in the modified failure assessment diagram for structural integrity assessment of gas pipelines damaged by scratches.

ACKNOWLEDGEMENTS Professor Yu.G. Matvienko and Professor G. Pluvinage acknowledge the support of the NATO (Grant CBP.NR.NRCLG 982800). This work was partly made when Yu.G. Matvienko was a Visiting

Professor at National School of Engineering of Metz (ENIM) and University of Metz (UPVM), France.

References

1. Rapport de l'enquête MH-2-95. Fissuration par corrosion sous tension des oléoducs et des gazoducs canadiens, Office National d'Energie, 1996.
2. Zerbst U, Ainsworth RA, Schwalbe K-H. Basic principles of analytical flaw assessment methods. Int J Press Vessels Piping 2000; 77: 855–867.
3. Webster S, Bannister A. Structural integrity assessment procedure for Europe – of the SINTAP programme overview. Eng Fract Mech 2000; 67: 481–514.
4. Ainsworth RA, Bannister AC, Zerbst U. An overview of the European flaw assessment procedure SINTAP and its validation. Int J Press Vessels Piping 2000; 77: 869–876.
5. SINTAP: Structural integrity assessment procedure. Final Revision, EU-Project BE 95-1462, Brite-Euram Programme, Brussels, 1999.
6. Adib-Ramezani H, Jeong J, Pluvinage G. Structural integrity evaluation of X52 gas pipes subjected to the external corrosion defects using the SINTAP procedure. Int J Press Vessels Piping 2006; 83: 420–432.
7. Adib H, Jallouf S, Schmitt C, Carmasol A, Pluvinage G. Evaluation of the effect of corrosion defects on the structural integrity of X52 gas pipelines using the SINTAP procedure and notch theory. Int J Press Vessels Piping 2007; 84: 123–131.
8. Pluvinage G. Fracture and fatigue emanating from stress concentrators. Dordrecht: Kluwer; 2003.
9. Taylor D, Cornetti P, Pugno N. The fracture mechanics of finite crack extension. Eng Fract Mech 2005; 72: 1021–1038.
10. Kim JH, Kim DH, Moon SI. Evoluation of static and dynamic fracture toughness using apparent fracture toughness of notched specimen. Mater Sci Eng A 2004: 381–384.
11. FITNET Consortium, FITNET: European FITNET for Service Network, EU's Framework 5, Proposal No. GTC-2001-43049, Contract No. G1RT-CT-2001-05071.
12. Kim YS, Matvienko YuG, Jeong HC. Development of experimental procedure based on the load separation method to measure the fracture toughness of Zr-2.5Nb tubes. Key Eng Mater 2007; 345–346: 449–452.
13. Paris PC, Ernst H, Turner CE. A J-integral approach to development of η-factors. In: Fracture Toughness: Twelfth Conference, ASTM STP 700, American Society for Testing and Materials, 1980: 338–351.
14. Sharobeam MH, Landes JD. The load separation criterion and methodology in ductile fracture mechanics. Int J Fract 1991; 47: 81–104.
15. Cassanelli AN, Cocco R, de Vedia LA. Separability property and η_{pl} factor in ASTM A387-Gr22 steel plate. Eng Fract Mech 2003; 70: 1131–1142.
16. Wainstein J, de Vedia LA, Cassanelli AN. A study to estimate crack length using the separability parameter S_{pb} in steels. Eng Fract Mech 2003; 70: 2489–2496.
17. Matvienko YuG. Separable functions in load separation for the η_{pl} and η_{plCMOD} plastic factors estimation. Int J Fract 2004; 129: 265–278.
18. Akourri O, Louah M, Kifani A, Gilgert G, Pluvinage G. The effect of notch radius on fracture toughness J_{IC}. Eng Fract Mech 2000; 65: 491–505.

SAFE AND RELIABLE DESIGN METHODS FOR METALLIC COMPONENTS AND STRUCTURES DESIGN METHODS

G. PLUVINAGE
Laboratoire de Fiabilité Mécanique
ENIM-UPVM, Metz, 57045, France

Abstract Tools available for structure design are various and are different according to the fact that the risk is associated with an eventual presence of defect and possibility of brittle fracture or a ductile failure. The following design tools are described: (1) allowable stress and safety factor, (2) linear fracture mechanics and safety factors, (3) crack driving force, (4) failure assessment diagram, (5) allowable strain, (6) critical gross strain, (7) strain based design.

Keywords: Design tools, safety factors, failure assessment diagram, strain based design

1. Introduction

Structural engineering has existed since humans first started to construct their own structures. It became a more defined and formalized profession with the emergence of the architecture profession as distinct from the engineering profession during the industrial revolution in the late nineteenth century. Until then, the architect and the structural engineer were often one and the same – the master builder. Only with the understanding of structural theories that emerged during the nineteenth and twentieth centuries did the professional structural engineer come into existence.

On 24 May 1847, the Dee Bridge collapsed as a train passed over it, with the loss of five lives. It was designed by Robert Stephenson, using cast iron girders reinforced with wrought iron struts. The bridge collapse was subject to one of the first formal inquiries into a structural failure. The result of the enquiry was that the design of the structure was fundamentally flawed, as the wrought iron did not reinforce the cast iron at all, and due to repeated bending it suffered a brittle fracture due to fatigue.

The role of a structural engineer today involves a significant understanding of both static and dynamic loading, and the structures that are available to resist them. Structural engineers often specialize in particular fields, such as bridge engineering, building engineering, pipeline engineering, industrial structures or special structures such as vehicles or aircraft. Tools available for structure design are various and are different according to the fact that the risk is associated with an eventual presence of defect and possibility of brittle fracture or ductile failure. Table 1 provides a classification of the existing tools.

In this paper, these different structure design tools are described with in each case an example for the design of pipelines.

TABLE 1. Classification of structure design tools.

Brittle fracture	Material fracture	Admissible stress and safety factor
	Emanating from defect	Linear fracture mechanics and safety factors
Elastoplastic fracture	Material fracture	Admissible stress and safety factor
	Emanating from defects	Crack driving force Failure assessment diagram
Ductile failure	Material failure	Admissible strain
	Emanating from defect	Critical gross strain
	Instability	Strain based design

2. Safety factor and design factor

The early methods for design against the risk of failure were based on the concept of ppermissible stress design (in USA construction more commonly called allowable strength design) which is a design philosophy used by civil engineers. The designer ensures that the stresses developed in a structure due to service loads do not exceed the strength limit. But the knowledge of the service loads is not sufficient: it is necessary to envisage an unsuited use such as: imprudence of the user, accidental overload, failure of a component, unforeseen external event etc. One uses for that a safety factor usually noted f_s. The method is usually determined by ensuring that permissible stresses σ_{ad} remain within the limits through the use of safety factor and strength limit is generally the yield stress Re for conservative reasons.

$$\sigma_{ad} = \frac{Re}{f_s} \tag{1}$$

The most ancient information about the use of safety factor is given in the code of Hammurabi. (Codex Hammurabi), the best preserved ancient law code, was created. 1760 BC (middle chronology) in ancient Babylon. It was enacted by the sixth Babylonian king, Hammurabi. At the top of a basalt stele is a bas-relief image of a Babylonian god (either Marduk or Shamash), with the king of Babylon presenting himself to the God, with his right hand raised to his mouth as a mark of respect. The text covers the bottom portion with the laws written in cuneiform script. It contains a list of crimes and their various punishments, as well as settlements for common disputes and guidelines for citizen conduct. It is mentioned that an architect who carried out a house which is ploughed up on its occupants and cause their death, is condemned to the capital punishment. In addition, it is noted that when *"when a stone is necessary to build a palace, the architect has to plane to use two stones"*.

The use of a safety factor of 2 was also mentioned by Buffon (1744). Georges-Louis Leclerc. Comte de Buffon (September 7, 1707–April 16, 1788) was a French naturalist, mathematician, biologist, cosmologist and author. He has study the strength of wood for civil and military construction. We can read: *"thus in buildings design for a long life duration, one should give to wood at most only half the load which can make it break, It is only in pressing cases and for in buildings for construction which should not last, as when it is necessary to make a Bridge to pass an army, or a scaffold to help or attack a City, which one can hazarder give to wood two thirds of his load"*.

We can note that a safety factor for 1.5 is authorized in special and emergency cases. Safety factor (f_s) can mean either the fraction of structural capability over that required, or a multiplier applied to the maximum expected load (force, torque, bending moment or a combination) to which a component or assembly will be subjected. The two senses of the term are completely different in that the first is a measure of the reliability of a particular design, while the second is a requirement imposed by law, standard, specification, contract or custom. Careful engineers refer to the first sense as a safety factor, or, to be explicit, a realized factor of safety, and the second sense as a design factor (DF), but usage is inconsistent and confusing, so engineers need to be aware of both. The safety factor is given to the engineer as a requirement. The design factor is calculated by the engineer. The safety factors are defined by the "state of art" for each field, possibly codified in standards. It is equal to or higher than 1, and is as much higher as the system is badly defined, and than service loads are badly controlled. Typical values of safety factor are given in Table 2.

TABLE 2. Typical safety factors.

Safety factor (f_s)	Applied structures loads	Structure stresses	Material behavior	Observations
$1 \leq f_s \leq 2$	Regular and well known	Known	Known after test	Operation without jolt
$2 \leq f_s \leq 3$	Regular and known	Relatively known	known	Usual operation with light shocks and moderate overloads
$3 \leq f_s \leq 4$	Not well known	Not well known	No tests	
	Uncertain	Unknown and uncertain	Not very known	

Appropriate safety factors are based on several considerations. Prime considerations are the accuracy of load, strength, and wear estimates, the consequences of engineering failure, and the cost of over engineering the component to achieve that safety factor. For example, components whose failure could result in substantial financial loss, serious injury or death usually can use a safety factor of 4 or higher (often 10). Non-critical components generally might have a design factor of 2. Risk analysis, failure mode and effects analysis and other tools are commonly used.

Buildings commonly use a factor of safety of 2.0 for each structural member. The value for buildings is relatively low because the loads are well understood and most structures are redundant. Pressure vessels use 3.5 to 4.0, automobiles use 3.0, and aircraft and spacecraft use 1.4 to 3.0 depending on the materials. Ductile, metallic materials use the lower value while brittle materials use the higher values. The field of aerospace engineering uses generally lower design factors because the costs associated with structural weight are high. This low design factor is why aerospace parts and materials are subject to more stringent quality control.

Many codes require the use of a Margin of Safety (M.S.) to describe the ratio of the strength of the structure to the requirements.

$$\text{Design Factor} = \text{Failure Load}/\text{Design Load}$$
$$\text{Margin of Safety} = [\text{Failure Load}/(\text{Design Load}*f_s)] - 1 \qquad (2)$$

For a successful design, the design factor must always equal or exceed the required safety factor and the safety margin is greater than zero. The Safety Margin is sometimes, but infrequently, used as a percentage, i.e., a 0.50% M.S versus a 50% M.S. When a structure meets all requirements it is said to have a "positive margin". A measure of strength frequently used in Europe is the Reserve Factor (RF). With the strength and applied loads expressed in the same units, the reserve factor is defined as:

DESIGN METHODS

$$RF = \text{Proof Strength/Proof Load}$$
$$RF = \text{Ultimate Strength/Applied Load} \quad (3)$$

The applied loads have safety factors including applied safety factors. The use of a safety factor does not imply that a structure or design is "safe". Many quality assurance, engineering design, manufacturing, installation, and end-use factors may influence whether or not a structure is safe in any particular situation. The use of safety factor has been extended for determination of admissible defect size a_{ad} for brittle materials using linear fracture mechanics. For that, three safety factors are used:

f_s^K the safety factor on fracture toughness

f_s^a the safety factor on applied stress

f_s^a the safety factor on defect size

The admissible defect size is then given by the following relationship:

$$f_s^a a_{ad} = \frac{1}{\pi} \left(\frac{K_{Ic}}{f_s^K} \right)^2 \left(\frac{1}{f_s^\sigma \sigma_{g,ap}} \right)^2 \quad (4)$$

where K_{IC} is the material fracture toughness and $\sigma_{g,\,ap}$ is the applied gross stress. According to the degree of knowledge of the material properties and applied loads, the following safety factors are used (Table 3):

TABLE 3. Safety factors used for determination of admissible defect size a_{ad} for brittle materials using linear fracture mechanics.

Safety factors	f_s^K	f_s^a	f_s^σ
Case 1	1.0	2	1.0
Case 2	1.0	1.1	1.2
Case 3	1.2	1.4	1.4

Safety factor on fracture toughness, defect size and applied loads are also defined in Failure Assessment Diagram (FAD) as we can see later.

The values of classical safety factor are codified in codes such as Eurocode 3. However, the actual trend is to adapt the safety factor according to the degree of uncertainties of the material properties in order to avoid over conservatism. The design material properties are defined as some percentile of the material resistance distribution (The mean values is generally used for metals and alloy but for wood the admissible stress is defined as the fifth percentile of the distribution).

When using a probabilistic approach to design, the designer no longer thinks of each variable as a single value or number. Instead, each variable is viewed as a probability distribution. The main characteristics of these distributions are:

$$\bar{x} = \sum_{i=1}^{n} x_i / n, \; s_x^2 = \sum_{i=1}^{n} (x_i - \bar{x})^2 / (n-1), \; s_x^3 A_x = \sum_{i=1}^{n} (x_i - \bar{x})^3 / n \quad (5)$$

where $\bar{x}$ and s_x are mean and standard deviation of a random variable x and A_x is the asymmetry parameter of the distribution. Generally, after statistical data processing, it appears that the mean and the standard deviation of a random variable x are constant and independent of the distribution shape. For this reason, we will present the density probability functions versus coefficient of variation.

$$c_{V,x} = s_x / \bar{x} \quad (6)$$

The coefficient of variation is an excellent indicator of homogeneity of the sample. This one will be declared homogeneous if $c_{V,x} < 1/3$. Concerning the properties of materials, if the mechanical tests were carried out carefully, the coefficient of variation is an excellent indicator of the process quality. Thus, the production of a low carbon steel led to a coefficient of variation $c_{V,x} = 0{,}1$ with Rm the ultimate resistance (Pluvinage and Sapounov (2007)). From a general point of view, one can estimate that for structural components the concept of continuous medium is hardly applicable for $c_{V,x} < 0{,}2$. We will rewrite the probability density f functions of distribution by introducing into the corresponding relations the coefficient of variation. For normal distribution (most widespread), the density of probability is expressed with two parameters as:

$$p(x) = \frac{1}{\sqrt{2\pi}\, s_x} \exp\left[-\frac{(x/s_x - 1/c_{V,x})^2}{2}\right] \quad (7)$$

For the lognormal distribution with $y = \lg x$, we have by analogy with the formula (7):

$$p(y) = \frac{1}{\sqrt{2\pi}\, s_y} \exp\left[-\frac{(y/s_y - 1/c_{V,y})^2}{2}\right] \quad (8)$$

For the Weibull distribution, it is expressed by:

$$p(x) = cmx^{m-1} \exp(-cx^m) \tag{9}$$

where m is the Weibull's modulus and c the normalisation factor. The mean, standard deviation and asymmetry are given by:

$$\bar{x} = c^{-1/m}\Gamma(1+1/m) \quad s_x^2 = c^{-2/m}\left[\Gamma(1+2/m) - \Gamma^2(1+1/m)\right]$$

$$A_x = \frac{\Gamma(1+3/m) - 3\Gamma(1+1/m)\Gamma(1+2/m) + 2\Gamma^3(1+1/m)}{\left[\Gamma(1+2/m) - \Gamma^2(1+1/m)\right]^{3/2}} \tag{10}$$

The coefficient of variation is the given by:

$$c_{V,x} = \sqrt{\frac{\Gamma(1+2/m)}{\Gamma^2(1+1/m)} - 1} \tag{11}$$

where Γ is the symbol of the Gamma–Euler function. The Weibull's modulus can be estimate by the following empirical relationship:

$$m = c_{V,x}^{-1.09} \tag{12}$$

From this perspective, probabilistic design predicts the flow of variability through a system. By considering this flow, a designer can make adjustments to reduce the flow of random variability and improve quality. Proponents of the approach contend that many quality problems can be predicted and rectified during the early design stages and at a much reduced cost. The safety factor is then defined as the ratio of the ultimate strength which corresponds to the mean value of the strength distribution over the admissible stress. The admissible stress is the failure stress associated with a low and conventional probability of failure P_f^* (10^{-4} or 10^{-6} if there is human life risk). If the ultimate strength follows the Weibull's distribution, the probability of failure is given by the following relationship:

$$p_f^* = \exp-\left[\frac{\Gamma(1+1/m)}{f_s}\right]^m \tag{13}$$

and the safety factor is:

$$f_s = \Gamma(1+1/m) \bigg/ \left[Ln\frac{1}{P_s^*}\right]^{1/m} \tag{14}$$

We can note that the safety factor increases considerably when the Weibull's modulus decreases i.e. the scatter of material strength increases.

3. Transition temperature

The concept of brittle–ductile transition temperature was discovered during the Second World War, because of the rupture of liberty ships at sea. The ignorance of this phenomenon is also one of the causes of the shipwreck of Titanic. At low temperatures where the impact energy required for fracture is less, a faceted surface of cleaved planes of ferrite is observed, indicating brittle fracture. Using 50% shear fracture area as a reference point, this would occur in ASTM A36 at −3°C, while for the Titanic steel, this value would occur at 49°C in the longitudinal direction and at 59°C in the transverse direction. At elevated temperatures, the impact-energy values for the longitudinal Titanic steel is substantially greater than the transverse specimens, as shown in Figure 1. The difference between the longitudinal and transverse shear fracture percent from the Titanic is much smaller.

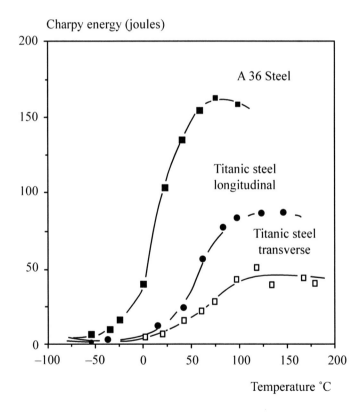

Figure 1. Charpy impact energy versus temperature for longitudinal and transverse Titanic specimens and ASTM A36 steel. (after www.tms.org.)

3.1. SPECIFIC IMPACT ENERGY TRANSITION TEMPERATURE

The use of the Charpy test is a cheap way to determine the transition temperature from brittle fracture to ductile failure (Pluvinage 2003). This transition temperature can be defined on three manners:

- For a given level of resilience for example 27 Da J/cm^2
- At the mid level between lower shelf and upper shelf region
- At 50% of cleavage fracture surface

This transition temperature is not intrinsic of the material as we can see in Figure 2. In this figure, we compare the work done for fracture-temperature curve obtained on the same steel but with a V notch in the first case and with a U notch in the second case. Tests have been performed on a St 52-3 steel (Pluvinage 2003). We can note that for Charpy V specimen, the transition temperature for brittle to ductile fracture is shifted from higher temperature and the level of the ductile plateau is decreasing.

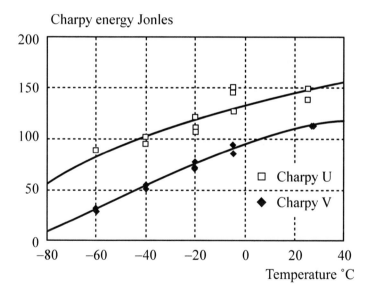

Figure 2. Charpy energy versus temperature for a Charpy U and V specimens. (Steel St 52-3.)

3.2. FRACTURE TOUGHNESS K_{IC} TRANSITION TEMPERATURE

The facture toughness K_{IC} evolution versus temperature for brittle and quasi brittle fracture can be modelled through plasticity–temperature relationship because fracture process needs a preliminary yielding. Plasticity is a thermally activated phenomenon which follows Arrhenius law. As shown in Figure 3 relative to a pipe steel, fracture toughness versus temperature is given by:

$$K_{Ic} = K_{min} + A.\exp(-BT) \quad (15)$$

K_{min} is the fracture toughness threshold, A and B are constants and T temperature in kelvin. According to the Krasowsky and Pissarenko model (Krasowsky and Pluvinage (1993)), $K\mu$ is connected to the yield stress at 0 K and characteristic distance X_c by the following relationship:

$$K_{min} = Re^0 \sqrt{\pi X_c} \quad (16)$$

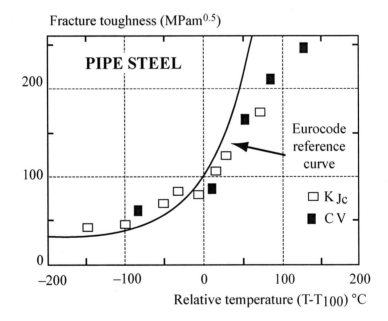

Figure 3. Evolution of fracture toughness with temperature for a pipe steel.

DESIGN METHODS

This evolution is now made in a statistical way in US standard ASTM E1921-08ae1 (2007). It consists in fact in three curves: the median curve and curves for lower and upper limits. The median curve is given by a universal representation:

$$K_{Jc}(50\%) = 30 + 70 \exp[0.0019(T - T_{100})] \qquad (17)$$

in which T_{100} is the temperature associated with the fracture toughness value of 100 MPa$\sqrt{m}$ (It is not possible to speak about abrupt transition because with have a continuous evolution of fracture toughness). Let us notice that the fracture toughness threshold value of 30 MPa$\sqrt{m}$ is a high value considering experimental values found for low temperature tests. The reference temperature T_0 can be evaluated by the following methods:

- Method of single temperature (ASTM E1921 (2007))
- Multi temperature method
- Method of single temperature associated with a statistical distribution of fracture toughness according to the three parameters Weibull's model (Wallim 1990):

$$P_f = 1 - \exp\left[-\left(\frac{K_J - K_{min}}{K_o - K_{min}}\right)^4\right] \qquad (18)$$

K_{min} is a fracture toughness threshold and K_0 a scale factor.

$$K_o = \left[\sum_{i=1}^{N}\left(K_{J(i)} - K_{min}\right)^4 \Big/ r - 0{,}3008\right]^{Y4} + K_{min} \qquad (19)$$

r is the number of uncensored results and N total number of results. The fracture toughness mean value is given by:

$$K_J(50) = \left(K_o - K_{min}\right)\left[L_n(2)\right]^{Y4} + K_{min} \qquad (20)$$

and the T_{100} temperature:

$$T_{100} = T - \frac{1}{0{,}019} L_n\left[\frac{K_{J(50\%)} - 30}{70}\right] \qquad (21)$$

The T_0 standard deviation estimation is given by:

$$d_s(T_o) = \beta/\sqrt{N} \qquad (22)$$

with β confidence interval for the reference temperature. For the pipe steel of Figure 3 the T_{100} temperature has been estimated to $T_{100} = -51°$. The curve $K_{IC} = f(T - T_{100})$ has been plotted on the same graph that the curve $K_{mat} = f(T - T_{100})$ given by Eurocode 3 (2005). It can be noticed that results obtained from experiments are close to the standard $K_{mat} = f(T - T_{100})$ curve but only below the transition temperature. At higher temperature, the Eurocode K_{mat} curve overestimates the fracture toughness of the material.

Charpy V energy has been measured at different temperatures and the transition temperature $T_{27\,J}$ has been estimated to $T_{27\,J} = -129°C$. Charpy V energy has then been converted into K_{IC} using the Sanz' correlation (Sanz 1980):

$$K_{Ic} = 19\sqrt{KCV} \qquad (23)$$

where K_{CV} is the specific impact energy in J/cm². The obtained values have been plotted versus the temperature $(T - T_{100})$ on the same diagram. The concept of design presented in the last version of Eurocode 3 part 2 (2005) considers the brittle assessment in the transition zone and consists of a comparison of the fracture toughness resistance K_{mat} of the material with the applied stress intensity factor $K_{I,eq}$.

$$K_{mat} = \frac{K_{I,eq}}{(k_{R6} - \omega)} \qquad (24)$$

k_{R6} is the value of the non-dimensional stress intensity factor according to the CEGB-R6 method option 2 (1998) and ω is a plasticity correction according to the same method. $K_{I,eq}$ is the equivalent stress intensity factor of a structural detail taking into account the design stress σ_d and the design defect value a_d.

$$K_{I,eq} = M_K \cdot \sigma_d \cdot \sqrt{\pi \cdot a_d} \cdot F_\sigma(a/W) \qquad (25)$$

where $F_\sigma(a/w)$ is the geometrical correction factor and M_K is the correction factor to take into account the value of the stress concentration factor. The relationship between K_{mat} and the design temperature take into account the Wallim's correlation (Wallim 1990) between the transition temperature T_{100} defined at a conventional level of 100 MPa√m of fracture toughness and the transition temperature defined for Charpy energy $T_{27\,J}$ at conventional level of 27 J of fracture energy.

DESIGN METHODS

$$T_{100} = T_{27J} - 18°C \qquad (26)$$

Design against brittle fracture considers that the material exhibits at service temperature, a sufficient ductility to prevent cleavage initiation and sudden fracture with an important elastic energy release. Concretely, this means that the service temperature T_s is higher than the transition temperature:

$$T_s > T_t \qquad (27)$$

The service temperature is conventionally defined by codes or laws according to the country where the structure or the component is build or installed. For examples, in France, a law published in July 1974 indicates that the service temperature in France is $-20°C$. Recommendations of AIEA for nuclear waste containers indicate that the service temperature for these components is $-40°C$.

A Fracture Mechanics based design ensuring that the stress intensity factor for design is lower than admissible fracture toughness and the fracture toughness is greater than 100 MPa$\sqrt{m}$ (i.e. in accordance with the rules, the service temperature defined is above the reference temperature). This additional criterion is written:

$$T_s > RT_i + \Delta T \qquad (28)$$

where ΔT is uncertainty on reference temperature. This reference temperature RT_i differs according to codes (RT_{NDT} or RT_{T100}):

$RT_{NDT} = T_{NDT}$
$RT_{T100} = T_{100} + 19.4°C$
$RT_{NDT} \cong RT_{T00}$ ASME code case No 629, ASME (2005)

This method based on a long practice of the engineers is very conservative and does not allow appreciating the safety coefficient. As it was announced, the transition temperature approach is not a computing design method. It aims is to make sure that design cannot be possibly carried out in a bad way with a brittle material.

Transition temperature can be defined also on the graph critical crack opening displacement versus temperature. In this case, this temperature is defined when the plastic zone size is equal to thickness B (see Figure 4).

$$\delta_{c,Tt} = 2\pi \varepsilon_y B \qquad (29)$$

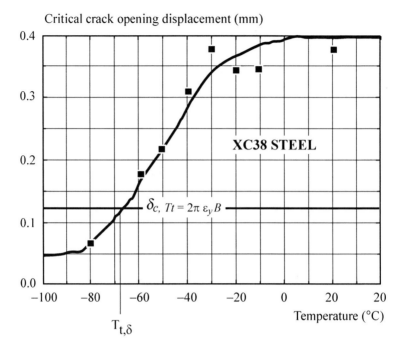

Figure 4. Definition of the critical crack opening displacement transition temperature $T_{t,\delta}$.

4. Crack driving force methods

4.1. THE CRACK DRIVING FORCE

The work done until fracture U_c, as the area of the curve load–displacement until fracture initiation is a direct measurement of fracture resistance. In order to take into account the geometry of the component, it is more convenient to use the specific fracture energy U_s to refer to material fracture resistance:

$$U_S = \frac{U_C}{Bb} \qquad (30)$$

where B is the thickness and b the ligament size. This specific energy is the sum of energy for fracture initiation and plastic dissipation in the process zone. According to Griffith, fracture resistance R is defined as:

$$R = \frac{dU_\Gamma}{B \cdot da} \tag{31}$$

where a is the crack length and U_Γ is the surface energy, which includes the plastic dissipation in the plastic zone.

For energetic equilibrium, the fracture resistance is counterbalanced by the crack driving force (CDF). From the above mentioned considerations, it has been seen that:

$$R = CDF = -\frac{1}{B}\frac{\partial U_c}{\partial a} \tag{32}$$

It is assumed according to Turner (1979) that the crack driving force is proportional to the specific fracture energy:

$$R = CDF = \eta U_s = \eta \frac{U_c}{Bb} \tag{33}$$

The crack driving force is then energy dissipated for creation of increment of crack length with units force per length (N/m). Several fracture parameters can expressed the crack driving force, the elastic strain energy release rate G; the stress intensity factor K, the J integral and the crack opening displacement. For a linear elastic situation, these parameters are equivalent and connected by the following relationship (for elastic behaviour and plane stress conditions).

$$K^2 = G \cdot E = Re \cdot \delta \cdot E \tag{34}$$

with E the Young's modulus.

4.2. PRINCIPLE OF THE CDF DESIGN

CDF Design is based on the CDF = $f(L_r)$ curve where L_r is the loading parameter. L_r is defined as the ratio of the applied load L and the limit load L_y or applied pressure p and the limit pressure p_y or the ratio of the applied net stress and the material yield stress Re:

$$L_r = \frac{F}{F_y} = \frac{\sigma_n}{Re} = \frac{p}{p_y} \tag{35}$$

In the CDF route, fracture assessment is made in two steps, the crack tip loading and the material fracture toughness determinations. The CDF curve which relates the crack driving force in terms of J or CTOD with the applied load is assumed to be a geometry independent function which only depends on the material behaviour. Failure is predicted when the crack driving force exceeds the fracture toughness (Figure 5).

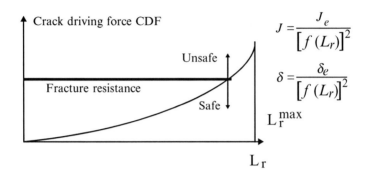

Figure 5. CDF design concept.

The basic equations of the CDF concept are:
for J integral as the CDF:

$$J = \frac{J_e}{[f(L_r)]^2} \qquad (36)$$

with

$$J_e = \frac{K^2}{E'} \quad E' = E(1-v^2) \qquad (37)$$

for COD as the CDF:

$$\delta = \frac{\delta_e}{[f(L_r)]^2} \qquad (38)$$

with

$$\delta_e = \frac{K^2}{Re.E'} \qquad (39)$$

4.3. J CRACK DRIVING FORCE

Turner has proposed (Turner 1979) a J design curve called Engineering J method (ENJ). For that, the J integral value is divided by the stress energy release rate for an applied net stress equal to the yield stress and the displacement is divided by the displacement for applied net stress also equal

DESIGN METHODS

to the yield stress. The relationship has been also established for the ratio of the net stress over the ligament and the same net stress at yielding. The ENJ consists of two parts:

$$\frac{JE}{Re^2 a} = 2\pi \cdot \left(\frac{d}{dy}\right)^2 \quad for \quad \left(\frac{d}{dy}\right) < 2.0 \tag{40}$$

$$\frac{JE}{Re^2 a} = 20 \cdot \left[\left(\frac{d}{dy}\right) - 0.75\right] \quad for \quad \left(\frac{d}{dy}\right) \geq 2.0 \tag{41}$$

$$\frac{JE}{Re^2 a} = 1.254\pi \cdot \left(\frac{\varepsilon_N}{\varepsilon_{N,Y}}\right)^2 \quad for \quad \left(\frac{\varepsilon_N}{\varepsilon_{N,Y}}\right) < 1.0 \tag{42}$$

$$\frac{JE}{Re^2 a} = 1.254\pi \cdot \left[\left(\frac{\varepsilon_N}{\varepsilon_{N,Y}}\right) - 1\right] \quad for \quad \left(\frac{\varepsilon_N}{\varepsilon_{N,Y}}\right) \geq 1.0 \tag{43}$$

4.4. COD CRACK DRIVING FORCE

Burdekin and Dawes (1971) has proposed a COD design curve made into two parts:

$$\Phi_\delta = \left(\frac{\varepsilon_g}{\varepsilon_{g,y}}\right)^2 \quad for \ \varepsilon_g / \varepsilon_{g,y} \leq 0.5 \tag{44}$$

$$\Phi_\delta = \left(\frac{\varepsilon_g}{\varepsilon_{g,y}}\right) - 0.25 \quad for \ \varepsilon_g / \varepsilon_{g,y} > 0.5 \tag{45}$$

where ε_g is gross stain and $\varepsilon_{g,y}$ the gross strain at yielding, Φ_δ is the non dimensional crack opening displacement.

$$\Phi_\delta = \frac{\delta}{2\pi \varepsilon_y a} \tag{46}$$

δ is COD and ε_y the yield strain. The COD design curve is obtained as an upper bound of experimental results.

4.5. ENGINEERING TREATMENT MODEL

This method has been proposed by (Schwalbe (1991)) is particularly used for welded joints. The crack driving force for welded joint in term of COD is given by:

$$\delta = \frac{\pi a \, \text{Re}_B}{2M^3 E} \cdot \left(\frac{\varepsilon g}{\varepsilon g, y, B}\right) \left[2M^2 \left(\frac{\varepsilon g}{\varepsilon g, y, B}\right)^2\right] \quad \text{for } 0 < F < F_{y,B} \quad (47)$$

$$\delta = \frac{\pi a \, \text{Re}_B}{E} \cdot \frac{1}{M} \left(\frac{\varepsilon g}{\varepsilon g, y, B}\right)^{2n_B} \left[1 + \frac{0.5}{M^2}\left(\frac{\varepsilon g}{\varepsilon g, y, B}\right)^{2n_B}\right] \quad (48)$$

for $F_{y,B} < F < F_{y,w}$

$$\delta = \frac{1.5 \pi a \, \text{Re}_B}{E} \cdot M^{(1-1/n_w)} \left(\frac{\varepsilon g}{\varepsilon g, y, B}\right)^{n_B/n_w} \quad (49)$$

for $F > F_{y,w}$

The subscript w and B indicates respectively the base and the weld metal and M is the mismatch factor.

$$M = \frac{\text{Re}_w}{\text{Re}_B} \quad (50)$$

5. Failure assessment diagram

The basic fracture mechanics relationship associates three parameters: defect size a, applied gross stress σ_g and fracture toughness R into a failure criterion expressed by the following equation.

$$F(\sigma_g, a, R) = 0 \quad (51)$$

Another presentation of the failure criterion can be made using a two parameters relationship:

$$k_r = f(L_r) \quad (52)$$

where k_r is the non dimensional crack driving force and L_r the non dimensional load.

$$k_r = \frac{K_{ap}}{K_{Ic}} = \sqrt{\frac{J_{ap}}{J_{Ic}}} = \sqrt{\frac{\delta_{ap}}{\delta_c}}; \quad L_r = \frac{F}{F_c} \quad (53)$$

where K_{ap}, J_{ap} and δ_{ap} are the applied stress intensity factor, J Integral or COD, K_{IC}, J_{IC} and δ_c the fracture toughness in given conditions of constraint, F applied load and F_C critical load.

Initially, the relationship $k_r = f(L_r)$ was obtained from a plasticity correction given by the Dugdale's model. This approach allows considering any failure conditions from purely brittle fracture to plastic collapse derived from the brittle one by a correction.

$$k_r = \left[\sqrt{\frac{8}{\pi^2} Ln \left(\frac{1}{\cos\left(\frac{\pi L_r}{2}\right)} \right)} \right]^{0.5} \quad (54)$$

In the past several failure assessment curves have been proposed (Newmann, Nureg, tangent method etc). Actually, as engineering tools, relationship (52) has been established from full scale tests and using the lower bound of results. There are several relationships given by different codes Actually the following methods are used EPRI (Kumar et al. 1981) in US, R6 (R6 1998) in UK, RCC MR (Moulin et al. 1998) in France and SINTAP (SINTAP 1998) in EU.

In all methods, it is assumed that the applied J integral consists of an elastic part and a plastic part $J_{ap} = J_{el} + J_{pl}$. J_{el} is obtained from classical linear fracture mechanics. Getting J_{pl} depends on methods and is based on the assumption that J_{pl} is get from J_{el} multiply by a correction factor

- In EPRI method EPRI (Kumar et al. 1981), this correction is given by the following relationship:

$$J_{pl} = J_{el} \cdot \left(\frac{E \varepsilon_{ref}}{\sigma_{ref}} + \phi \right) \quad (55)$$

ϕ is a plastic zone correction,

$$\sigma_{ref} = \frac{F}{F_0} \sigma_0 \quad (56)$$

where F_0 is the reference load and σ_0 the flow stress. ε_{ref} is the corresponding strain on the Ramber–Osgood relationship.
- For the R6 route R6 (R6 1998):

$$J_{pl} = \frac{J_{el}}{(k_r - \rho)^2} \quad (57)$$

ρ is a plasticity correction
- For RCC-MR A16 method RCC-MR (Moulin et al. 1998):

$$J_{pl} = J_{el} \cdot K_{A16} \quad (58)$$

where K_{A16} is the correction factor.

The failure assessment curve is similar in the different methods. The SINTAP procedure (SINTAP 1998) can be generally simplified to several distinct levels according to the no yield point elongation assumptions. The used mathematical expressions of the SINTAP procedure with the aforementioned assumption for "level 1" are presented below:

$$f(L_r) = \left[1 + \frac{L_r^2}{2}\right]^{-\frac{1}{2}} \left[0.3 + 0.7 \times e^{(-0.6 \times L_r^6)}\right], \quad (59)$$

$$\text{for } 0 \leq L_r \leq 1 \text{ where } L_r^{max} = 1 + \left(\frac{150}{\sigma_Y}\right)^{2.5}$$

where, $f(L_r)$, L_r, L_r^{max}, σ_Y, μ, E, σ_U, N, ε_{ref} and σ_{ref} are interpolating function, non dimensional loading or stress based parameter, maximum value of non dimensional loading or stress based parameter, yield stress, first correction factor, modulus of elasticity, ultimate stress, second correction factor, reference strain and reference stress, respectively.

In the FAD method, a failure curve or "interpolating curve" is used to assess the failure zone, safe zone and safety factor. In Figure 6, a typical failure assessment diagram is illustrated. In this FAD, an assessment curve, interpolating brittle fracture to plastic collapse separates the safe zone to the failure zone. As a consequence of L_r and K_r definition, the loading path OC is linear when load increases from 0 to critical load F_c. Under service conditions, a defect in a given material is represented by an assessment point B of coordinates L_r^* and k_r^*. If this assessment point in inside the safe zone, no failure occurs, if the assessment point is on the assessment curves or above critical conditions are reached. The safety factor associated with the defect situation is simply defined by relationship:

$$F_s = \frac{OC}{OB} \quad (60)$$

The criticality of the situation is generally given by comparing the obtained safety factor to the conventional value of $f_s = 2$.

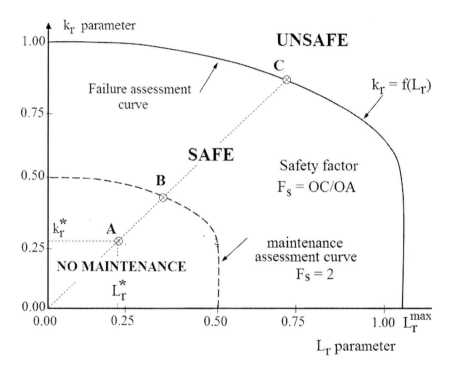

Figure 6. Presentation of Failure Assessment Diagram (FAD) which presents the evolution of non-dimensional stress intensity parameter versus non-dimensional loading or stress based parameter including an assessment point A (L_r^*, k_r^*).

However, The SINTAP procedure has been established for crack like defects and in the present study, we assess defect such as gouges which represents the majority of incidents defects for gas pipes. In this case a modify Notch Failure Assessment Diagram (NFAD) (Pluvinage 2007). The parameters of this NFAD are:

$$k_{\rho,r} = \frac{K_{\rho,ap}}{K_{\rho,c}}; \quad L_r = \frac{F}{F_c} \quad (61)$$

where $K_{p,\,ap}$ is the applied notch stress intensity factor obtained from the Volumetric Method (VM) and $K_{p,\,C}$ the critical notch stress intensity factor measured here on roman tile specimen. One notes that the notch fracture toughness is notch radius dependant. The used of non dimensional parameter $k_{p,\,r}$ allows to use the same assessment curves for nay notch radius. In our study, the SINTAP failure assessment curve is also used and the safety factor keeps the same definition.

6. Probabilistic defect assessment method

Monte Carlo (MC) is a simple method that uses the fact that the failure probability integral can be interpreted as a mean value in a stochastic experiment. An estimate is therefore given by averaging a suitably large number of independent outcomes (simulations) of this experiment. The basic building block of this sampling is the generation of random numbers from a uniform distribution (between 0 and 1). This random number can be used to generate a value of the desired random variable with a given distribution. Using the cumulative distribution function $F(X)$, the random variable would then be given as:

$$X = F_X^{-1}(u) \tag{62}$$

Then, to calculate the probability of failure, a multi-dimensional integral has to be evaluated:

$$P_F = \Pr[g(X) < 0] = \int_{g(X) < 0} f_X(x) dx \tag{63}$$

where $g(X)$ is a limit state function and $f_x(x)$ is a known joint probability density function of the random vector X.

To calculate the failure probability, one performs N deterministic simulations and determines whether the component analysed has failed (i.e. if $g(X) < 0$) after every simulation. For a count of the number of failures, N_F, an estimate of the mean probability of failure is:

$$P_F = \frac{N_F}{N} \tag{64}$$

The advantage with MCS is that it is robust, easy to implement into a computer program and for a sample size $N \rightarrow \infty$, the estimated probability converges to the exact result. Another advantage is that MCS works with any distribution of the random variables. There is no restriction on the limit state functions. However, MCS is rather inefficient, when calculating very low failure probabilities, since most of the contribution to P_F is in a limited part of the integration interval transform method This integral is very hard (impossible) to evaluate, by numerical integration, if there are many random parameters.

First and second-order reliability methods are general methods of structural reliability theory (Bjerager 1989). These methods are based on linear (first-order) and quadratic (second-order) approximations of the limit state surface g(x) = 0 tangent to the closest point of the surface to the origin of the space. The determination of this point involves nonlinear programming. The FORM/SORM algorithms involve several steps:

- In the first step, the space of uncertain parameters x is transformed into a new N-dimensional space u consisting of independent standard Gaussian variables. The original limit state g(x) = 0 then becomes mapped onto the new limit state $g_u(u) = 0$ in the u space.
- In the second step, the point on the limit state $g_u(U) = 0$ having the shortest distance to the origin of u space is determined by using an appropriate nonlinear optimization algorithm. This point is referred to as the design or β-point, and has a distance β_{HL} to the origin of the u space.
- In the third step, the limit state $g_u(u) = 0$ is approximated by a surface tangent to it at the design point. Let such limit states be $g_L(u) = 0$ and $g_Q(u) = 0$, which correspond to approximating surfaces of hyperplane (linear or first-order) and hyperparaboloid (quadratic or second-order), respectively.

The probability of failure P_F is thus approximated by $\Pr[g_L < 0]$ in FORM and $P_r[g_Q(u)<0]$ in SORM. These first- and second-order estimates $P_{F,1}$ and $P_{F,2}$ are given by:

$$P_{F,1} = \phi(-\beta_{HL})$$

$$P_{F,2} \approx \phi(-\beta_{HL})\prod_{i=1}^{N-1}(1-\kappa_i\beta_{HL})^{-1/2},$$

Where (65)

$$\phi(u) = \frac{1}{\sqrt{2\pi}}\int_{-\infty}^{u}\exp(-\frac{1}{2}\xi^2)d\xi$$

$\Phi(u)$ is the cumulative distribution function of a standard Gaussian random variable, and κ_i's are the principal curvatures of the limit state surface at the design point. FORM/SORM are analytical probability computation methods. Each input random variable and the performance function g(x) must be continuous.

Within the SINTAP procedure (SINTAP 1998), the following parameters are treated as random parameters:

- Fracture toughness
- Yield strength
- Ultimate tensile strength
- Defects distribution
- Pressure distribution

These random parameters are treated as not being correlated with one another. The parameters can follow a normal, log-normal, Weibull or some special distributions (for the defects).

The coefficient of variation CV_x is an excellent indicator of the homogeneity of the analyzed unit (Pluvinage 2007). This one will be declared homogeneous if $CV < 1/3$, concerning the properties of materials, if the mechanical tests were carried out carefully, the coefficient of variation is an excellent indicator of the manufactures quality, thus, the manufacture of low carbon steel leads to a coefficient of variation $CV = 0.1$, for ultimate strength, yield stress and fracture toughness. The pressure distribution obeys to the same coefficient of variation. We note that for exponential distribution the coefficient of variation is necessary taken as unit. The presentation of the method will be arrived out with the value of coefficient of variation.

6.1. FRACTURE TOUGHNESS DISTRIBUTION

The fracture toughness is assumed as Weibull's distribution. The Weibull's probability density function has the following form:

$$f(K_{IC}) = C \times m \times K_{\rho,c}^{m-1} \exp\left(-C \times K_{\rho,c}^{m}\right) \tag{66}$$

where C (scale) and m (shape) are the Weibull's distribution parameters. μ (mean) and σ (standard deviation) are the input data into the program and are related to the Weibull's parameters as follows:

$$\mu = \frac{C^{-1}}{m * \Gamma\left(1+\frac{1}{m}\right)} \qquad \sigma = C^{\frac{-2}{m}}\left[\Gamma\left(1+\frac{2}{m}\right) - \Gamma^2\left(1+\frac{1}{m}\right)\right] \tag{67}$$

where $\Gamma(Z)$ is the gamma function, defined by the following integral:

$$\Gamma(Z) = \int_0^\infty t^{Z-1} e^{-t} dt \tag{68}$$

This non-linear equation system is solved using a globally convergent method with line search and an approximate Jacobian matrix.

DESIGN METHODS

6.2. YIELD, ULTIMATE STRENGTH AND LOAD DISTRIBUTION

Yield strength ultimate, tensile strength and load can be mainly assumed as a normal distribution. The normal probability density function has the following form:

$$F(X) = \frac{1}{\sigma\sqrt{2\pi}} \exp\left[-\frac{1}{2}\left(\frac{X-\mu}{\sigma}\right)\right] \tag{69}$$

6.3. DEFECT SIZE DISTRIBUTION

The exponential distribution generally governs for defect size analysis. Consequently, the probability density function has the following form:

$$F(X) = \lambda \exp(-\lambda a) \tag{70}$$

where λ is the exponential distribution parameter. μ mean and the standard deviation, σ is related to λ as below:

$$\mu = \sigma = \frac{1}{\lambda} \tag{71}$$

An example of application of Monte Carlo and FORM-SORM method is given in Figure 7 in the case of a longitudinal defect in a pipe submitted to a 70 bar internal pressure (Jallouf et al 2005).

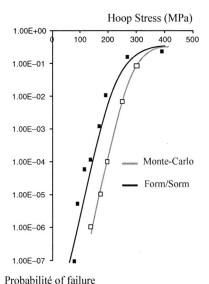

Figure 7. Evolution of the probability of failure of the gas pipe exhibiting a gouge defect with the hoop stress induced by internal pressure.

7. Limit analysis

To set the background for plastic limit analysis, it is helpful to consider in general terms the behavior of an elastic-plastic solid or structure subjected to mechanical loading. The solution to an internally-pressurized elastic-perfectly plastic cylinder provides a representative example. The solution shows that the vessel first yields at a critical internal pressure

$$p_a = \frac{Y}{\sqrt{3}}\left(1 - \frac{a^2}{b^2}\right)$$
$$p_y = \frac{\text{Re}}{\sqrt{3}}\left[1 - \left(\frac{a}{b}\right)^2\right] \quad (72)$$

The material first reaches yield at the interior of the cylinder $r = a$.

7.1. BOUNDING THEOREMS OF LIMIT ANALYSIS

The limit analysis method allows to compute two bound of the plastic collapse load which frame the real limit load P_L. This method is based on two principles:

The principle of virtual work
The Drucker's or minimum energy dissipation principle

This method uses the concept of statically and cinematically admissible stress fields (Save et al. 1997). The following assumptions are used:

The material behavior is elastic perfectly plastic.
The limit load is connecting with a totally plastic ligament.
The material is ductile until failure.
There is no instabilities.
The load path is proportional to a λ factor.

We consider a body of volume V loaded by external forces P_i, submitted to surface loads T_i and displacements u_i.

7.1.1. Lower bound theorem

If σ_{ij} is the real stress field (associated with a λ_{real} factor) and σ'_{ij} a statically admissible stress field (associated with a λ_- factor), the double application of the principle of virtual works gives:

$$\lambda_{-}\left|\int_V P_i u_i dV + \int_S T_i u_i dS\right| = \int_V \sigma'_{ij}\varepsilon_{ij}dV \qquad (73)$$

$$\lambda_{real}\left|\int_V P_i u_i dV + \int_S T_i u_i dS\right| = \int_V \sigma_{ij}\varepsilon_{ij}dV \qquad (74)$$

$$(\lambda_{real} - \lambda_{-})\left|\int_V P_i \cdot u_i \cdot dV + \int_S T_i \cdot u_i \cdot dS\right| = \int_V \sigma_{ij}\varepsilon_{ij}dV - \int_V \sigma'_{ij}\varepsilon_{ij}dV \geq 0 \qquad (75)$$

The second term is always positive according to the principle of virtual works then:

$$(\lambda_{real} - \lambda_{-}) \geq 0 \text{ and } \lambda_{real} > \lambda_{-} \qquad (76)$$

7.1.2. Upper Bound Theorem

We consider the real stress field σ_{ij} which is according to the Drucker's principle normal to the yield surface and a cinematically admissible stress σ''_{ij} non normal to the yield surface associated with a λ_+ factor). The Drucker's principle implies that the product A is always positive:

$$A = \overrightarrow{(\sigma_{ij} - \sigma''_{ij})} \wedge \vec{\varepsilon}_{ij} \geq 0 \qquad (77)$$

$$(\lambda_+ - \lambda_{real})\left|\int_V P_i u_i dV + \int_S T_i u_i dS\right| = \int_V \sigma''_{ij}\varepsilon_{ij}dV - \int_V \sigma_{ij}\varepsilon_{ij}dV \geq 0 \qquad (78)$$

Then
$$(\lambda_+ - \lambda_{real}) \geq 0 \text{ and } \lambda_+ > \lambda_{real} \qquad (79)$$

Combination of static and cinematic theorem gives:

$$\lambda_+ \geq \lambda_{real} \geq \lambda_{-} \qquad (80)$$

So, we can choose any collapse mechanism (statically or cinematically admissible), and use it to estimate a safety factor. The actual safety factor is likely to be lower than our estimate (it will be equal if we guessed right). This method is evidently inherently unsafe, since we overestimate the safety factor but in practice one can guess the collapse mechanism quite well, and so with practice you can get excellent estimates.

7.2. APPLICATION OF LIMIT ANALYSIS TO CORRODED PIPES

ASME B31G, ASME (1984) is a code for evaluating the remaining strength of corroded pipelines. It is a supplement to the ASME B31 code for pressure piping. The code was developed in the late 1960s and early 1970s at Battelle Memorial Institute and provides a semi-empirical procedure for the assessment of corroded pipes. Based on an extensive series of full-scale tests on corroded pipe sections, it was concluded that pipeline steels have adequate toughness and the toughness is not a significant factor. The failure of blunt corrosion flaws is controlled by their size and the flow or yield stress of the material. The input parameters include pipe outer diameter (D) and wall thickness (t), the specified minimum yield strength (σ_Y), the maximum allowable operating pressure (MAOP), longitudinal corrosion extent (L_c) and defect depth (d). According to the ASME B31G code, a failure equation for corroded pipelines was proposed by means of data of burst experiments and expressed with consideration of two conditions below:

First, the maximum hoop stress cannot exceed the yield strength of the material ($\sigma_{\theta\theta} \leq \sigma_Y$),
Second, relatively short corrosion is projected, on the shape of a parabola and long corrosion is projected on the shape of a rectangle.

The failure pressure equation for the corroded pipeline is classified by parabola and rectangle.
For parabolic defects:

$$P_f = \frac{2(1.1\sigma_Y) \times t}{D} \left[\frac{1-(2/3)\times(d/t)}{1-(2/3)\times(d/t)/M} \right] \qquad (81)$$

where $M = \sqrt{1 + 0.8 \left(\frac{L}{D}\right)^2 \left(\frac{D}{t}\right)}$, $\sqrt{0.8 \left(\frac{L}{D}\right)^2 \left(\frac{D}{t}\right)} \leq 4$

For rectangular defects:

$$P_f = \frac{2(1.1\sigma_Y) \times t}{D} \left[1 - (d/t)\right] \qquad (82)$$

where $M = \infty$ $\sqrt{0.8 \left(\frac{L}{D}\right)^2 \left(\frac{D}{t}\right)} > 4$

where, P_f, D, d, t, M, σ_Y and L are the failure pressure, outer diameter, maximum corrosion depth, wall thickness, bulging factor, yield stress and longitudinal corrosion defect length, respectively.

Due to some problems associated with the definition of flow stress $\sigma_f = 1.1 \times \sigma_Y$ and the bulging factor, a new flow stress was proposed as:

$$\sigma_f = 1.1 \times \sigma_Y + 69 \ (\text{MPa}) \tag{83}$$

The modified ASME B31G including this new modified flow stress and bulging factor is a follows:

$$P_f = \frac{2(1.1\sigma_Y + 69) \times t}{D} \left[\frac{1 - 0.85 \times (d/t)}{1 - 0.85 \times (d/t)/M} \right] \tag{84}$$

8. Strain based design

Strain-based design (Denys 2007) generally refers to pipeline designs expected to have longitudinal strains greater than 0.5%. Strain-based design encompasses both applied strain demand and strain resistance. At least two limit states are associated with strain-based design: tensile failure and compressive buckling. Strain-based design in recent years has been driven primarily by the need to construct pipelines in the arctic regions, deep-water offshore and other areas with high probability of large ground movements. Cases of in-service plastic strain were also observed through the history of pipeline usage due to soil movement on unstable slopes, mining subsidence, and seismic loadings. Confidence developed from the resistance of steel pipelines to these loadings and the understanding of pipe behaviour compared to known strains in installation and test has allowed pipeline designers to include strain-based design for in-service plastic strain.

Tensile failure in the presence of defect is caused by plastic instability. Two kinds of instabilities can be developed:

- For defect less than a critical value, plastic instability with plastic band at 55° of the direction of loading (plate instabilities) are associated with large plastic strain.
- For defect larger than a critical value, instability by striction associated with small plastic strain is developed.

The transition of these two types of instabilities described in Figure 8 has been exploited to develop a critical defect size failure criterion initially associated with a 1% critical gross strain but now with a reduced 0.5% strain.

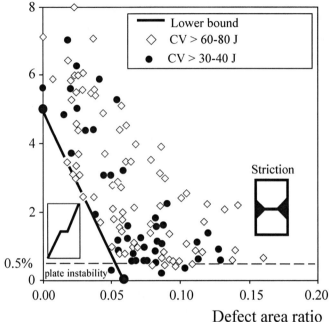

Figure 8. Critical gross strain versus defect ratio for tensile failure in presence of defect and associated with plastic instability. (Denys 2007).

The strain capacity is affected by material properties, geometry/defect features and interaction between pipes and their surrounding environment. The applied strains usually have large uncertainties. The methods for assessing tensile failure resistance of pipelines by engineering critical assessment become fewer when the plastic strain exceeds 0.005 (0.5%) and fewer still as the strain increases to 0.02 (2%) or more. These engineering critical assessment methods are used to demonstrate the sizes and types of imperfections that can remain in pipes and welds for high-strain service.

In compression, the failure modes relate to several varieties of buckling. The entire length of pipeline segment can buckle like an Euler beam, either vertically or horizontally. Alternatively or in combination with these modes, a pipeline may buckle a local area of the pipe wall. Internal pressure increases resistance to local buckling because the tensile hoop stress creates helps for the pipe to resist the diametrical changes that occur locally at buckling. External pressure reduces the resistance to local buckling. It can also create the possibility of a propagating buckle, a local buckle that extends along the pipe leaving a section of collapsed pipe. Yield strength should be recognized as an important parameter for assessment of the risk of local buckling, particularly for materials with a strong change of slope in the stress–strain curve near the yield point. Higher yield strength correlates to a lower critical strain for local buckling (Figure 9).

Several codes have provisions that apply to strain-based design of pipelines. The first limited residual longitudinal strain to below 0.002 (0.2%) for areas that are not reeled or pulled through a J-tube or have similar displacement-loading conditions imposed. This limit was for global strain as local strain was limited to 0.02 (2%) at areas of variable stiffness. Permanent curvature methods, such as reeling or J-tube installation could have 0.02 (2%) bending strain or 1% with bending and straightening.

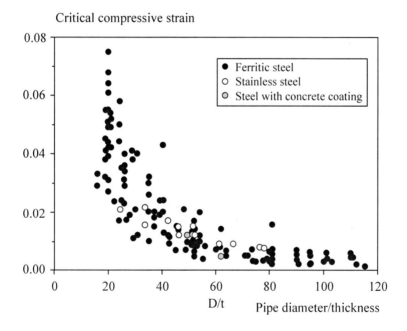

Figure 9. Critical buckling strain for pipe in bending. (Denys, 2007).

The safety factor used for the engineering critical assessment of the Northstar pipeline can be used as an example for those cases where very high safety levels are required for strain-based design. There a safety factor of 3 on strain was applied, somewhat higher than the safety factor on buckling from the compression side of the pipe in bending. The use of the "strain based" design criterion in offshore pipeline technology has widely increased in the last years since a general consensus has been developed about the fact that, in many circumstances, it is more valid than a "stress based" design criterion.

9. Conclusion

Several methods related to failure or fracture mechanisms have been presented. Theses methods are precisely described in the frame of codes and procedures with sometimes large difference according to nationality of the codes.

Mainly stress based design, International codes and regulations clearly state that the "strain based" design criterion is applicable only at displacement controlled conditions to be identified as the system conditions in which the structural response is primarily governed by imposed geometric displacements.

References

American National Standard Institute (ANSI)/American Society of Mechanical Engineers (ASME), 1984. Manual for determining strength of corroded pipelines, ASME B31G.

American Society for Mechanical Engineering (ASME 2005), Boiler and Pressure Vessel Code (Code), Code Case N-629, Use of Fracture Toughness Test Data to establish Reference Temperature for Pressure Retaining materials of Section III.

ASTM E1921-08ae1, 2007 Standard Test Method for Determination of Reference Temperature, T_o, for Ferritic Steels in the Transition Range ICS Number Code 77.040.106.

Bjerager P., 1989, Methods for structural reliability computations, in Computational Mechanics of Probabilistic and Reliability Analysis, W. K. Liu and T. B. Belytschko (Eds), Elmepress International, Lausanne, Switzerland.

Buffon G.L., 1744Expériences sur la force du bois, Second Mémoire, in *Mémoires de mathématique et de Physique, tirés des registres de l'Académie Royale des Sciences*, Imprimerie Royale, Paris, pp. 292–334.

Burdekin F.M. and Dawes M.G., 1971, Practical use of linear elastic and yielding fracture mechanics with particular reference to pressure vessels, Institution of Mechanical Engineers, pp. 28–37. In: Practical application of fracture mechanics to pressure-vessel technology: a conference held at the Institution of Electrical Engineers, Savoy Place, London, 3–5 May 1971.

Denys R., 2007, Interaction between material properties, inspection accuracy and defect acceptance levels in strain based pipeline design, in Safety, Reliability and Risks Associated with Water, Oil and Gas Pipelines, G. Pluvinage and M. Elwany (Eds), Springer, Dordrecht, pp. 1–22.

Eurocode 3, 2005, Design of steel structures. General rules and rules for buildings Division 1, Class 1.

http://www.tms.org/pubs/journals/JOM/9801.html: Felkins, K., Leighly, H.P., Jr. and Jankovic, A, 1988, The Royal Mail Ship Titanic: Did a Metallurgical Failure Cause a Night to Remember?JOM, 50(1), pp. 12–18.

Jallouf S., Pluvinage G., Carmasol A., Milović L. and Sedmak S., 2005, Determination of safety margin and reliability factor of boiler tube with surface crack, Structural Integrity and Life, vol 5, no 3, pp. 131–142.

Krasowsky A. and Pluvinage G., 1993, Structural parameters governing fracture toughness of Engineering Materials. Physico Chemical Mechanics of Materials, vol 29, Mai–Juin, no 3, pp. 106–113.

Kumar V., German M.D. and Shih C.F., 1981, An Engineering approach of elastic plastic fracture mechanics, EPRI, NP1931, Res.Pr 1237-1.

Moulin D., Drubray B. and Nedelec M. 1998, Méthode pratique de calcul de J, Annexe du RCC-MR : Méthode J_s.

Pluvinage G., 2003, Un siècle d'essai Charpy, de la résistance vive à la rupture à la mécanique de rupture d'entaille, Revue Mécanique et Industries, vol 4, no 3, pp. 197–212.

Pluvinage G., 2007, General approaches of pipeline defect assessment, Safety, Reliability and Risks Associated with Water, Oil and Gas Pipelines, G. Pluvinage and M. Elwany (Eds), Springer, Dordrecht, pp. 1–22.

Pluvinage G. and Sapounov V., 2007, Conception fiabiliste de la sécurité des matériaux composites, Revue des Sciences et de la technologie, Juin, no 16, pp. 6–15.

R6, 1998, Assessment of the integrity of structures containing defects, Nuclear electric procedure R/H/R6, Revision 3.

Sanz G., 1980, Essai de mise au point d'une méthode quantitative de choix de qualités d'acier vis à vis du risque de rupture fragile, Revue de Métallurgie, CIT Juillet, pp. 621–642.

Save M.A., Massonnet, C.E. and de Saxce, G., 1997, Plastic Limit Analysis of Plates, Shells and Disks North-Holland, Amsterdam, 6019972 pp. ISBN 0-444-89479-9.

Schwalbe K.H. and Cornec A., 1991, The engineering treatment model (ETM) and its practical application, Fatigue and Fracture of Engineering Materials and Structures, vol 14, pp. 405–412.

SINTAP: Structural Integrity Assessment Procedure, Final Report E-U project BE95-1462 Brite Euram Programme Brussels (1999).

Turner C.E., 1979, Methods for post yield fracture safety assessment, in Post-Yield Fracture Mechanics, Applied Science Publishers, London, pp. 23–210.

Wallim K., 1990, Methodology for selecting Charpy Toughness. Criteria for thin high strength steels, Jernkontorets Forskning, VTT manufacturing Technology Finland, Part I, II, III, Nr 4013/89, TO 40-05–06-31.

DEVELOPMENT AND APPLICATION OF CRACK PARAMETERS

ALEKSANDAR SEDMAK[*]
University of Belgrade, Faculty of Mechanical Engineering,
Kraljice Marije 16, 11000 Belgrade, Serbia

LJUBICA MILOVIĆ
University of Belgrade, Faculty of Technology and
Metallurgy, Karnegijeva 4, 11000 Belgrade, Serbia

JASMINA LOZANOVIĆ
University of Belgrade, Faculty of Mechanical Engineering,
Innovation Center, Kraljice Marije 16, 11000 Belgrade,
Serbia

Abstract The development, definition and implementation of crack parameters (stress intensity factor, crack tip open displacement, J integral, final stretch zone) are presented. They are now used for structural integrity and residual life assessment and fracture analysis of cracked components in elastic and plastic range, what is illustrated by typical examples.

Keywords: Crack parameters, development, testing, application, structural integrity

1. Introduction

1. Most structures fail when the stress of applied material reaches certain critical level. In the first theory on fracture[1] this stress limit was considered as elastic and an inherent material property.
2. The above theory was soon questioned by the measured fluctuation in fracture strength as high as an order of magnitude. Fracture mechanism was not clear, since understanding of the fracture phenomena was missing.
3. The breakthrough came in 1920 in a paper by Griffith.[2] He proposed the energy-balance concept of fracture in his classic paper based on the principle of energy conservation laws of mechanics and thermodynamics.

---

[*] E-mail: asedmak@mas.bg.ac.yu

Griffith used the results of Inglis[3] on the stress concentration around an elliptical hole from the theory of elasticity and established the base for introduction of new discipline, Fracture Mechanics.

Linear elastic fracture mechanics (LEFM) is valid only as long non-linear material deformation is confined to a small region surrounding the crack tip. The basic crack parameter in LEFM is stress intensity factor (SIF). In many materials, it is not possible to characterize material fracture behaviour by LEFM, so an alternative fracture mechanics model is required.

Elastic-plastic fracture mechanics (EPFM) applies to materials that exhibit time-independent, nonlinear behaviour (i.e. plastic deformation). Three EPFM parameters are shortly presented: crack-tip opening displacement (CTOD), the J contour integral and final stretch zone (FSZ). These parameters describe crack-tip conditions in elastic-plastic materials, and each can be used as a fracture criterion. Critical values of CTOD or J give nearly size-independent measures of fracture toughness, even for relatively large amount of crack-lip plasticity.

4. Anyhow, the development and intensive implementation of nano materials and structures in all facilities, requiring reliable, save and secure operation, have shown that fracture mechanism requires further investigation, specifically at micro and nano scale. Most probably, for that reason, the story about fracture parameters is far from being finished.

2. Linear elastic fracture mechanics

2.1. STRESS CONCENTRATION, PLANE STRAIN AND PLANE STRESS STATE

Applying Mathematical Theory of Elasticity, Kirsch[4] calculated stress concentration by factor 3 around circle in infinite plate, indicating that tensile strength could be locally reached although remote stress is well bellow yield point. Using the same approach, Inglis[3] defined normal stress in cross section for elliptic hole in an infinite plate (Figure 1) as:

$$\sigma_\theta = \sigma_y = \sigma_{max} = \sigma\left(1 + \frac{2a}{b}\right) = \sigma\left[1 + 2\left(\frac{a}{\rho}\right)^{1/2}\right] \quad (1)$$

where a stands for ellipse major axis, b for minor axis, σ is normal stress in remote section, ρ ellipse root radius ($\rho = b^2/a$). If minor axis $b \to 0$ ($\rho \to 0$) normal stress tend to infinity ($\sigma_y \to \infty$), and in elastically deformed material the condition for fracture is fulfilled, e.g. the component will be broken in brittle manner just after load is applied. Only cracked materials of expressed brittle behaviour will fracture in this way, not the ductile material.

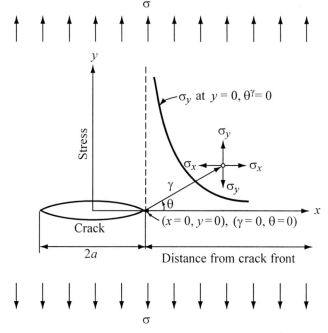

Figure 1. Stress distribution around elliptical hole.[5]

Since the maximum stress, reached at the crack tip can cause local yielding of ductile material, the plastic zone will develop ahead the crack tip and the effect of stress concentration will be reduced, crack will not grow.

Post fracture examination of surfaces can sometimes reveal a great deal about the type of material behavior. This is true not only of the fracture of cracked or notched body but also of a smooth sample – e.g. in a tensile test (Figure 2). The fracture types in tensile test are ductile cup-cone (Figure 2a) or brittle of flat surface (Figure 2b).

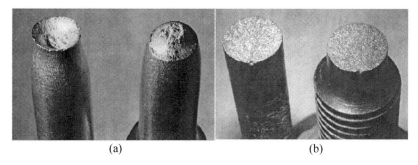

Figure 2. Smooth specimen after tensile test. (a) Ductile fracture. (b) Brittle fracture.

2.2. GRIFFITH'S CRACK

The theoretical fracture strength of a solid is of the order of $E/10$ (E is elasticity modulus), but the strengths of crystal metals in practice tend to be lower than this value by two orders of magnitude. Griffith first suggested reasons for this discrepancy between predicted and actual values and by developing his arguments the present methods of measuring a material's fracture toughness is evolved.[5]

To understand the fracture stress let us suppose a sample with crack-like defect (Figure 1), of length $2a$ and negligible width ($2b \approx 0$), normal to the applied stress, σ. Griffith's accurate calculation for potential energy U gives:

$$U = -\frac{1}{2} \cdot \frac{\sigma^2 \pi a^2}{E} \qquad \frac{\partial U}{\partial a} = -\frac{\sigma^2 \pi a}{E} \qquad \text{in plane stress} \qquad (2)$$

$$U = -\frac{1}{2} \cdot \frac{\sigma^2 \pi a^2}{E}(1-v^2) \qquad \frac{\partial U}{\partial a} = -\frac{\sigma^2 \pi a}{E}(1-v^2) \quad \text{in plane strain} \quad (3)$$

Here v is Poisson's ratio. These values of $\partial U/\partial a$ represent potential energy decrease when a crack extends by an infinitesimal amount δa under constant load.

The Griffith criterion for fracturing a body with crack of half length a may be visualised, as in Figure 3a, by drawing the way in which energy changes with crack length. In plane stress, for example, the total energy, W, consisting of potential energy, U, and surface energy, $S = 2\gamma$, is given by

$$W = U + S = \frac{1}{2} \cdot \frac{\sigma^2 \pi a^2}{E} + 2\gamma a \qquad (4)$$

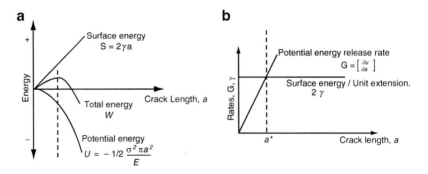

Figure 3. Variation of energy (a) and energy rates (b) with crack length (a* – critical length).[5]

The maximum in the total energy curve is given by $\partial W/\partial a = 0$, i.e.

$$\frac{\sigma^2 \pi a}{E} = 2\gamma \qquad (5)$$

This situation is depicted in Figure 3b by the intersection of the lines for $\sigma^2 \pi a/E$ and 2γ. The positive value of slope ($\partial U/\partial a$) is conventionally defined as the strain energy release rate (with respect to crack length) with the symbol G; it is the potential energy release rate for crack extension under constant load for unit thickness. For crack length a, fracture stress σ_c is:

$$\sigma_c = \sqrt{\frac{2E\gamma}{\pi a}} \text{ (in plane stress)} \quad \sigma_c = \sqrt{\frac{2E\gamma}{\pi(1-v^2)a}} \text{ (in plane strain)} \qquad (6)$$

Formulae 5 and 6 relate external load (remote stress σ and fracture stress σ_c) with crack size (length a) via material property (fracture surface energy γ). This is pre-condition for definition of material crack parameters.

2.3. STRESS INTENSITY FACTOR

The modes of displacement of crack tip, which govern crack growth in x–z plane, are given in Figure 4.

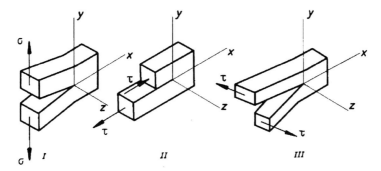

Figure 4. Basic modes of crack growth: I – opening; II – sliding; III – tearing.

Mode I – crack growth by opening (Figure 4I) is defined by opening of surface cracks symmetrically with respect to initial crack plane.
Mode II – crack growth by sliding (in-plane shear along x axes – Figure 4II) in which crack surfaces glide over one another in opposite directions.
Mode III – crack growth by tearing (anti-plane shear – Figure 4III). This is local movement in which crack surfaces are shifted along crack front in a way that points, positioned in the same vertical plane are moved in different directions and took places in different vertical planes after fracture.

The fracture mode I is the most dangerous for component security. The relationships for the stress components $\sigma(\sigma_x, \sigma_y, \tau_{xy})$ and crack of length a in polar coordinates (r, θ) (Figure 1) ahead the front of are[5]:

$$\sigma_x = K_I \frac{\cos\frac{\theta}{2}}{\sqrt{2\pi r}}\left(1 - \sin\frac{\theta}{2}\sin\frac{3\theta}{2}\right) \quad \sigma_y = K_I \frac{\cos\frac{\theta}{2}}{\sqrt{2\pi r}}\left(1 + \sin\frac{\theta}{2}\sin\frac{3\theta}{2}\right)$$

$$\tau_{xy} = K_I \frac{\cos\frac{\theta}{2}}{\sqrt{2r\pi}} \sin\frac{\theta}{2}\cos\frac{3\theta}{2} \tag{7}$$

by implementing new parameter *stress intensity factor* (SIF) (K_I).

Displacement components, u in x direction and v in y direction, are

$$u = \frac{K_I}{\mu}\sqrt{\frac{r}{2\pi}}\cos\frac{\theta}{2}\left(\frac{\kappa - 1}{2} + \sin^2\frac{\theta}{2}\right) \quad v = \frac{K_I}{\mu}\sqrt{\frac{r}{2\pi}}\sin\frac{\theta}{2}\left(\frac{\kappa + 1}{2} + \cos^2\frac{\theta}{2}\right) \tag{8}$$

Stress intensity factor K_I is expressed in units of stress·length$^{1/2}$, measured in terms of the opening mode is indicated with subscript '*I*' (Figure 4). Its critical value for plane strain condition is material characteristic, *plane strain fracture toughness* K_{Ic}. The distinction between K_{Ic} and K_I is important, and is comparable to the distinction between strength and stress. To determine a K_{Ic} value, a cracked specimen of suitable size is quasi statically loaded until the crack becomes unstable and extends abruptly. The ratio of K_I to the applied load is a function of specimen design and dimensions which is evaluated by stress analysis. The K_I value corresponding to the load at which unstable crack extension is observed is the K_{Ic} value, determined in the test. This material property is a function of temperature and strain rate; they are specified in standards for plane strain fracture toughness determination.[6]

The effective toughness of a material is expected to be less than its K_{Ic} level for any practical conditions. It has been found that the K_{Ic} level in general is independent on specimen design and sizes when standard specifications for valid testing are met. So, stress intensity factor K_I for crack length a and its critical value K_{Ic} for critical flaw size a_c at applied normal stress σ are:

$$K_I = \sigma\sqrt{\pi a} \quad K_{Ic} = \sigma\sqrt{\pi a_c} \tag{9}$$

Stress state in loaded component depends on its thickness (Figure 5), and plane strain condition is a basic requirement for fracture toughness testing.[6]

In this case again, when the distance from crack $r \to 0$, stress tends to infinity, whereas displacements tend to zero. For plane strain, stress state is close to hydrostatic pressure, and these are conditions for brittle fracture.

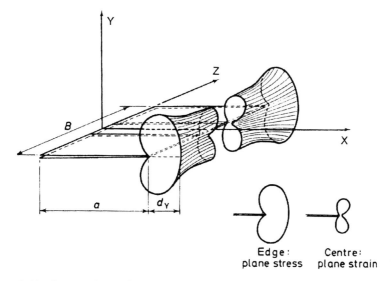

Figure 5. Plastic zone ahead of crack in a plate of finite thickness. At the edges of the plate, for ($z = \pm B/2$) the stress state is close to plane stress. In the centre of a sufficiently thick plate the stress state approximates to plane strain.

Irwin had demonstrated that all materials exhibit also in plane stress state critical value of stress intensity factor, K_c, which is not material characteristic, but depends on thickness. When the critical value of stress intensity factor is achieved, crack can initiate to grow. Accordingly, the condition representing the situation for existence of non growing crack is

$$K < K_c \qquad (10)$$

For better understanding of the SIF nature the relation between Griffith's solution and critical stress intensity factor K_c is derived calculating the energy, released in crack growth process from initial its length a to the value $a + \delta a$.[7]

$$G_{Ic} = \frac{K_{Ic}^2}{E}(1-v^2) \text{ in plane strain} \quad G_I = \frac{K_I^2}{E} \text{ in plane stress} \qquad (11)$$

with G as new defined term "crack driving force" or "energy release rate".

Critical value of crack driving force, G_c, is much greater than the material surface energy 2γ. This suggests that the energy release γ_p in sample is to a large extent dissipated by plastic flow around the crack tip, so:

$$\sigma_c = \sqrt{\frac{E(2\gamma + \gamma_p)}{\pi a}} \qquad (12)$$

It was found experimentally that ($\gamma_p + 2\gamma$) is much greater than 2γ, so

$$\sigma_c = \sqrt{\frac{E\gamma_p}{\pi a}} \quad (13)$$

and values of γ_p could then be determined directly from the fracture stress of specimen containing crack of known length.

By introducing in plane stress the "notional" or an equivalent elastic crack, proportional by coefficient α, the plastic zone of total extent

$$2r_Y = \frac{\alpha}{\pi} \frac{K^2}{\sigma_{YS}^2} \quad (14)$$

contributes to crack of half-length $(a + r_Y)$, where σ_{YS} is yield stress. The failure stress σ_c is then given by:

$$\sigma_c = \sqrt{\frac{EG_c}{\pi(a+r_Y)}} = \sqrt{\frac{EG_c}{\pi\left(a + \sigma_c^2 a / 2\sigma_{YS}^2\right)}} \quad (15)$$

or, in terms of the critical stress intensity factor:

$$K_c = \sigma_c \sqrt{\pi a \left(1 + \frac{\sigma_c^2}{\sigma_{YS}^2}\right)} \quad (16)$$

3. Elastic-plastic fracture mechanics

3.1. CRACK–TIP OPENING DISPLACEMENT

The schedule of events on the crack tip (Figure 6) in sample exposed to increasing load can help in understanding of crack behavior in elastic-plastic condition.[8]

When plastic zone of significant size is established in the crack tip region, it is not more possible to describe stress and strain fields by a single parameter, as it was the case with critical stress intensity factor K_{Ic}. The specimen will behave as its compliance is greater than that corresponding to crack size due to the effect of blunted crack, surrounded by elastically deformed material (Figure 6III). This effect is expressed by value r_Y – Eq. (14).

Plastic zone is a region ahead of the crack tip, in which material behaviour is not linear: it is small if surrounded by singular stress field, described by stress intensity factor K_I (Figure 6III). Plastic zone is large if it is surrounded by material of behaviour which can not be described by K_I value.

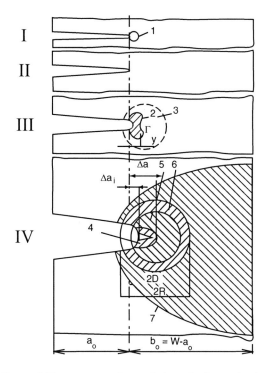

Figure 6. The events on the crack tip under increasing load.

I – Initial stage (1 – plastic zone, visible after unloading)
II – Stage of relaxation (relaxed residual stress around crack tip)
III – Blunting of a crack and small plastically deformed zone (2); 3 is the region in which material behavior is described by K_I value
IV – Stable crack growth (4 – zone of elastic unloading; 5 – process zone; 6 – HRR zone; 7 – zone of large plastic deformation, for net section yielding)

The term "net section yielding" comprises yielding spread across the ligament plane, and "full scale yielding" denotes yielding spread over the total specimen. Radius R (Figure 6IV) determines the extension of HRR zone, named according to the material behaviour model of Hutchinson–Rice–Rosengren (HRR), in which the state of stress and strain is described by parameter J_I (path independent J integral), and in which no relaxation occurs with uniform growth of J integral. Radius D defines the process zone (fracture process), in which free crack surfaces can be formed from crack tip during crack extension, caused the relaxation of elastic-plastic material with uniform growth of J integral. Radius D is usually small and comparable to the initial crack opening in the vicinity of crack tip (Figure 6IV).

Crack-tip opening displacement (CTOD) as a parameter has been involved by Wells[9] based on an approximate analysis, in which it was related to the stress intensity factor in the limit of small-scale yielding (Figure 7).

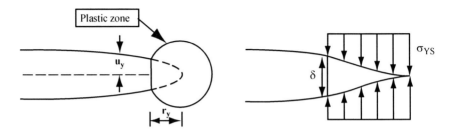

Figure 7. Estimation of CTOD from the displacement u_y with Irwin correction r, (left), Dugdale model (right).

Considering the crack with a small plastic zone, as illustrated in Figures 6 and 7, Irwin[10] postulated in Eq. (14) that crack-tip plasticity enables that the crack behaves as it was slightly longer. Thus, one can estimate the CTOD by solving for the displacement at the physical crack tip, v in Eq. (8), assuming an effective crack length of $a + r_Y$ from Eq. (14)

$$u_y = \frac{\kappa+1}{2\mu} K_I \sqrt{\frac{r_Y}{2\pi}} = \frac{4}{E'} K_I \sqrt{\frac{r_Y}{2\pi}} \qquad (17)$$

where E' is effective Young's modulus. With the value $\alpha = \frac{1}{2}$, as Irwin accepted in Eq. (14), crack-tip opening displacement δ is obtained as

$$\delta = 2u_y = \frac{4}{\pi} \frac{K_I^2}{\sigma_{YS} E} \qquad (18)$$

Thus, in the limit of small-scale yielding, CTOD is related to K, K_{Ic} and also to G. Wells postulated that CTOD is an appropriate crack-tip-characterising parameter when LEFM is no longer valid. This assumption was shown to be correct several years later when a unique relationship between CTOD and the J integral has been established.

Using the strip-yield model of Dugdale[11] provided an alternate means for analysing CTOD, which confirmed the former approach of Wells, as:

$$\delta = \frac{8\sigma_{YS} a}{\pi E} \ln \sec\left(\frac{\pi}{2} \frac{\sigma}{\sigma_{YS}}\right) \qquad (19)$$

3.2. RICE'S CONTOUR J INTEGRAL

By definition J integral is given in following form:

$$J = \oint_\Gamma \left(Wdy - T_i \frac{\partial u_i}{\partial x} ds \right) \qquad (20)$$

with $W = \int \sigma_{ij} d\varepsilon_{ij}$ – strain energy density; Γ – integration path; ds – element of segment length; $T_i = \sigma_{ij} n_j$ – traction vector on the contour; u_i – displacement vector, Figure 8.

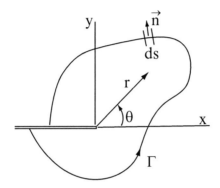

Figure 8. Arbitrary contour path for J integral calculation.

Rice has shown that J integral is path independent if necessary conditions are fulfilled. This is the prerequisite for its calculation along properly selected path, because its value is the same for the contours close to the crack tip, outside plastic zone, as well as for path along specimen sides.

J integral can also be presented as the energy, released on crack tip for unit area crack growth, Bda, by following expression

$$JBda = B \oint_\Gamma Wdyda - B \oint_\Gamma T_i \frac{\partial u_i}{\partial x} ds da \qquad (21)$$

where B is specimen thickness. The member $B \oint Wdyda$ denotes the energy obtained (and released) along the contour Γ for crack increase, da, supposing non-linear elasticity. Second member represents the work of traction forces on contour displacement for crack extension da. The value $JBda$ is total energy at crack tip available for crack growth Δa, equal to the value G:

$$J = G = \frac{K^2}{E'} \quad E' = E \text{ for plane stress;} \quad E' = \frac{E}{1-v^2} \text{ for plane strain} \qquad (22)$$

In plastic region, W is not strain energy density, being dissipated inside the material, so J is not the energy at the crack, available for crack growth.

In order to assure the existence of J singular field around the crack tip, some requirements have to be fulfilled. It is possible to see in Figure 9 that ahead the crack tip there is a region in which material is significantly deformed, with occurrence of voids, slip lines and other forms beyond the continuum mechanics application. This region is called "fracture process zone", and its magnitude could be small compared to body dimensions in plane. In order to fulfil plane strain condition process zone must be small compared to specimen dimensions, e.g. its thickness. These conditions are experimentally based, and general condition for dimensions is applied in the form:

$$B, b, a < 25 \frac{J}{\sigma_o} \qquad (23)$$

where $b = W - a$ stands for specimen ligament.

The behaviour of elastic-plastic material at stable crack growth is described by $J-\Delta a$ relationship (J–R curve), where Δa is crack extension (Figure 10).

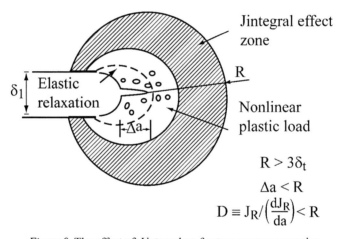

Figure 9. The effect of J integral on fracture process zone size.

In brittle fracture no energy is spent for crack growth, and corresponding relation for $J-\Delta a$ is given by horizontal straight line, that intersects the ordinate at the level J_{Ic}. This level corresponds to critical energy consummation for crack initiation. For ductile material, in initial stage of load increase and energy release, deformation is expressed by increase of existing crack opening, but not with its extension. This corresponds to very steep line in $J-\Delta a$ relationship, representing crack tip blunting stage (Figure 11). In critical point the slope of the line is changed, due to next crack opening increase connected with crack extension and its length increase. Point of deflation in the initial stage is taken as J_{Ic}, so it is possible to calculate, based on this value, the value of critical stress intensity factor K_{Ic}.

In the moment when material can't suffer plastic deformation any more, crack will start to grow (point B in Figure 11). The end of plastic deformation can be characterized by final stretch zone (Figure 12).

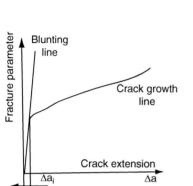

Figure 10. J integral crack growth resistance curve (J–R curve).

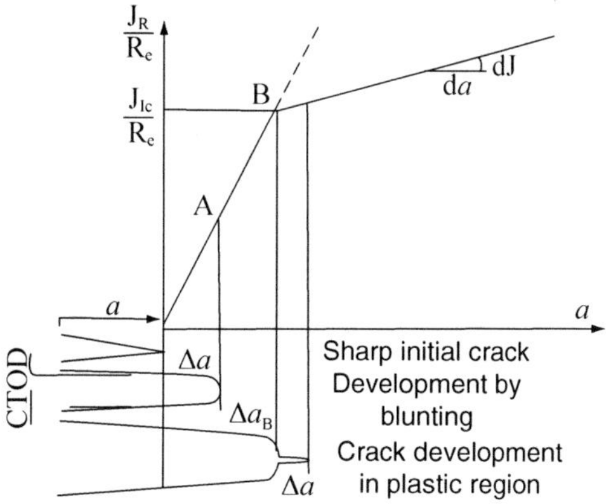

Figure 11. Blunting and stable crack growth processes, marked on R curve.

3.3. FINAL STRETCH ZONE

Apparent crack growth during its tip blunting, Δa_B, Figure 11, can be identified by final stretch (FSZ), Figure 12. In simplified way FSZ width, ω, can be expressed via crack tip opening displacement, δ_t, according Figure 12a, as:

$$\omega = \delta_t/\sqrt{2} \tag{24}$$

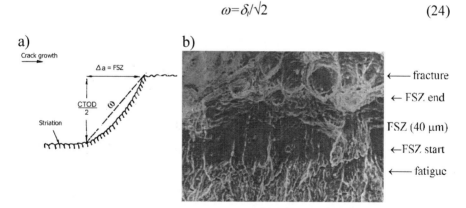

Figure 12. Final stretch zone: (a) scheme, (b) SEM fractograph.[12]

Applying strip-yield model for yield stress level ($\sigma_y = \sigma_{YS}$), simple relation between J integral and crack tip opening displacement, δ_t, is found.[12]

$$J = m\sigma_{YS}\delta_t \tag{25}$$

with coefficient m depending on stress state.

Since apparent crack extension is $2\Delta a_B = \delta_t$, Figure 11, it follows:

$$J = \sqrt{2}m\omega\sigma_Y = 2m\Delta a_B\sigma_Y = 2\Delta a_B\sigma_Y \tag{26}$$

for flow strength σ_Y (average of yield and tensile strengths) and plane stress ($m = 1$), since in plane strain there is no blunting.

4. Integrals describing crack resistance at elevated temperature

Under conditions of steady-state creep the material exhibits nonlinear viscous flows described by uniaxial constitutive equation

$$\dot{\varepsilon} = B\sigma^n \tag{27}$$

where B and n are material parameters. In that case J integral can be replaced by C^* integral, defined as

$$C^* = \int_\Gamma W(\dot{\varepsilon})dy - \mathbf{T}\frac{\partial \dot{\mathbf{u}}}{\partial x}ds \tag{28}$$

where $W(\dot{\varepsilon}) = \int_0^{\varepsilon}\sigma d\dot{\varepsilon}$ is the strain energy rate density, and symbol "$\dot{u}$" stands for the time derivative. Under the same conditions as in the case of J integral, C^* integral is path independent, has the physical meaning of crack driving force and can be used as the loading parameter for the strength of crack-tip fields.

Having in mind the complete analogy between the J integral and C^* integral, it is obvious that the latter can be obtained from expressions of the first one, given for hardening materials described by Ramberg–Osgood relation, and using B from Eq. (28):

$$C^* = Bbg_1\left(\frac{a}{W}\right)h_1\left(\frac{a}{W},n\right)\left(\frac{P\sigma_y}{P_o}\right)^{n+1} \tag{29}$$

where b is the ligament, $b = W - a$, W is the width of component, $\varepsilon_y = \sigma_y/E$, g_1 and h_1 are dimensionless functions depending on component geometry and are given in handbook,[13] P_o is the limit load.

If the steady-state conditions are not reached, other fracture parameters besides C^* integral should be used. Two of the most popular are C_t parameter and $C(t)$ integral. For the same constitutive equation, $C(t)$ integral is defined as an integral expression:

$$C(t) = \lim_{\Gamma \to 0} \int_{\Gamma} \frac{n}{n+1} \sigma \dot{\varepsilon} dy - \mathbf{T} \frac{\partial \dot{\mathbf{u}}}{\partial x} ds \qquad (30)$$

whereas C_t is defined as follows

$$C_t = -\frac{\partial u^*}{\partial a} \qquad (31)$$

where u^* is the instantaneous power dissipation rate.

5. Application of crack parameters for structural integrity assessment

Various crack parameters enable the application in different circumstances, depending on responsibility of a structure and severity of loading condition and environment. Several typical examples are presented here to get closer insight of way in how to use them and in results that can be obtained.

Interesting application refers to the pressure vessels of the same design, produced of the same steel by welding, with some differences in size and dimensions, shown in Figure 13. After about 30 years of service, in regular inspection many defects in welded joints had been detected by NDT.[14]

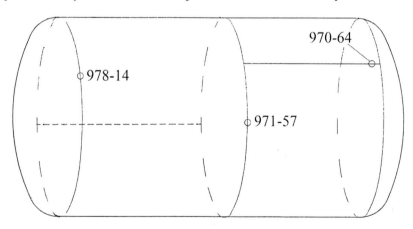

Figure 13. Pressure vessel with crack-like defects in welded joints.

Three crack-like defects are selected for structural integrity assessment based on evaluated significance. The assessment could be performed only by involved simplifications and assumptions, which have to be conservative for safe acceptance of results and conclusions by inspection authority.

Defect 970-64 is represented as a part-through crack. Data:

- Pressure vessel geometry (thickness $t = 50$ mm, diameter $D = 2150$ mm)
- High strength low alloy steel: yield stress-$R_{eh} = 500$ MPa, tensile strength – $R_M = 650$ MPa; plane strain fracture toughness – $K_{Ic} = 1,580$ MPa$\sqrt{\text{mm}}$
- Crack geometry in the root of longitudinal weld (50 mm from the circular weld, away from connections length 60 mm, depth 2 mm)
- Loading (maximum pressure $p = 8.1$ MPa, assumed maximum value of residual stress transverse to the weld $\sigma_R = 200$ MPa).

Defect 978-14, also represented as a part-through crack, was found in a second pressure vessel. The data different from former pressure vessel are:

- Pressure vessel geometry (thickness $t = 42$ mm, diameter $D = 1,958$ mm)
- Crack geometry in circular weld with the bottom cover – lack of penetration in the root (length 25 mm, depth 2 mm, away from connections)

Defect 971-57, lack of fusion, represented as through crack 10 mm long in central circular weld of the second vessel, away from connections.

The values are measured or obtained by non-destructive testing, except assumed residual stress based on experience with similar structure, and K_{Ic}, taken from references for the welded join of the same steel.[15]

5.1. THE USE OF SINGLE PARAMETER

Basically, single crack parameter used for structural integrity assessment can be only used to compare applied and critical values (Eqs. (9) and (10)) for stress intensity factor). In the same sense CTOD and J integral can be used. This application is presented for the pressure vessels given in Figure 13.

The significance of defect 970-64 is assessed by stress intensity factor, with conservative assumptions for missing data. It was taken (Figure 14 left) that crack extends along whole cylindrical part. Curvature effect is neglected since $2t/D = 2 \times 50/2,150 \approx 0.05$.

Using Eq. (9) SIF is calculated with normal stress σ consisting of applied plus residual stress, $\sigma = (pR/t) + \sigma_R$, for detected crack half-length of size $a = 1$ mm, and vessel radius $R = D/2 = 1,075$ mm, so

$$K_I = \sigma\sqrt{\pi a} = [(pR/t) + \sigma_R]\sqrt{\pi a} = 663 \text{ MPa}\sqrt{\text{mm}},$$

what is significantly less than $K_{Ic} = 1,580$ MPa$\sqrt{\text{mm}}$ for applied steel.

If one takes $a = 2$ mm as conservative value regarding NDT result, K_I is still lower compared to K_{Ic} (937 < 1,580 MPa$\sqrt{\text{mm}}$). In this way the resistance against brittle fracture is documented by stress intensity factor.

CRACK PARAMETERS

In the same way for **defect 978-14** is obtained $K_I = 515$ MPa√mm.

The **defect 971-57** is analysed as a through crack 10 mm long in tensile panel (Figure 14 right), of diameter $D = 1,958$ mm, wall thickness $t = 42$ mm, assuming residual stress transverse to the weld $\sigma_R = 175$ MPa, so
$K_I = \sigma\sqrt{\pi a} = [(pR/t)+\sigma_R]\sqrt{\pi a} = 1,039$ MPa√mm $< K_{Ic}$ (1,580 MPa√mm).
For $2a = 20$ mm the value $K_I = 1,465$ MPa√mm is still lower than K_{Ic}.

It is to say that all the assumptions involved (crack size and position, the level of residual stress, geometry of vessels) are on the safe side, so these defects are not danger in regard to brittle fracture.

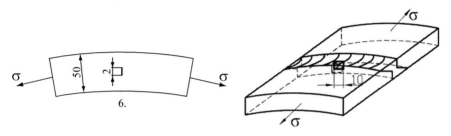

Defect 970-64 as a part-through crack   Defect 971-57 as a through crack

Figure 14. Simplification and assumption for significance assessment of defects.

5.2. TWO PARAMETERS APPROACH

Failure assessment diagram (FAD), Figure 15, is the most used form of two parameters approach. Detailed description can be found elsewhere.[16,17]

Brittle behaviour of material is presented on the ordinate, and plastic behaviour on the abscisa. Normalized coordinates, $S_r = \sigma/\sigma_c$ and $K_r = K_I/K_{Ic}$, are used, where σ and K_I are applied values, σ_c and K_{Ic} are critical values of stress and stress intensity factor, respectively. They are related as

$$K_r = S_r \left[\frac{8}{\pi^2}\ln\sec\left(\frac{\pi}{2}S_r\right)\right]^{-\frac{1}{2}} \quad (32)$$

For perfectly ductile material the structure will fail by plastic colapse at $S_r = 1$ ($K_r = 0$), whereas failure of fully brittle material took place at $K_r = 1$ ($S_r = 0$). In other cases of mixed-mode failure the values K_r and S_r are less than 1, and both should be applied in FAD to locate corresponding point in safe or in unsafe region (Figure 15).

The approach can be illustrated for failure assessment of pressure vessels (Figure 13). Coordinates K_r are defined as: 0.59 for defect 970-64, 0.45 for 978-14 and 0.92 for 971-57. For coordinates S_r only primary stress, caused by pressure, should taken, and secondary stress is neglected.[17]

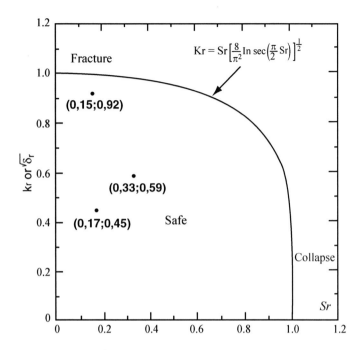

Figure 15. Failure assessment diagram (FAD) applied to pressure vessels from Figure 13.

Stress σ_n in net cross-section for the defect 970-64 is $\sigma_n = 1.08pR/t$, with coefficient 1.08 for section weakened by crack 4 mm long at 50 mm thickness $S_r = \sigma_n/\sigma_F = 2(1.08pR/t)/(R_{eH} + R_M) = 0.33$ accepting flow stress σ_F, average of yield stress and tensile strength $(R_{eH} + R_M)/2$ as critical value of coordinate S_r. The effect of lid is neglected since changes reflects to torical segment, but not to spherical part of the lid.

Stress σ_n for the defect 978-14 is $\sigma_n = 1.05pR/t = 95$ MPa, with coefficient 1.05 for crack length 2 mm in thickness 42 mm, so
$S_r = \sigma_n/\sigma_F = 95/575 = 0.17$.

Stress σ_n for the defect 971-57 is $\sigma_n = pR/2t = 87$ MPa, and
$S_r = \sigma_n/\sigma_F = 87/575 = 0.15$.

Corresponding points (0.33; 0.59) (0.17; 0.45) i (0.15; 0.92) are drawn in Figure 15 and all of them are in safe region of FAD.

5.3. STABLE CRACK GROWTH PREDICTION BY *J* INTEGRAL

Failure assessment could be performed also by *J* integral, comparing applied crack driving force, expressed by *J* integral according to Eq. (22) and material crack resistance in the form of J–R curve. This approach offers more insight in loaded structure behaviour, including not only critical situation but also how the crack will grow in stable manner. Since it offers much better assessment, with numerous advantages, it became popular, although it is not simple in use.

Crack driving force in cylindrical shell can be calculated by different model, e.g. Ratwani, Erdogan and Irwin (REI),[18] and King[19]. J–R curves can be obtained by testing methods from different documents,[16,17] also by others, e.g. ASTM E1152, using convenient specimens, e.g. C(T) or SEN(B).

More details about this procedure can be found in the paper,[20] published in ARW NATO "Safety, Reliability and Risks Associated with Water, Oil and Gas Pipelines". As an illustration, diagram for structural integrity assessment is presented in Figure 16. Upon the special request J integral approach was applied to the welded penstock of Hydroelectric Pumping-up Power Plant "Bajina Bašta", produced of weldable high-strength low-alloyed (HSLA) of tensile strength of min. 800 MPa and yield stress above 700 MPa. The experience with this class of steel was not sufficient, because undermatching was required and brittle fracture possible, having in mind the wall thickness of 47 mm and eventual constraint. For that in experiments both, parent and weld metals were included.

Set of crack driving forces was calculated applying REI model. Crack length $c = 90$ mm was constant in the analysis, and shell parameter was calculated as $\lambda = 0.52$.[20] This is presented in $\sqrt{J}$ term in Figure 16 for different values of $pR/WR_{p0,2}$ ratio, up to 0.93 (R is vessel radius, W wall thickness), obtained for selected pressures p, using measured yield strength $R_{p0,2}$, for parent metal (PM) 755MPa and for weld metal (WM) 722 MPa.

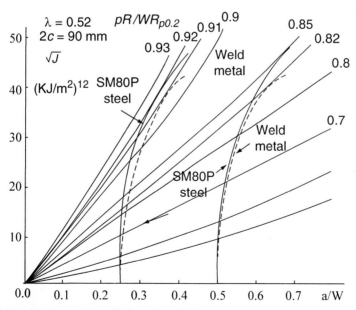

Figure 16. Residual strength prediction of cracked model, using crack driving force calculated according REI model and J–R curves obtained by SEN (B) in standard test.

Resistance curves for applied steel SM80P and SAW weld metal had been obtained by single edge notched bend specimens SEN(B), 22.5∗45∗180 mm (ASTM E1152). They are adapted to the same plot with crack driving force curves in Figure 16. The analysis has shown that for the assumed ratio $a/W = 0.25$ (crack depth $a = 11.75$ mm), crack will grow in a stable manner for 3.75 mm in WM and 4.25 mm in PM, and critical pressure values for fast fracture will be 144 and 155 bar, respectively. For $a/W = 0.5$ ($a = 23.5$ mm) corresponding values are 6.1 mm and 104 bar for WM, 8.5 mm and 140 bar for PM.

Additionally performed experiments of HAZ demonstrated satisfactory behaviour of virtual crack in pressure vessel.[20] Next investigation enabled to understand better the effect of HAZ heterogeneity in microstructure and mechanical properties.[15]

It is to emphasize that presented J integral application is well beyond the common use in assessment of existing crack significance,[21] because closer insight into welded joint behaviour is obtained, with the analysis of crack size and growth effects, in different microstructures of PM and WM, and also HAZ. Additional benefit is J integral direct measurement method (D. Read), used in this analysis for the verification of applied model conservatism.[20]

5.4. LONG TERM LOADING AT ELEVATED TEMPERATURE

High-pressure steam turbine rotor had been analysed by C^* integral regarding brittle fracture prediction.[22] For that, the stress analysis was performed by finite elements taking into account non-stationary and stationary temperature fields, centrifugal force and steam pressure, indicating some critical points (Figure 17). For the most critical point, K5, one can estimate the residual life as 3.1×10^5 h for the case of an edge crack in a large component, loaded by remote tension ($\sigma^\infty = 39$ MPa) and the following data in Eqs. (27) and (29):

$$B = 7.41 \times 10^{-3}, n = 0.72$$

with the critical crack depth value for brittle fracture

$$a_c^{BF} = 34 \text{ mm}$$

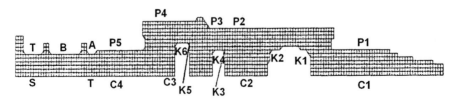

Figure 17. Critical stress points on high-pressure steam turbine rotor for brittle fracture.

6. Discussion and conclusion

It is clear form presented crack parameters and their application that three aspects have to be involved: theoretical, numerical modelling and experimental. The scope of necessary knowledge and skill is large. In theoretical sense at least mechanics, mathematics, physics, material science, theory of elasticity, theory of plasticity are inevitable. Examples, briefly presented in numerical modelling, show that not only basic knowledge, but also skill and experience are required to solve complicated problems of stress and strain analysis, including simplifications which must be introduced when exact solution is not possible. Finally, all obtained solutions and results must be verified by experiments, far from being simple, especially in the case of full scale testing.[20] This is also to add the complex standard testing of cracked specimens, demanding in performance and result analysis.

Fracture mechanics and its parameters significantly contributed to improve structures by crack significance and residual life assessment, but also in design, including materials properties improvement and development of new materials.

References

1. R. W. Nichols, *The Use of Fracture Mechanics as an Engineering Tool*, The 1984 ICF Honour Lecture. Sixth International Conference on Fracture, ICF 6, New Delhi, India, 1983.
2. A. A. Griffith, *The phenomena of ruptures and flow in solids*. Phil. Trans. R. Soc. London. A, 221: 163–198, 1920.
3. C. E. Inglis, *Stresses in a plate due to the presence of cracks and sharp corners*. Proc. Inst. Naval Arch. 55: 219–241, 1913.
4. G. Kirsch, *Die Theorie der Elastizität und die Bedürfnisse der Festigkeitslehre*. (*Theory of elasticity and needs of strength of materials*), Zeitshrift des Vereines deutscher Ingenieure, 42, 797–807, 1898.
5. J. F. Knott, Fundamentals of Fracture Mechanics, London: Butterworths, 1973.
6. ASTM E399-87, Standard Test Method for Plane-Strain Fracture Toughness of Metallic Materials, Annual Book of ASTM Standards, Vol. 04.01. p. 522. 1986.
7. T. L. Anderson, Fracture Mechanics, CRC, Taylor & Francis Group, LLC, 2005.
8. S. Sedmak, Z. Burzic, *Fracture Mechanics Standard Testing*, From Fracture Mechanics to Structural Integrity Assessment, IFMASS 8 Monograph, pp. 95–122, TMF – DIVK, Belgrade, 2004.
9. A. A. Wells, *Application of fracture mechanics at and beyond general yielding*. Br. Welding J. 11: 563–570, 1963.
10. G. R. Irwin, *Plastic Zone Near a Crack and Fracture Toughness*. Proceedings of Seventh Sagamore Research Conference on Mechanics and Metals Behavior of Sheet Material. Vol. 4, 463–478, Racquette Lake, NY, 1960.

11. D. S. Dugdale, *Yielding of steel sheets containing slits.* J. Mech. Phys. Solids. 8: 100–104, 1960.
12. K. Gerić, *Merenje kritične zone razvlačenja.* (*Measurement of final stretch zone*). Eksperimentalne i numeričke metode mehanike loma u oceni integriteta konstrukcija IFMASS 7, pp. 271–279, TMF-JSZ-GOŠA, 2000.
13. H. Tada, P. C. Paris, G. R. Irwin, The Stress Analysis of Crack Handbook. Del Research Corporation, Hellertown, PA, 1973.
14. A. Sedmak, Primena mehanike loma na integritet konstrukcija. (The Use of Fracture Mechanics in Structural Integrity), Mašinski fakultet, Beograd, 2003.
15. K. Gerić, Prsline u zavarenim spojevima (Cracks in welded joints). FTN Izdavaštvo, Novi Sad, 2005.
16. British Standard BS7910 Guide on Methods for Assessing the Acceptability of flaws in Fusion Welded Structures, 1999.
17. Structural INTegrity Assessment Procedures for European Industry (SINTAP Procedure), BRITE European project, Final version: November 1999.
18. M. M. Ratwani, F. Erdogan, G. R. Irwin, Fracture Propagation in Cylindrical Shell Containing an Initial Flaw, Lehigh University, Bethlehem, 1974.
19. R. B. King, *Elastic-plastic analysis of surface flaws using a simplified line-spring model.* Eng. Fract. Mech. 18: 217–231, 1983.
20. S. Sedmak, A. Sedmak, Welded penstock, produced of high strength steel and application of fracture mechanics parameters to structural integrity assessment. G. Pluvinage and M.H. Elwany (eds.) ARW NATO Safety, Reliability and Risks Associated with Water, Oil and Gas Pipelines, Springer Verlag, pp. 271–286, 2008.
21. S. Sedmak, B. Petrovski, A. Sedmak, Z. Lukačević, *The study of crack significance in spherical storage tanks by J resistance curve,* International Conference on Weld Failures, TWI, Welding Institute, London, 1988.
22. A. Sedmak, S. Sedmak, *Critical crack assessment procedure for high pressure steam turbine rotors.* Fatigue Fract. Eng. Mater. Struct., 18(9): 923–934, 1995.

WELDED JOINTS BEHAVIOUR IN SERVICE WITH SPECIAL REFERENCE TO PRESSURE EQUIPMENT

STOJAN SEDMAK[*]
Faculty of Technology and Metallurgy, University of
Belgrade, Karnegijeva 4, 11000 Belgrade, Serbia

KATARINA GERIĆ
Faculty of Technical Sciences, Trg Dositeja Obradovića 6,
University of Novi Sad, 21000 Novi Sad, Serbia

ZIJAH BURZIĆ
Military Technical Institute, Ranka Ristanovića 1,
11000 Belgrade, Serbia

VENCISLAV GRABULOV
Institute for Material Testing, Bulevar voj. Mišića 43,
11000 Belgrade, Serbia

RADOMIR JOVIČIĆ
Faculty of Mechanical Engineering, University of Belgrade,
Kraljice Marije 16, 11000 Belgrade, Serbia

Abstract Quality and structural integrity of welded pressure equipment are defined by The Pressure Equipment Directive (97/23/EC). The problem is complex, due to welded joint imperfections, matching effect, heterogeneity of microstructure and material properties. Equipment can be fit for service with defects unacceptable by codes, what should be proved by fracture mechanics approach, especially for cracks in the heat-affected-zone.

Keywords: Welded joint, steel, pressure equipment, structural integrity

---

[*] E-mail: sedmak@divk.org.yu

1. Introduction

In spite of precaution measures involved, security of welded structures can not be guaranteed, and their failures occurred in different situation under different circumstances also in responsible welded structures, such as bridges or pressure equipment, being in some cases catastrophic.[1]

International Institute of Welding (IIW),[2] conceived in 1947 and founded in 1948 by 13 countries, is the largest worldwide network for welding and joining technologies, with 51 member countries acting to improve the global quality of life. The IIW objectives are exchanges of scientific and technical information on welding research and education, including the formulation of international standards for welding. In this way, IIW enables to prepare the base for quality and service security of welded structures.

By introduction of ISO 9000 series standards for quality assurance the welding has been defined as "special process". In the case of welding, the quality can not be verified on the product, as it is usual practice, but has to be built-in in the product. This approach is dictated by the nature of fabrication in welding. For that, quality requirements for welding are defined (ISO 3834), with approval testing of welders (EN 287), qualification of specified welding procedures for metallic materials (EN 288) and acceptability criteria "Guidance on quality levels for imperfections" (ISO 5817). Anyhow, the quality of welded joint can be endangered: (1) by defects induced during manufacturing and service and (2) by inevitable heterogeneity in microstructure and mechanical properties, corresponding to the nature of welding process and applied heating – cooling cycles of material. It is possible to avoid failure of welded structure only if defects and heterogeneity are under strict control. Many codes and rules are defined for that purpose, one of them is The Pressure Equipment Directive (97/23/EC).[3]

2. Effect of The Pressure Equipment Directive on the security

At the first glance one can conclude that The Pressure Equipment Directive (97/23/EC) defined everything regarding the pressure equipment security. More detailed examination will reveal that the irregularities in welded joints in the form of defects and heterogeneity have an important effect, which can not be controlled in all situations in service, thus contributing to failure. In order to emanate substantial aspects of this effect, it is necessary to find out what is covered by 97/23/EC and supplementary documents.

The requirements for design, manufacture, testing, marking, labelling, instructions and materials of pressure equipment, where the hazard exists, are mandatory and must be met before products may be placed on the market in the European Community, see compulsory Essential safety requirements

(ESRs) (Annex I of 97/23/EC). In that sense, "Pressure equipment must be designed, manufactured and checked in such a way as to ensure its safety when put into service in accordance with the manufacturer's instructions, or in reasonably foreseeable conditions" with the hazard treated according to its significance (Guidelines: 8/15,8/7-97/23/EC).

The pressure equipment must be properly designed taking all relevant factors into account in order to ensure that the equipment will be safe throughout its intended life. The design must incorporate safety coefficients using comprehensive methods which are known to adopt safety margins against all relevant failure modes, and designed for loadings expected in its intended use for foreseeable operating conditions. Internal/external pressure, ambient and operational temperatures, static pressure and mass of contents in operating and test conditions are most important factors. For special products it is necessary to account with traffic, wind, earthquake loading, reaction forces and moments resulting from the supports, attachments or piping, corrosion and erosion, fatigue, decomposition of fluids.

Calculation method of pressure containment and other loading aspects includes allowable stresses, limited regarding to reasonably foreseeable failure modes under operating conditions. To this end, safety factors must be applied to eliminate fully any uncertainty arising out of manufacture, actual operational conditions, stresses, calculation models and the properties and behaviour of the material. This can be achieved applying design by formula, by analysis, by fracture mechanics, or combining these approaches.

Material characteristics to be considered, where applicable, include:

- Yield strength, 0.2% or 1.0% proof strength at calculation temperature
- Tensile strength
- Time-dependent strength, i.e. creep strength
- Fatigue data
- Young's modulus (modulus of elasticity)
- Appropriate amount of plastic strain
- Impact strength and toughness
- Fracture toughness
- Appropriate joint factors, applied to the material properties depending, for example, on the type of non-destructive testing, the materials joined and the operating conditions envisaged (e.g. corrosion, creep, fatigue)

Pressure equipment must be designed and constructed so that all necessary examinations to ensure safety can be carried out, including proof pressure test. Means of determining the internal condition of the equipment must be available, where it is necessary to ensure the continued safety of the equipment, such as access openings allowing physical access to the inside.

Next important stage is manufacturing. The manufacturer must ensure the competent execution of the provisions set out at the design stage.

Preparation of the component parts must not give rise to defects or cracks or changes in the mechanical characteristics likely to be detrimental to the safety of the pressure equipment. Permanent joints and adjacent zones must be free of any surface or internal detrimental defects. The properties of permanent joints must meet the minimum properties specified for the materials to be joined unless other relevant property values are specifically taken into account in the design calculations. For pressure equipment, permanent joining of components which contribute to the pressure resistance must be carried out by suitably qualified personnel (EN 287), according to suitable operating procedures (EN 288). For higher quality categories (II, III and IV),[3] operating procedures and personnel must be approved by a competent third party which may be a notified body or a third-party organization.

For pressure equipment, non-destructive tests of permanent joints must be carried out by suitable qualified personnel, and for categories III and IV, the personnel must be approved by a third-party organization.

Pressure equipment must be subjected to assessment through final inspection, proof test and inspection of safety devices. Proof test is hydrostatic pressure corresponding to the maximum operating loading of pressure equipment in service, multiplied by prescribed coefficient.

Materials used for the manufacture of pressure equipment must be suitable for such application during the scheduled lifetime. Welding consumables and other joining materials need fulfil the relevant requirements in an appropriate way, both individually and in a joined structure. That means (a) appropriate properties for all operating conditions and for all test conditions, with sufficient ductility and toughness, but also being capable to prevent brittle-type fracture, when necessary; (b) sufficient chemical resistance to the fluid contained in the pressure equipment; (c) not be affected by ageing; (d) be suitable for the intended processing procedures.

The pressure equipment manufacturer must define the values necessary for the design calculations and the characteristics of the materials and their treatment, provide in technical documentation elements related to compliance with the materials specifications of the Directive 97/23/EC and take measures to ensure that the material used conforms with the specification. If a material manufacturer has an appropriate quality-assurance system, its issued certificates are presumed the conformity with the requirements.

Detailed quantitative requirements for certain pressure equipment given in corresponding provisions 97/23/EC can be applied as a general rule.

Introduction of Directive 97/23/EC was justifiable and beneficial. Increased quality level of produced and repaired welded equipment improved the situation in all aspects. This is important for extended use of oil and gas for energy supply. The implementation of 97/23/EC required changes and improvements of practice in pressure equipment design, manufacturing and service.[4] It helps to solve many problems in use of pressure equipment.[5]

3. Heterogeneity of welded joint

Selection of material play an important role in design and use of pressure equipment. Material has to be of convenient weldability, assuring required properties of welded joint.

Two microalloyed steels (NIOVAL 47 and NIOMOL 490K) of the same strength level, but different in chemical composition and manufactured by different procedures are selected here to present the significance of heterogeneity of welded joint.

Three welded joint constituents are parent metal (PM), weld metal (WM) and heat-affected-zone (HAZ). Parent metal is used structural steel, of defined uniform microstructure and mechanical properties, according to chemical composition and manufacturing procedure. Weld metal is obtained by crystallization of parent metal and electrode material mixture, molten by induced welding heat. Its microstructure is cast, less uniform compared to PM, but both, PM and WM have more or less homogeneous microstructure. By proper selection of welding procedure and electrode it is possible to obtain evenmatching of welded joint, when WM strength is very close to the strength of PM. General welding practice is to apply overmatching, when WM is superior in strength compared to PM. For structural steels of yield strength above 700 MPa, high-strength low-alloy (HSLA) steel, undermatching is recommended in order to avoid cold cracks, that means the strength of WM is designed to be lower compared to PM.[6,7] Heat-affected-zone, disposed between WM and PM, is characterized by heterogeneous microstructure and corresponding localized material properties, being exposed in each welding pass to transformation at the temperature from melting point to 500°C. The boundary between HAZ and PM, so-called "fusion line", is of special importance due to its inhomogeneity.

It is possible to conclude that welded joint is heterogeneous on the global level due to matching effect, the difference in WM and PM strength. Depending on chemical composition and steel strength, heterogeneity at local level can be obtained in HAZ, with various microstructures and mechanical properties.

3.1. MICROSTRUCTURAL HETEROGENEITY

Frequent failures of spherical storage tanks produced of NIOVAL 47 steel (Steelworks "Jesenice"), and also produced of the same class steels worldwide, due to cracking in HAZ required detailed analysis.[8,9]

In order to analyse HAZ microstructure and properties, experiments were performed with simulated and welded samples, produced by shielded metal arc welding (SMAW) of 13 mm thick plates of NIOVAL 47.[10]

For expereminetal work used normalized steel NIOVAL 47 ("A" in next text), of older generation, with 0.2%C, microalloyed by 0.18%V, intended for pressure equipment application, also at low temperatures, exhibited yield strength of 547 MPa, high tensile strength (738 MPa) and elongation (28.6%). Steel microstructure is fine grained ferrite–pearlite, size 11 after JUS C.A3.004, with banded pearlite (Figure 1).

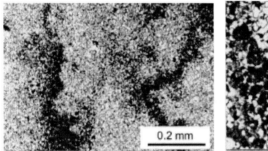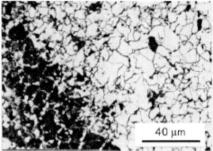

Figure 1. Microstructure of NIOVAL 47 steel at different magnifications.

In simulation of HAZ, samples (Figure 2) had been exposed to different temperatures for 15 s as $\Delta t_{8/5}$ cooling time, typical for microstructural transformations: 1,350°C – formation of coarse grains (CG), 1,100°C – fine grains region (FG), 950°C – intercritically heated zone (IZ) of fine grains above A_{c3}, 850°C – subcritically heated zone (SZ), partial transformation between A_{c1} and A_{c3} temperatures. Some samples had been obtained by two successive simulations: first at 1,350°C, followed by 750°C or 650°C.[11,12]

Roughly presented heterogeneous microstructures in Figure 3 correspond to one pass welded joint, with no clear boundary between different microstructures. The second and next passes have a beneficial effect, thanks to induced heat, producing unchanged coarse zone (NGZ), zone with maximum temperature above A_{c3} and bellow 1,200°C (FGZ, Figure 4), between A_{c1} i A_{c3} (intercritical – IGZ, Figure 4), and bellow A_{c1} (subcritical – SGZ). Sample, welded in four passes (Figure 4), has microstructures presented in Figures 5 and 6.

Coarse microstructure in HAZ of steel A corresponding to 1,350°C, is prone to brittle behaviour due to martensite–austenite (MA) constituent.[10,11]

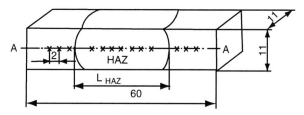

Figure 2. Simulated sample and positions (x) for hardness measurement.

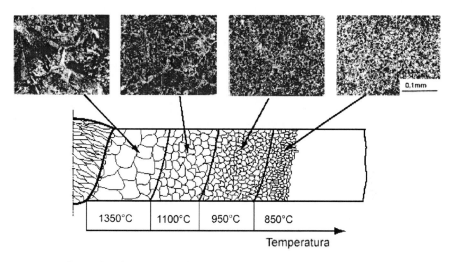

Figure 3. Microstructures in simulated samples of NIOVAL 47 steel.

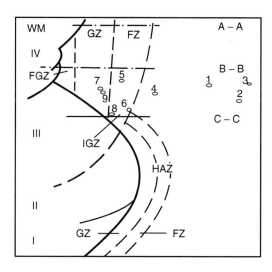

Figure 4. Scheme of sample, welded in four passes (I to IV), with sections of analysed microstructures.

Fractures in spherical storage tanks developed from initial cracks by stress corrosion mechanisms, with final leakage experienced in some cases.[8,9] However, local process industry was not in position to eliminate failed equipment, but used them after repairing.[13]

Based on the experiance with steel A, Steelworks "Jesenice" developed new quenched and tempered microalloyed steel with low carbon content (C = 0.1%), NIOMOL 490K ("B" in next text), applicable up to −60°C. Extended experiments by testing samples of simulated HAZ and welded joint have been performed.[14] Welded joint was produced by SMAW of steel B, 16 mm thick (575 MPa yield strength, 700 MPa tensile strength, 28.1% elongation), and properly qualified (EN 288-3). Passes sequence of welded joint, together with macrograph, are given in Figure 7. Fine grained and uniform ferrite–pearlite microstructure of steel B is not much degradated by welding, and microstructure of all welded joint constituents is beneficial, Figure 8.

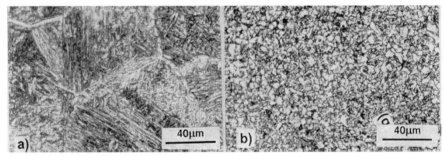

Figure 5. HAZ microstructure of coarse grain zone GZ (left), and fine grain zone FZ (right), close to upper surface of welded joint sample, section A-A in Figure 4.

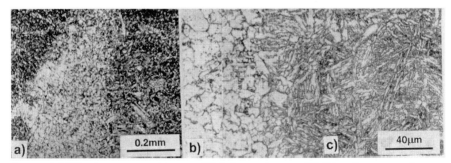

Figure 6. FGZ microstructure (a), magnified fine grain zone FZ (b) and coarse grain zone GZ (c), section B-B in Figure 4, positioned 2 mm bellow upper surface.

Comparison of microstructures of steels A and B and of their HAZes indicate substantial difference. Steel B is characterised by fine grains in all regions of welded joint (Figure 8), and HAZ of steel A close to fusion line has coarse grains region (Figures 3 and 5), prone to brittle fracture. The reason is (1) carbon content (0.2% in steel A, 0.1% in steel B) and manufacturing procedure (steel A is normalized, steel B is quenched and tempered).

WELDED JOINTS IN PRESSURE EQUIPMENT 239

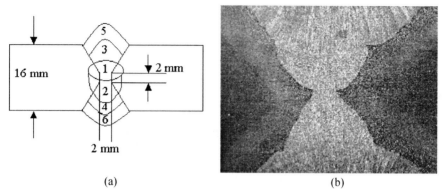

Figure 7. Experimental welded joint: sequence of passes (a), macrograph (b).

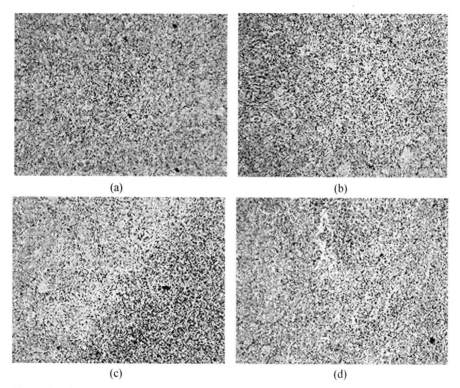

Figure 8. Microstructures of welded joint regions NIOMOL 490K steel (200×). (a) Parent-metal. (b) Heat-affected-zone. (c) Fusion line region. (d) Weld metal.

3.2. HETEROGENIETY OF MECHANICAL PROPERTIES

In design, material is considered as homogeneous and uniform, the differences in welded joint material properties are taken into account by coefficients from codes and rules. The basic approach in design of welded structure is to assure sufficient strength and to exclude the occurrence of plastic yielding, allowing only elastic deformation of components, considering also the hazard of brittle fracture. Different loadings and service condition, producing fatigue, creep or corrosion are treated in special ways. In order to avoid plastic deformation designers accept safety factor, e.g. 1.5 relative to yield strength. It is also general practice to neglect the crack existence in design stage. However, in service condition welded structures can fail, and they do.[1] By acceptability criteria in standard ISO 5817 cracks, as most severe imperfection in welded joint, are not allowed (except cracks in the crater of low quality welded joints and microcracks less than 1 mm^2 cross section). It is not possible to detect reliably cracks of small size by available equipment for nondestructive testing, what means that pressure equipment can operate in real conditions with some imperfections, not revealed during inspection. With this in mind, the full description of welded joint and welded structure requires the knowledge of tensile and toughness properties, but also crack resistance of material, including HAZ.

Hardness measurement (HV5), tensile test (Φ4.5 mm specimen), Charpy V impact test and fracture mechanics test were performed using corresponding specimens, produced from simulated samples (Figure 2).

The results obtained testing steel A and steel B are compared by hardness distribution across welded joints (Figure 9). This approach can be applied since indentation size presents both strength and toughness, as shown in Figure 10. The material in welded joint is continuous, but variation of mecha-

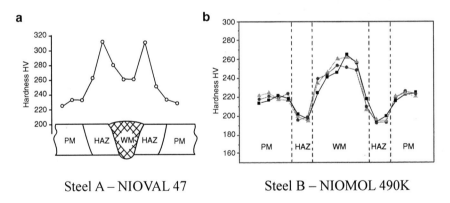

Figure 9. Hardness distribution across welded joint.

Figure 10. The size of hardness measurement indents in different regions.

nical properties (strength, ductility, toughness, hardness) affects its response to applied loading.

Testing results of steel A (Figure 11, Table 1) indicated the HAZ regions of very high strength. Brittle behaviour had been found for the specimens treated at 1,350°C (A1) and 1,100°C (A2). This is attributed to M-A constituent in martensite microstructure. Increased strength and reduced ductility and toughness compared to PM exhibited also samples A3 (950°C) and A4 (850°C). High hardness of simulated samples of steel A

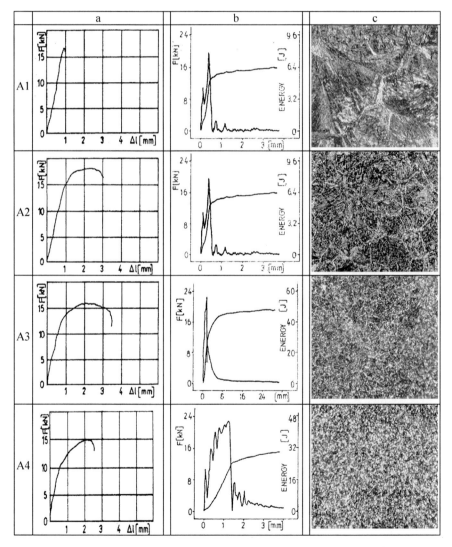

Figure 11. Diagrams of tensile test (a) and instrumented impact (b) test, and microstructures of samples (c) of steel A, simulated at different temperatures (Table 1).[10]

corresponds to 0.2%C content. The values of impact toughness correlate well with tensile properties and hardness. Beneficial effect of subsequent welding passes could be recognized for specimens A5 and A6. Detrimental effect of welding on both, microstructure and mechanical properties is clear.

TABLE 1. Test results of simulated samples testing of NIOVAL 47 steel (A).

Sp	Simulation temperature, °C	Yield strength, MPa	Tensile strength, MPa	Elongation, %	Impact toughness, J	Hardness, HV5
A1	1,350	1,101	1,101	–	7.7	480
A2	1,100	943	1,189	12.6	10.6	418
A3	950	818	1,036	18.0	46.2	353
A4	850	660	936	11.8	31.6	338
A5	1,350/750	815	889	7.3	57.5	364
A6	1,350/650	948	1,035	12.6	25.2	395

With inevitable stress concentration in welded joint, initial plastic deformation and stress redistribution has to be taken into account,[7,15] as general behaviour of welded structure. After unloading, plastically deformed material would behave as elastic in next reloading, up to prior maximum achieved loading level.

3.3. HETEROGENEITY OF CRACK RESISTANCE PARAMETERS

Standard fracture mechanics testing had been performed applying a single specimen compliance technique for J – integral evaluation (ASTM E1737). Charpy size specimens with V notch, produced from simulated samples, with measuring region 20 mm long in central part of 60 mm HAZ length (Figure 2), or specimens sized from available welded joint samples, had been fatigue pre-cracked on Crackthronik pulsator, by variable loading, R = 0.1 ratio, to produce 1 mm long pre-crack in notch root for 80,000 cycles.

Crack resistance of steel B welded joint is not critical, due to pretty uniform microstructures (Figure 8). This is confirmed by J integral standard testing.[14] Plane strain fracture toughness, K_{Ic}, about 136 MPa m$^{1/2}$ for PM, 127 MPa m$^{1/2}$ for WM and 101 MPa m$^{1/2}$ for HAZ, are of acceptable scatter.

This is not the case of steel A. Results from Table 1 and Figure 9 show that brittle behaviour can be expected for specimens of highest hardness (A1, A2, A6), while the specimens A3, A4, A5 are more tough. For brittle specimens plain strain fracture toughness, K_{Ic}, is applied and for other specimens crack opening displacement, δ, was measured for all cases. Brittle behaviour of samples A1 and A2 is confirmed by low values of crack parameters, samples A3, A4 and A5 exhibited better crack resistance, and the value of 82 MPa·m$^{1/2}$ for specimen A6 is higher than K_{Ic}, since testing condition were not fulfilled (Table 2). Crack resistance of welded joint specimens are measured by crack opening displacement, δ (Table 3). Crack tip position corresponds to the number in Figure 4.

TABLE 2. Crack opening displacement and fracture toughness for steel A samples.

Parameter	Sample	A1	A2	A3	A4	A5	A6
Crack opening displacement, δ_c mm		0.007	0.008	0.164^a	0.130^a	0.095	0.017
Plane strain fracture toughness, K_{Ic} MPa m$^{1/2}$		60	48	–	–		(82)

^{a}Crack opening displacement, δ_u, achieved before fracture.

TABLE 3. Crack tip opening displacement values for steel A welded joint specimens.

Distance from fusion line	mm	3.1	2.6	1	0.4	0.32	0
Crack opening displacement, δ	mm	0.56	0.42	0.45	0.39	0.41	0.21

4. Analysis of cracks in welded pressure vessel with defects

Defect free manufacturing of welded structure is costly and probably impossible,[16] resulting in approach of defect acceptance criteria, developed based on experiments and experience. This approach is known as the fitness-for-service (FFS) assessment. In service, repairs are necessary, in order to save integrity and security of equipment. In many cases they have been requested after several years of service, when operation became critical or irregularity are discovered during inspection. In that case it is not possible to follow completely the requirements from the Directive 97/23/EC, but rather the FFS approach should be applied.

In a real welded structure stress concentration caused by inevitable geometrical changes can produce local plastic strain, and the strain hardening capacity could be partly exhausted, affecting the behaviour of crack and structure. Fracture mechanics can provide an approximate value of used parameter when welded joint is tested, especially for the crack in HAZ.

Hydrostatic pressure test of pressure equipment is accepted as final proof for the safety of a welded structure before vessel would be accepted for service, giving realistic answer of loaded welded joints. Some misuse of proof pressure test leaded to crack growth in spherical storage tanks.[9] It happened that small cracks in inner welded joint stayed undetected by applied non-destructive testing, and they had been opened during proof test by applied pressure, higher than working pressure. These cracks could be developed in next service due to stress corrosion, up to the leakage.

The analysis presented here is performed on the storage tank designed for liquefied CO_2 (Figure 12). It is a pressure vessel produced of microalloyed steel NIOVAL 47, steel A, 14 mm thick, 12.5 m^3 in volume, 1,600 mm in diameter, 7,180 mm long; classified in class II, working pressure 30 bar, proof pressure 39 bar, minimum temperature −55°C. High alloyed austenite steel X7CrNiNb18.10 (in next text steel "C") has been selected for manhole flange and outlet pipe connection (weldolet), requiring austenite–ferrite welded joint. Consumable INOX 29/9 was used for welding.

During regular periodical inspection two irregularities had been detected, requiring repair.[17] First one was leakage on manhole flange – austenite steel, the second were cracks in ferrite steel A (Figure 13).

Careful examination during pressure proof test confirmed the existence of two tinny cracks, passed through wall of flange, but also a net of part-through cracks in the same region, close to welded joint. In last inspection some drops have shown leakage through these cracks, but in previous inspection wet surface was attributed to condensation due to temperature

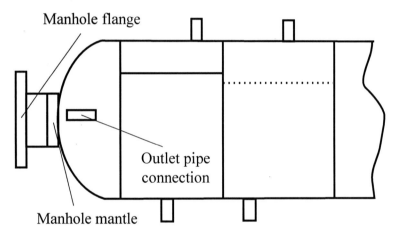

Figure 12. Storage tank for liquefied CO_2.

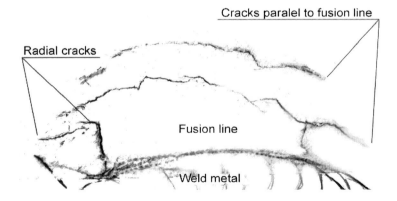

Figure 13. Cracks detected in ferrite steel A.

effect. The solution was to cut the flange, reduce its length to the size with no defects, and re-welded using original procedure, but qualified by an analysis of produced samples.

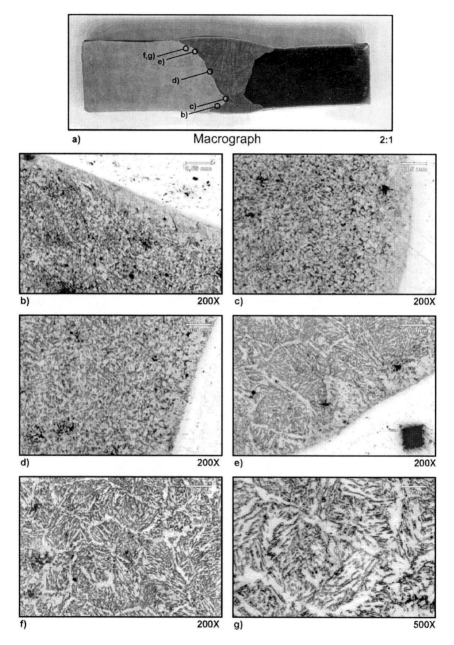

Figure 14. Macrograph of welded joint (a): PM steel A (left), WM steel 29/9 (middle), PM steel C (right). Microstructures of locations (b) to (g).

The cracks in steel A had to be grind out and new austenite connection of diameter adopted to new situation should be welded, applying the same welding procedure and same consumable, INOX 29/9.

In order to confirm FFS of repaired pressure vessel in eventual presence of cracks, similar to those detected in final NDT control, additional test had been made with samples, used also for qualification of welding specification procedure. Micrograph of one sample is presented in Figure 14a, followed by microstructures of indicated locations (b to g) on the side of steel A as PM. Three-material body can be easily recognized: PM steel A – ferrite (left), WM of steel 29/9 – austenite (middle), PM steel C – austenite (right), with the comment that the HAZes on both sides represent also multi-material body due to heterogeneity of microstructure, what was neglected in this analysis.

Tensile test specimen and test diagrams, required for the analysis, are given in Figure 15, and tensile properties are listed in Table 4.

It is necessary to understand behaviour of considered three-material body in tensile test (Figure 15b), since it is similar to the behaviour of real welded joint in pressure vessel. Plastic strain will initiate, Figure 15b, point A and develop till the point B, but only in austenitic steel C of the lowest yield strength, with expressed ductility and strain hardening (Figure 15d). In point B yield strength level of steel A (Figure 15f) is achieved, and further tensile behaviour of steel A causes the redistribution of deformation, which develops in both, steel C and steel A, but faster in steel A up to the point D. Point D presents the tensile strength of steel A (Figure 15f) and final fracture took place in steel A (Figure 15a). At the fracture, neglectible plastic strain occurred in WM, as can be seen from distribution of contraction across welded joint (Figure 15c), since WM yield strength level (Figure 15e) is slightly lower than steel A tensile strength (Figure 15f), and WM tensile strength is much higher (Table 4).

TABLE 4. Properties of three-material body constituents.

Material	Yield strength, MPa	Tensile strength, MPa	Elongation, %	Young modulus, GPa
PM steel C	315	600	36	193
PM steel A	435	555	25	207
WM	545	755	35	193

In one experimental sample hot cracks were detected in WM (Figure 16). This type of cracks can not be completely excluded in real welded joint of flange, but according described tensile behaviour these cracks are not significant. In overmatched WM due to only elastic deformation and high ductility of WM critical driving force for crack initiation can not be achieved.

WELDED JOINTS IN PRESSURE EQUIPMENT 247

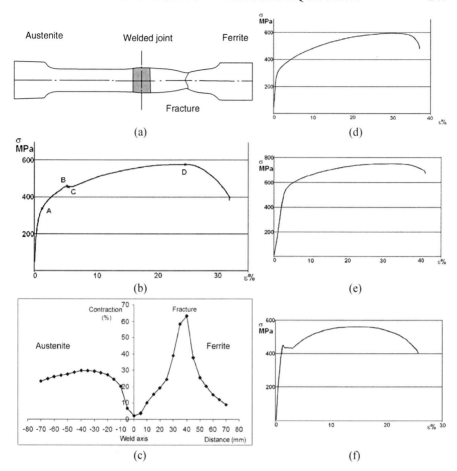

Figure 15. Tensile test specimen and diagrams obtained in tensile tests. (a) Fractures welded joint specimen. (b) Stress–strain curve of welded joint. (c) Distribution of contraction across welded joint. (d) Stress–strain curve of PM steel C. (e) Stress–strain curve of WM. (f) Stress–strain curve of PM steel A.

Crack shielding effect in overmatched welded joint is already reported.[15] Anyhow, more precise control during next periodical inspection is required.

Next problem were cracks in HAZ of steel A (Figure 13). Pressure vessel was in service with these cracks for certain time period and they did not propagate, but following ISO 5817 requirements they were repaired. In another pressure vessel shallow similar microcracks were detected, in HAZ of steel A, 1.8 mm long (Figure 17). Dark upper region with crack is part of steel A, etched by NITAL, so the lower part, austenite, is not etched.

Based on gathered experience they could be neglected, but more evidence for FFS was required. For that fracture mechanics analysis is performed by J integral.[15] Crack driving force is determined applying King's

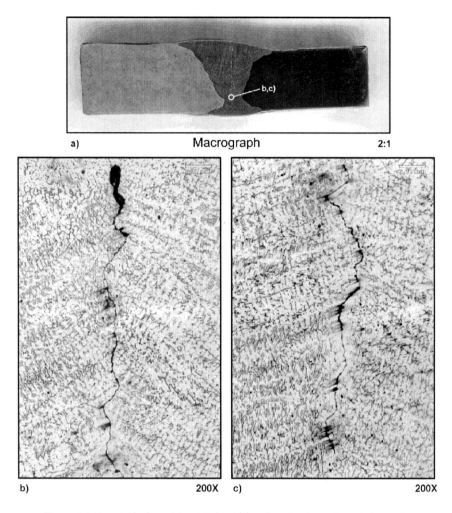

Figure 16. Hot cracks in weld metal dendritic microstructure of welded samples.

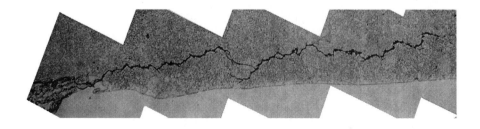

Figure 17. Microcracks in HAZ of NIOVAL 47 (steel A).

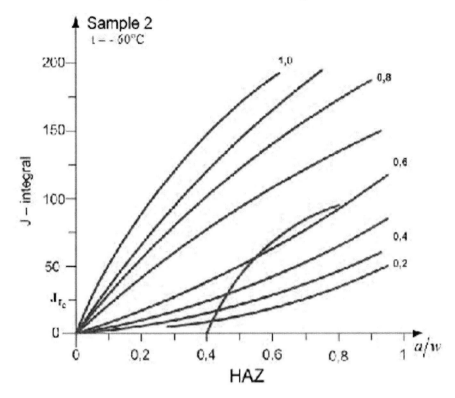

Figure 18. Crack driving forces curves (lines marked by 0.2 to 1.0) and J–R curve for steel C HAZ material have shown that crical crack depth is 7.5 mm.[17]

model, J resistance curve is designed based on some simplifications and assumptions regarding the properties of material in HAZ,[10,11] and applied for analysis (Figure 18). It was found that conservative prediction of critical crack depth is about 7.5 mm, enabled the conclusion that the crack of detected size can be neglected.[17]

5. Discussion and conclusion

Significant efforts of International Institute of Welding (IIW) and necessity to reduce or eliminate frequent failures by increasing security lead to development of codes, rules and directives in how to manage production of welded structures.

Final achieved results are ISO 9000 standard series together with The Pressure Equipment Directive (97/23/EC), covering the aspects of welded structures quality accepted for service in Europe. The last document deals with new welded product aimed for European Community market, but there is still the problem with defective equipment in exploitation. In that sense welding has a special significance because it represents unavoidable procedure in maintenance and repair of equipment.

On the other hand, it is not possible to exclude the irregularity in welded joints, so in some cases clear and strict requirement must be defined. Fittness-for-service approach enables the use of equipment with defect after a conservative assessment regarding security, safety and life. In that context behaviour of welded structures in service attracts special consideration, having in mind inavoidable irregularities and heterogeneity of welded joint.

The comparison of two steels of similar strength level, NIOVAL 47 and NIOMOL 490K showed superior properties of new generation. Nowadays much better steel properties are offered by manufacturers, contributing to easier welding in production and much safer final welded structures.

However, there is still in service equipment made of steels of older generation, for decades, and even more than century, which must be maintained and repaired. For that intensive research followed by practical solution in service are welcome, as it is presented here in the case of pressure vessel. Most important findings of that research are connected with service behaviour of welded joints, based on facts that cracks, due to heterogeneity and service condition, can be present even they are not detected, and can develop up to critical size. The assessment of crack significance is an important achievement of Fracture mechanics and its application is now accepted world-wide. High stress levels for initiation of stable crack growth suggest the possibility the welded structure can work safely even with relatively large surface cracks. Also the integrity of heterogeneous welded joints must not be affected by the presence of surface cracks when overmatching with crack shielding effect is applied. The latest conclusion holds at low temperatures, as well. It has also be shown that in welded joints displacement in plastic range has an imortant role, allowing to exploit strain hardening capacity of welded joint constituents in preventing crack to grow.

Some efforts are also done to improve the control of equipment from the very beginning of its service. The idea is to establish initial state of material and welded joints, and reveal the changes and deviations by inspection after properly defined time period and in-service monitoring, allowing an action to prevent failure.[18,19]

References

1. S. Sedmak, Crack problems in structures, in monograph The Challenge of Materials and Weldments – Structural Integrity and Life Assessments, edited by S. Sedmak, Z. Radaković, J. Lozanović (Faculty of Mechanical Engineering – MF, Society for Structural Integrity and Life – DIVK, Faculty of Technology and Metallurgy – TMF, Institute GOŠA, Beograd, 2008), pp. 3–32.

2. International Institute of Welding, http://www.iiw-iis.org/
3. The Pressure Equipment Directive (97/23/EC), http://ec.europa.eu/enterprise/pressure_equipment/ped/index_en.html
4. A. Bređan, J. Kurai, Evropska direktiva za opremu pod pritiskom (PED) i integritet konstrukcija (European Pressure Equipment Directive (PED) and Structure Integrity), Structural Integrity and Life (IVK), Vol. 3, No 1, 2003, pp. 31–42.
5. A. Bređan, J. Kurai, Zahtevi, praksa i dileme pri tehničkom nadzoru nad opremom u eksploataciji (Requirements, practise and surveillance of equipment in exploitation), IVK, Vol. 1, No 1, 2001, pp. 19–22.
6. K. Satoh, and M. Toyoda, Joint Strength of Heavy Plates with Lower Strength Weld Metal, AWS (Translation from Japanese), Welding Journal, 54, 311S (1975).
7. S. Sedmak, A. Radović, and Lj. Nedelković, The strength of welds in HSLA steel after initial plastic deformation, in Mechanical Behaviour of Materials, edited by K. J. Miller and R. F. Smith (Pergamon Press, Oxford/New York, 1979), Vol. 3, pp. 435–446.
8. I. Hrivnjak, Reparaturno zavarivanje velikih sfernih rezervoara (Repair welding of huge spherical storage tanks) in monograph Service Cracks in Pressure Vessels and Storage Tanks (in Serbian), edited by S. Sedmak and A. Sedmak, (Faculty of Technology and Metallurgy, European Center for Peace and Development, Institute GOŠA, Beograd, 1994), pp. 328–351.
9. J. Kurai, B. Aleksić, Ispitivanje pritiskom kao uzročnik pojave prslina kod opreme pod pritiskom u eksploataciji (Proof pressure test as the cause of crack occurrence in pressure equipment in service), IVK, Vol. 3, No 2, 2003, pp. 65–72.
10. K. Geric, Pojava i rast prslina u zavarenim spojevima čelika povišene čvrstoće (The occurrence and growth of cracks in welded joints of high strength steels), (in Serbian), Ph.D. Thesis, Faculty of Technology and Metallurgy, University of Beograd (1997).
11. K. Gerić, Prsline u zavarenim spojevima (Cracks in welded joints), (in Serbian) the Monograph, FTN Izdavaštvo, Novi Sad (2005).
12. S. Sedmak, K. Gerić, V. Grabulov, Z. Burzić, A. Sedmak, Static and dynamic fracture mechanics parameters of material in the heat-affected-zone, 11th International Conference on Fracture, ICF 11, Turin (2005).
13. B. Aleksić, A. Fertilio, Popravka sferne posude za skladištenje VCM zapremine 2000 m3 (Repair of spherical vessel for VCM storage, 2000 m3 in volume), monograph listed here as 8, pp. 255–260.
14. M. Manjgo, Kriterijumi prihvatljivosti prslina u zavarenom spoju posuda pod pritiskom od mikrolegiranih celika (Acceptance criteria for cracks in welded joint of pressure vessel of microalloyed steels), (in Serbian), Ph.D. Thesis, Faculty of Mechanical Engineering, University of Beograd (2008).
15. B. Božić, S. Sedmak, B. Petrovski, and A. Sedmak, Crack growth resistance of weldment constituents in a real structure, Bulletin T. Cl de l'Academie serbe des Sciences at des Arts, Beograd, Classe des Sciences techniques. No 25 (1989), pp. 21–44.
16. R. W. Nichols, The Use of Fracture Mechanics as an Engineering Tool, The 1984 ICF Honour Lecture. Sixth International Conference on Fracture, ICF 6, New Delhi, India (1983).
17. R. Jovičić, Analiza uticaja prslina na integritet feritno – austenitnih zavarenih spojeva (Analysis of cracks effect on integrity of ferrite–austenite welded joints), (in Serbian), Ph. D. Thesis, Faculty of Mechanical Engineering, University of Beograd (2007).
18. J. Kurai, Z. Burzić, N. Garić, M. Zrilić, B. Aleksić, Initial stress state of boiler tubes for structural integrity assessment, IVK, Vol. 7, No 3, 2007, pp. 187–194.
19. N. Gubeljak, Application of Stereometric Measurement on Structural Integrity, IVK, Vol. 6, No 1–2, 2006, pp. 65–74.

SECURITY OF GAS PIPELINES

VLADIMIR STEVANOVIĆ[*]
*University of Belgrade, Faculty of Mechanical Engineering
Kraljice Marije 16, 11 120 Belgrade 35, Serbia*

Abstract Security of the natural gas supply strongly depends on the integrity of the transportation pipelines. The statistical evidence shows that the most probable cause of the break occurrence at natural gas pipeline is the external third party interference. A potential damage to the surrounding objects and violation of the people lives during the pipeline accident depends on the mass flow rate of the natural gas leakage from the break. The paper presents an efficient method for the prediction of the natural gas leakage rates in case of pipeline accidents, as well as for the prediction of transient gas dynamic forces that are generated in case of an unsteady fluid flow and a fluid discharge from the pressurized volume to the surrounding. The method application is demonstrated on the test cases of natural gas outflow from a high pressure main transportation gas pipeline and a break occurrence at the distribution pipeline. Obtained data are a necessary input to the safe design of pipeline structures and supports.

Keywords: Gas pipelines, security, accident, blowdown, fluid dynamic forces

1. Introduction

The share of natural gas in the world total primary energy supply is constantly increasing in the last decades. In 1973, 16% of the total primary energy supply was provided with natural gas, while in 2005 this share was 20.7%. Compared with oil and coal, which provided respectively 35% and 25.3% of the world total primary energy needs in 2005, natural gas is more valuable fossil fuel. The combustion of natural gas emits the least amount of products that pollute the environment. For instance, the energy of 1 GJ in

---

[*] E-mail: estevavl@eunet.yu

natural gas combustion is obtained with the emission of 56 kg of carbon dioxide, the gas most responsible for the greenhouse effects in the Earth atmosphere and corresponding global warming, while the same amount of energy is obtained with the emission of approximately 73 kg of CO_2 in crude oil combustion or between 95 and 106 kg of CO_2 in combustion of various kinds of coal. The heat content of natural gas is high, between 33 and 38 GJ per 1,000 S m^3. It is more evenly distributed over the planet than oil. Investment and operational costs for natural gas extraction are about the same as for oil. The world reserves of natural gas are about equal in magnitude to oil reserves. Also, gas will be available for the next midterm period.[1]

A feature of natural gas exploitation is that its transportation is particularly advantageous when transported by pipeline. A hundred of thousand kilometers of high pressure main transportation gas pipelines are in operation in North America, Europe and Asia for the gas transport from gas source fields to large consuming sites, as well as low pressure distribution pipelines in urban and industry areas. Design pressures of main transportation gas pipelines reaches 7 MPa and diameters of 1 m, while distribution pipelines are designed for pressures up to 0.1 MPa and diameters of approximately 100–200 mm. A rupture of pressurized natural gas pipeline could lead to accidents with gas ignition, formation of burning jets or explosions and catastrophic consequences to surrounding objects and population. A number of disasters caused by gas pipeline ruptures and gas leakages have been reported. Although the engineering community devotes a considerable attention to the security and safety of gas pipelines, and despite the applied safety measures and systems, the rates of natural gas pipeline accidents still remain in the same level during the last 20 years.[2]

A gas pipeline rupture is characterized with the critical gas outflow at the rupture hole, the transient flow within the pipeline, the intensive pressure waves propagation in the starting period after the rupture and the intensive fluid-dynamic forces that act on the pipeline's structure. Therefore, the accurate prediction of these gas dynamic conditions is extremely important for the proper design of the gas pipeline structure and prevention of further pipes' destructions and for the mitigation of accidents' consequences including fire, explosion and environmental pollution. The transient compressible gas flow is modeled with the mass, momentum and energy balance equations in the form of hyperbolic partial differential equations. These equations can be effectively solved numerically with the application of the method of characteristics, as it was presented by Wylie and Streeter[3] and Mahgerefteh et al.[4] Several researches have based their predictions of the gas leakage at the pipeline brake on the quasi-steady-state conditions, for instance Jo and Ahn,[5,6] Keith and Crowl[7] and Morris.[8] These simplified models are useful engineering tools for the quantitative risk assessment or estimation of the hazard areas associated with high pressure natural gas pipelines. But, since these models are derived from the assumption that steady-state conditions hold in the gas pipeline during the accident, they provide only the first approximation. More precise and reliable predictions are needed, which take into account the transient nature of the gas pipeline blowdown during the pipe rupture accident.

In this paper a modeling approach to the simulation and analyses of the natural gas pipeline accidents with pipe rupture and gas leakage is presented. The method is based on the numerical solution of the conservative equations of transient compressible fluid flow and appropriate boundary conditions for the prediction of gas critical leakage, junction of two or several pipes, the influence of the valve on the gas stream etc. Also, the module for the prediction of the transient-fluid dynamic forces is presented. The model is solved numerically by applying the method of characteristics. The developed models are incorporated in the computer code TEA-NGAS (Transient Evolution Analyses in Natural Gas), which is an extended version of the general purpose code TEA.[9,10] Regarding previous analytical and numerical investigations of the gas pipeline flows, here presented results also incorporate the calculation procedure and predictions of the transient fluid-dynamic forces. The dynamics of these forces in the natural gas pipelines have not been analyzed in the up-to-date literature, for instance, in textbook by Wylie and Streeter[3] or the recent paper by Mahgerefteh et al.[4] These forces have impact character and they are the highest loads that could be expected to act on the pipelines in case of ruptures. The prediction of their magnitudes is important for the design of the pipeline supports and structures in order to prevent further escalation of pipeline damage in case of pipe rupture accident events. The presented modelling approach provides more reliable data on natural gas leakage rates and pressure transients than simplified analytical models of Jo and Ahn,[5,6] Keith and Crowl[7] and Morris.[8]

2. Causes and consequences of natural gas pipeline accidents

The main causes of the natural pipelines integrity degradation have been classified in five categories[11,12,6] as presented in Table 1. Also, Table 1 shows the failure frequency of each category per year and per kilometer of pipeline length, the share of each category in the total failure rate, and the distribution of the rupture hole sizes. The parameters in Table 1 are derived from historical data of the failure rate of onshore natural gas-pipelines in Western Europe. The utilized data base is the experience of 1.5 million kilometer-years in eight countries. As shown in Table 1, the external interference by third party activity is the leading cause of major accidents related to medium or great holes. Construction defects and corrosion contribute mainly to the occurrence of small rupture holes, while a ground movement is the main cause of the great rupture holes occurrence.

Investigations of real accidents of natural gas pipelines show that the consequences are dominated by a few accident scenarios such as the explosion and the jet fire.[6] Some examples of gas pipeline accidents with catastrophic effects on people and property are presented here.

In 1994 at New Jersey, USA, an explosion of an underground natural gas pipeline was followed by a crater of approximately 50 m diameter and massive flames that could be seen more than 80 km away. The accident resulted in one death and 50 injuries. Subsequent investigations revealed that the pipeline had been damaged by excavation works. Probably, a mechanically induced crack grew to a size as a result of enhanced fatigue leading to material failure.

In Venezuela, an explosion of a natural gas pipeline occurred underneath a highway in 1993. The rupture occurred while a state telephone company was installing fiber optic cables. The result was 40 injured people and 50 dead.

One of the most severe chemical accidents that ever happened took place in 1989 at Siberia in Russia. It was reported that there had been a leakage for several days at the petroleum gas pipeline that supplied an industrial city from the industry plant. Instead of investigating the complaints, the responsible engineers had responded by increasing the pumping rate in order to maintain the required pressure in the pipeline. The leakage point was found about half a mile away from the side of a railway.

TABLE 1. Failure frequencies based on failure causes and hole size.[6]

Failure causes	Failure frequency (1/year km)	Percentage of total failure rate (%)	Percentage of different hole size (%)		
			Small	Medium	Great
External interference	3.0×10^{-4}	51	25	56	19
Construction defects	1.1×10^{-4}	19	69	25	6
Corrosion	8.1×10^{-5}	14	97	3	<1
Ground movement	3.6×10^{-5}	6	29	31	40
Others/unknown	5.4×10^{-5}	10	74	25	<1
Total failure rate	5.75×10^{-4}	100	48	39	13

Note: The hole sizes are defined as follows: the small hole, the hole size is lower than 2 cm; the medium hole, the hole size ranges from 2 cm up to the pipe diameter; the great hole, the full bore rupture or the hole size is greater than the pipe diameter.

Some hours later, two passenger trains traveling in opposite directions, approached the area. One train sparked off the cloud of gas and air mixture and the explosion was initiated, and subsequently two more explosions succeeded. A wall of fire was formatted with the width of 1 mile. The trains were derailed, while trees were flattened and windows were broken within a radius of 2.5 and 8 miles respectively. Totally 462 people died and 706 were injured.

3. Safety measures for natural gas pipeline rupture accidents

After the pipeline rupture and natural gas release to the atmosphere a gas jet fire or a gas cloud explosion could occur with devastating effects to surrounding. Sklavounos and Rigas[13] determined that the most dangerous scenario is the jet fire formation after the break. According to this scenario they determined the safety distances in the vicinity of natural gas pipelines. It was shown that the safety distance is more sensitive to pipeline size than operating pressure. The atmospheric conditions strongly influence the thermal effect of a jet fire event, as well as the distance that fuel gas travels from source to its lower flammable limit position. Sklavounos and Rigas[13] proposed diagrams for the prediction of safety distances as a function of pipeline diameter and operating pressure. According to these results, the safety distance ranges from 50 m in case of a low pressure distribution pipe at 3 bar and 90 mm diameter, to 900 m in case of a transportation pipeline at 50 bar and 900 mm diameter.

Jo and Ahn[5] investigated hazardous event of the worst accident case of the full-bore rupture of a high-pressure pipeline with horizontal gas release resulting in explosion and fire. The developed hazard model is based on an effective release rate model at steady-state for high pressure pipeline rupture, a jet dispersion model that relates the operating condition of the pipeline and the effective hole size to the contour of the lower flammable limit, as well as on a model that relates the rate of gas release to the heat intensity of the fire. A simple relation is derived between the hazard distance r_h and the gas release rate $\dot{m}_{leakage}$

$$r_h = 10.285\sqrt{\dot{m}_{leakage}} \qquad (1)$$

In order to prevent significant gas leakage to the surrounding, quick sliding valves could be employed. Disadvantages of these valves are that they are limited to diameters less or equal to 0.6 m, they require heavy fixing equipment and their cost is high (closing a DN 600 pipeline requires 45,000 €). Hence, an airbag is developed for closing of pipelines on explosions and leakages.[14] The basic solution principle is to use airbags similar to those utilized in cars. This new device is still under investigation, but achieved results indicate effectiveness and high reliability of the pipeline isolation in cases of a rupture occurrence, much faster closing (within 50 ms) compared to the sliding valve closing for 300 to 500 ms, and they are cheaper than sliding valves.

A number of safety conditions of the natural gas pipeline in case of a pipe rupture accident depend on the system's gas dynamics. As presented, a safety (hazard) distance from the natural gas pipeline in case of the pipe rupture accident depends on the mass discharge rate at the rupture. The discharge rate depends on the pressure distribution along the damaged pipeline. Transient gas flow induces gas dynamic forces that act on the pipeline structure and supports. Hence, a crucial point in the design and safety analyses of the natural gas pipeline is to obtain a reliable prediction of the natural gas pipeline dynamics under the pipe rupture accident.

4. Modelling approach to gas dynamics of natural gas pipeline rupture accidents

Natural gas pipelines are exposed to high loads in transient conditions caused by the large size pipe rupture and gas leakage or due to rapid action of control, isolated or relief valves. These transient gas flow conditions generate intensive pressure waves that propagate and superimpose within a gas pipeline or network, and additional fluid dynamic forces are generated. Besides the fluid dynamic force caused by the fluid transient flow, the reactive force is also exerted at the ruptured pipe, as well as the component due to the difference between the rupture and atmospheric pressures. In case of intensive disturbances, the amplitude of these forces reaches tens or even hundreds kilo Newton. This circumstance should be taken into account especially during the design of gas pipeline that crosses the bridge, in the urban cites and in the vicinity of other vital objects.

Transient flow of compressible natural gas in the pipeline is described with the one dimensional model based on the mass and momentum balance equations. The thermal effects of gas heating or cooling are described with the energy balance equation. These balance equations are partial differential equations of the hyperbolic type, and they can be written as follows

- Mass balance

$$\frac{\partial \rho}{\partial t} + \frac{\partial (\rho u)}{\partial x} = 0 \qquad (2)$$

- Momentum balance

$$\frac{\partial(\rho u)}{\partial t} + \frac{\partial(\rho u^2)}{\partial x} = -\frac{\partial p}{\partial x} - f\frac{\rho u|u|}{2D_H} - g\rho\sin\theta \qquad (3)$$

- Energy balance

$$\frac{Dh}{Dt} - \frac{1}{\rho}\frac{Dp}{Dt} - \frac{fu^2|u|}{2D_H} - \frac{\dot{q}}{\rho} = 0 \qquad (4)$$

where ρ is density, u is velocity, p is pressure, f is friction coefficient, g is gravity acceleration, θ is the pipeline inclination, h is enthalpy, $\dot{q}$ is volumetric heat flux, x is spatial coordinate and t is time.

The set of the balance Eqs. (2) to (4) is closed by the equation of state in the form of the modified equation of state of the ideal gas

$$\frac{p}{\rho} = zR_g T \qquad (5)$$

The molar concentration of methane CH_4 in natural gas is 98%, and the rest of 2% belongs to the secondary components, such as ethane, propane, nitrogen etc. Hence, the natural gas physical properties may be identified with the properties of methane.

This system of equations is solved for the prescribed initial and boundary conditions. Initial conditions determine the pressure, velocity and temperature (enthalpies) in the gas network in the initial time instant before the distribution action. Boundary conditions describe the action of safety, control and technical components, such as various kinds of valves, pipeline junctions, compressor units etc. Equations (2) to (4) are solved numerically by the method of characteristics, which is based on the physics of mechanical disturbance propagation in the form of pressure waves by the speed of sound and the enthalpy propagation by the fluid particle flow. The method of characteristics transforms the partial differential equations in the form of ordinary differential equations that holds along the characteristic paths determined by the pressure waves propagations C^+ and C^- and fluid particle movement C^P in the time–space coordinate system, as presented in Figure 1. Obtained ordinary differential equations are approximated with the finite difference equations that are solved explicitly for every new time step of integration. The time step of integration is determined by the Courant criterion.

$$\Delta t \leq \min\left(\frac{\Delta x}{c_j + |u_j|}\right) \qquad (6)$$

where j indicates a pipe within the network and Δx is the distance between two adjacent numerical nodes.

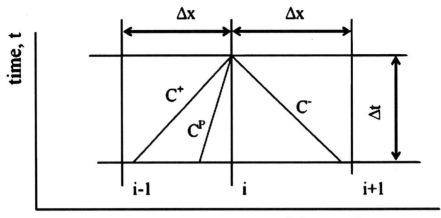

Figure 1. x–t plane and characteristics directions.

In order to simulate complex pipe networks and various transient scenarios, several models of boundary conditions (i.e. the flow channel discontinuities) are developed that can be linked in a modular way. The calculation of the flow parameters at the ends of a pipe must be done using additional hydraulic models. These additional equations describe the mass, momentum and energy balance at a point of discontinuity, and they replace the equations of the characteristics which do not belong to the physical domain of a pipe. The general form of these equations is

- Balance of mass

$$\Delta(\rho u A) = 0 \qquad (7)$$

- Balance of momentum

$$\Delta\left(\frac{\rho u^2}{2}\right) + \Delta p = \Delta M(t) \qquad (8)$$

- Balance of energy

$$\Delta\left(h + \frac{1}{2}u^2\right) = 0 \qquad (9)$$

where ΔM is the momentum change due to the local friction loss. It can be time dependant (for instance in the case of valve closure). The following boundary conditions are included in the code: a subcritical or critical leakage from a pipe, a closed end of a pipe, a pipe joining a reservoir, a junction of two or more pipes, a valve in a pipe etc.

4.1. FLUID DYNAMIC FORCES

The transient fluid dynamic force in a pipe which is bounded by other flow components, such as elbows or closed end, is caused by the propagation pressure waves. This transient wave force acts along the straight pipe axis and its positive direction is assumed to be opposite to the fluid flow direction. It is calculated by the following expression:[15]

$$F = \int_L \frac{d\dot{m}}{dt} dx \qquad (10)$$

The fluid force exerted on a pipe with an open end (expulsion from a pipe) includes both a wave force and a blowdown component associated with momentum expulsion and the difference between discharge and ambient pressure. It acts along the pipe axis and the force intensity is

$$F = \int_L \frac{d\dot{m}}{dt} dx + \left[(p_i - p_{atm}) + \rho_i u_i^2 \right] A \qquad (11)$$

The pipe wall is assumed to be rigid, which is a conservative assumption.
The developed models are incorporated in the TEA-NGAS computer code. The code is applied to the simulation and analyses of natural gas transient conditions caused by the pipe rupture.

5. Prediction of natural gas pipelines blowdown and related consequences

The gas dynamic phenomena and the forces exerted on the gas pipeline caused by the pipe rupture are simulated and analyzed with the developed numerical method and the code TEA-NGAS.

5.1. CASE 1: PRESSURE AND LEAKAGE FLOW RATE DURING THE GAS PIPELINE RUPTURE ACCIDENT

The blowdown of the long natural gas transportation pipeline due to the complete guillotine of the pipe (100% break) is simulated. The pipeline length is 100 km and the inner diameter is 0.87 m. The rupture occurs in 200 s at the pipeline end at 100 km from the gas inlet. Gas freely outflows to the atmosphere. Five minutes after the pipe break the valve at the gas inlet to the transportation pipeline is closed. Figure 2 shows mass flow rate from the pipeline to the atmosphere at the location of break. The rapid increase of gas flow is shown in the short interval after the break, while later on the mass flow gradually decreases. One hour after the break the leakage flow rate equals approximately one half of the initial flow rate to the consumers that existed before the rupture. The pressure distribution along the pipeline is shown in Figure 3. Due to the rupture occurrence, the gas pressure towards the break location decreases. Three hundred seconds after the break the valve at the long pipeline inlet starts closing, hence the gas inlet mass flow rate to the pipeline reaches zero and the pressure at the pipeline inlet decreases. Results show the critical gas outflow at the break. From the safety point of view, the important conclusion is that a considerable gas leakage exists for a long time period.

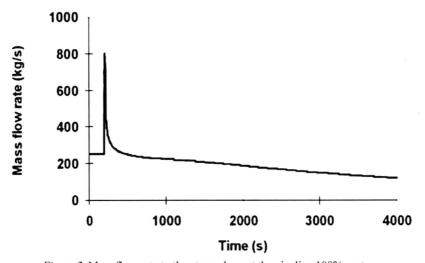

Figure 2. Mass flow rate to the atmosphere at the pipeline 100% rupture.

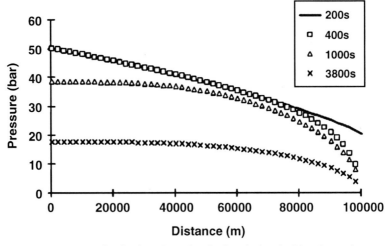

Figure 3. Pressure distribution along the pipeline during the blowdown phase.

5.2. CASE 2: TRANSIENT GAS DYNAMIC FORCES DURING THE NATURAL GAS BLOWDOWN

A scheme of the gas pipeline with the hypothetical break is depicted in Figure 4. The distribution gas pipeline is connected to the main transportation pipeline at the pressure of 4 MPa and with the diameter of 0.610 m. The distribution pipeline consists of two straight segments, the length of each one is 50 m and the diameter is 0.219 m. It is assumed that an instantaneous distribution pipeline break occurs at its end towards the consumer at 0 s. The calculated transient gas dynamic forces that act on the pipe 1, where the break occurs, and the pipe 2, which is connected to the main transportation pipeline, are shown in Figures 5 through 7 for the break sizes of 10%, 50% and 100% of the full pipeline cross section area, respectively. The rarefraction pressure wave propagates from the break towards the main transportation pipeline. At the junction with the main transportation pipeline, the rarefraction wave reflects as the compression wave and travels back to the break location, where it reflects as the rarefraction wave, and the periodic wave propagation continues till the wave attenuation. Due to the initial pressure of 4 MPa, the critical gas flow is established at the break location with the velocity of approximately 400 m/s. The transient gas dynamic forces act along the straight pipeline segments, where the positive force direction is opposite to the direction of gas flow. The following characteristic phenomena are observed:

- The pipe 1 at which the break occurred is loaded with a higher force than the pipe 2. The pipe 2 is loaded only with the transient force caused by the unsteady gas flow (i.e. pressure wave propagations), while the pipe 1 is additionally loaded by the reactive force due to the gas leakage and due to the difference between discharge critical pressure and atmospheric pressure.
- In case of lower break area (for instance 10% break, Fig. 5) the periodic character of the gas dynamic force is more pronounced due to the greater number of pressure wave reflections at the hole of the break and at the pipe 2 junction with the main transportation pipeline. In case of a large break (100% break, Figure 7), the gas dynamic force has an impulse character.
- The gas dynamic force intensity increases with the increase of the break area; in case of 100% break the total force in the pipe 1, where the break and gas leakage take place, has a value of approximately 150 kN immediately after the break occurrence.
- After the attenuation of the pressure wave propagations, only the gas dynamic force in the pipe 1 acts due to the gas leakage; in this later period the reactive force has quasi steady-state character.

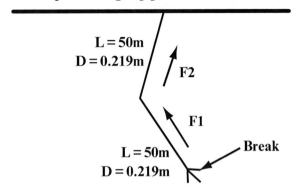

Figure 4. Scheme of the gas pipeline with the position of the break.

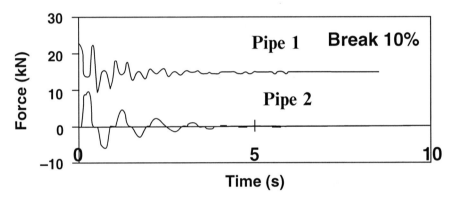

Figure 5. Gas dynamic force initiated by the 10% break.

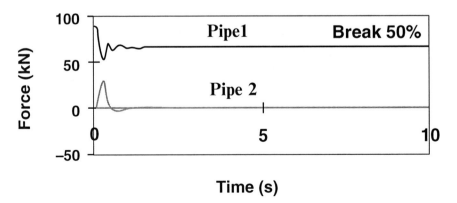

Figure 6. Gas dynamic force initiated by the 50% break.

The magnitudes of the predicted transient gas dynamic forces show that they exert the maximum dynamic loads on the pipeline structure. Hence, it is very important to predict these forces in order to design a safe gas pipeline structure and supports. An inadequate pipeline support could results in pipe whipping and further damage of the nearby objects.

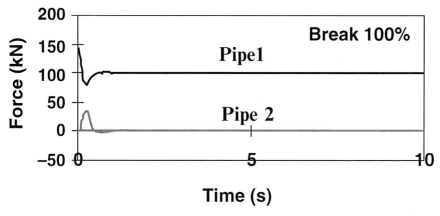

Figure 7. Gas dynamic force initiated by the 100% break.

6. Conclusion

The probabilistic failure frequency assessment of the gas pipeline ruptures show that these kinds of accidents can not be excluded during the exploitation period of both high pressure and large diameter transportation gas pipelines and low pressure distribution gas pipelines. The gas pipeline failure frequency rate of the order of 10^{-4}/year km of pipeline length is much higher than acceptable failure frequency of 10^{-6} for the accidents in energy plants. The consequences of the natural gas pipeline rupture accidents are strongly determined by the transient gas dynamics caused by the gas rupture leakage. Therefore, the model of one dimensional transient compressible gas flow and the numerical procedure based on the method of characteristics are presented with the aim of simulation and analyses of natural gas pipeline gas dynamics under pipe rupture accidents. The model is applied to the prediction of transients caused by the pipe break at the long high pressure transportation pipeline and at the lower pressure distribution pipelines. The obtained results show the dynamics of the gas discharge rate at the break, the pressure distribution along the pipeline during the pipe rupture accident and the gas dynamic forces induced by the transient gas flow and gas leakage. The obtained results are a necessary input for the design of the pipeline structure and supports, as well as for the determination of the protective measures in cases of the gas pipeline break, such as a prediction of the safety distance from the gas pipeline to surrounding objects and population, or application of the quick isolation devices in order to prevent gas leakage.

References

1. M. Kleinpeter, *Energy Planning and Policy* (Wiley, New York, 1995), pp. 39–40.
2. OPS, *Natural Gas Incident Yearly Summaries (1986–2004)*, US Department of Transportation, Office of Pipeline Safety, Washington, 2005.
3. E. B. Wylie and V. L. Streeter, *Fluid Transients* (McGraw-Hill, New York, 1978).
4. H. Mahgerefteh, O. O. Adeyemi, and Y. Rykov, Efficient numerical solution for highly transient flows, *Chemical Engineering Science* 61, 5049–5056 (2006).
5. Y. D. Jo and B. J. Ahn, Analysis of hazard areas associated with high-pressure natural-gas pipelines, *Journal of Loss Prevention in the Process Industries* 15, 179–188 (2002).
6. Y. D. Jo and B. J. Ahn, A method of quantitative risk assessment for transmission pipeline carrying natural gas, *Journal of Hazardous Materials* A123, 1–12 (2005).
7. J. M. Keith and D. A. Crowl, Estimating sonic gas flow rates in pipelines, *Journal of Loss Prevention in the Process Industries*, 18, 55–62 (2005).
8. S. D. Morris, Choked gas flow through pipeline restrictions: an explicit formula for the inlet Mach number, *Journal of Hazardous Materials* 50, 71–77 (1996).
9. V. Stevanovic and M. Studovic, Computer Code for the Simulation of Thermo-Hydraulic Transients in Thermal Power Systems, in *Proceedings of the 15th IAHR Symposium, Section on Hydraulic Machinery and Cavitation*, 2, Belgrade, paper L1, 1990.
10. V. Stevanovic, M. Studovic, and A. Bratic, Simulation and analysis of a main steam line transient with isolation valves closure and subsequent pipe break, *International Journal of Numerical Methods for Heat and Fluid Flow* 4(5), 387–398 (1994).
11. A. G. Papadakis, Major hazard pipelines: a comparative study of onshore transmission accidents, *Journal of Loss Prevention in the Process Industries* 12(1), 91–107 (1999).
12. G. B. DeWolf, Process safety management in the pipeline industry: parallels and differences between the pipeline integrity management (IMP) rule of the office of pipeline safety and the PSM/RMP approach for process facilities, *Journal of Hazardous Materials* 104, 169–192 (2003).
13. S. Sklavounos and F. Rigas, Estimation of safety distances in the vicinity of fuel gas pipelines, *Journal of Loss Prevention in the Process Industries* 19, 24–31 (2006).
14. N. Eisenreich, J. Neutz, F. Seiler, D. Hensel, M. Stancl, J. Tesitel, R. Price, S. Rushworth, F. Markert, I. Marcelles, P. Schwengler, Z. Dyduch, and K. Lebecki, Airbag for the closing of pipelines on explosions and leakages, *Journal of Loss Prevention in the Process Industries* 20, 589–598 (2007).
15. R. T. Lahey and F. J. Moody, *The Thermal-Hydraulics of a Boiling Water Nuclear Reactor* (ANS Monograph, 1984), American Nuclear Society, La Grange Park, IL, pp. 377–379.

SAFETY, RELIABILITY AND RISK. ENGINEERING AND ECONOMICAL ASPECTS

L. TÓTH, LENKEY BIRO GY
Bay Zoltán Institute for Logistics and Production Systems, Miskolctapolca, Hungary

Abstract   The risk-based approach (RBA) of the safety is able to create an equilibrium situation between the safety level and the investment value, i.e. RBA is only the tool for discussion between engineers and economists in safety issue. This will be illustrated through risk-based inspection (RBI) and reliability centred maintenance (RCM) in petrochemical industry.

Keywords: Safety, reliability, petrochemistry

1. Introduction

The guarantee of the safety issue of different industrial plants, systems, etc. is the task of the parliaments, which grant this task to different governmental bodies. These organisations formulate their requirements in national laws. On the basis of these documents different technical guides are issued for different industrial areas to make safe of the specific industrial systems. It is obvious that the required level of safety of the given system required own cost, which is the investment into the guarantee of the safety. The main question is always: how much investment is needed for the guarantee a given level of safety? It is generally true, that the answer strongly depends on the structure of the owners? If the plant's, organisation's owners are the governments (states), then the necessary investments in generally are much higher that at that privatised goods. It follows from the fact, those sources of the safety-issue investments are different: public (soft money) and private (hard money).

The basic words of the technical-economic life are the followings: **safety-reliability** and **risk**. It is absolutely true that these words are the driving forces for the activities made in privatized economical life. The

safety itself expresses the level of actual safety of system (structure, equipment, etc.) with a unit of %, i.e. it does not dealing with investment and it's cost items. In the expression of **reliability** are included all the tools are used for estimation of safety, i.e. all the knowledge, instruments, software, cost of the experts, etc. i.e. this item includes all the cost items are invested into the structural integrity assessment of the systems. Against the investment we are able to consider the **risk** level of the operating systems, i.e. the probability of failure of the system (having no any unit) times of the consequences, which can be expressed in cost item, in money.

If we speaking about invested cost items (reliability) and operational risk in the last analysis we are speaking about the amount of money, which is invested and risked. To define some kind of optimum is the basic task which looking for in the owner's group. By this approach the "invest of minimum and the profit of maximum" principle is defined.

Relating to structural integrity assessment of engineering components it need to be considered the

- Damage process takes place in materials during a given operation conditions.
- The existing discontinuities, flaws are in the structures and geometrical imperfections.
- The fields (stress–strain, temperature, magnetic, etc,) are raising in the structures during operation and simulated operation conditions.

This is shown in the Figure 1.

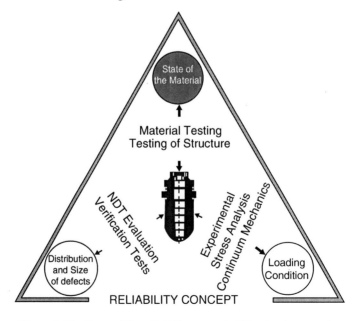

Figure 1. The items of the reliability concept of the structures, systems.

It is always very important question: who is responsible for the reliability, or the safety level of the system: the **economists**, or the **engineers**? During the last periods it can be observed that the role of the economists increased, i.e. in decision of the economical background of the safety, reliability the economists played more and more dominant role. This attitude has been strengthened in the privatisation process especially in the new EU-countries because of the new owners would like to receive the profit of the investments as quickly as possible. In the case of public (governmental) goods this ambition has a real background because at the earlier owners the public money had been invested into the assessment safety items of the systems. The public money was always "soft money", i.e. it's effectivity was always week. This is why the maintenance costs of different public plants were overestimated in general. Reduction of this cost item is the basic ambition of the new, private owners group. The main question is: how can it be done without increasing the level of possible environmental disasters, to expose more and more the human lives, etc. Only one solution can be used to find and define the cost effective way, i.e the risk based approach! This is only the tool for the communication of the economists and engineers as it is reflected on Figure 2.

Figure 2. Balance between the resources necessary for the determination of the level of safety/reliability and the operational risk.

2. The "house of the safety, reliability"

All the plants have different unites, equipment, elements, parts, as it is illustrated in Figure 3. The main question is: how can be built up the "house of the safety, reliability assessment" of the whole system?

If we would like to do it, it has to be considered some actions, the earlier operating experiences and it needs to be done decisions at different levels. These are summarized in Figure 4.

The first and deterministic step in decisions is the selection of the strategy, because it depends on the actual technical level, the goal of the supervision. That is why the strategies are depending on time, as it is illustrated in Figure 5.

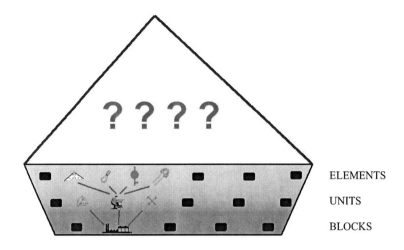

Figure 3. The basic structure of the engineering plants.

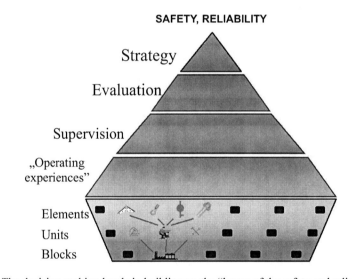

Figure 4. The decision-making levels in building up the "house of the safety and reliability".

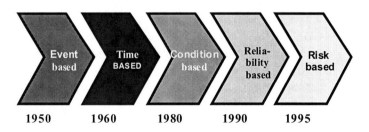

Figure 5. Development of the maintenance strategies.

It can exactly be seen in the Figure 5, that up to date strategy is the risk informed in harmony of the "cost effective" ones as it has been illustrated in Figure 1., i.e. according to the interface between the "engineering" and "economists" point of views.

Selecting the strategy, the second step is to analyse the system according to the functions and conditions of the consisting elements. The main task of the elements is on the one hand **to be safe**, and the other hand to **carry out its function**. The level of safety can be controlled by selected testing methods, control during periodical inspection considering the damage processes takes place in different parts of the selected equipment. In this case the following question needs to be answered:

- What kind of damage process can be realised in the supervised equipment?
- In which part of the equipment takes it place?
- What kind of testing procedure able to detect it?
- What kind of qualification is required from the specialists?
- How often needs to be controlled? etc.

These questions are summarized n Figure 6.

Figure 6. The basic questions in design of the periodical supervisions.

Selecting the testing procedures, performing the control some defects, imperfections could be detected. Their effects on reliability of the equipment have to be evaluated. For this either specialists or expert systems can be used, as the Figure 7 illustrates it.

Figure 7. The tools for evaluation of the detected flaws, imperfections.

In order to have a real picture about operating experiences of the system, structure it is obvious that the earlier damages, failures, unexpected situations have to be considered. These are available either in paper- or electronic forms, as it illustrates in Figure 8.

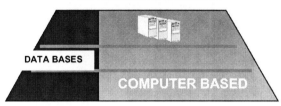

Figure 8. The earlier operating experiences, damages, failures of the systems, equipment.

Selecting the strategy, performing the most suitable testing procedures, detecting imperfections, evaluating them, analysing the importance of the selected equipment, estimating its probability of failure the equipment can be put into the risk-matrix. It is illustrated in Figure 9. Having its position it can be decided that the risk is acceptable or not. If yes, that it is not necessary to do anything, but if the answer is "yes", i.e. the risk value has to be decreased, than further investment is needed.

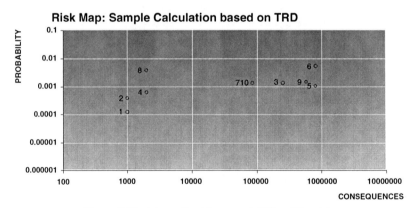

Figure 9. Decisions about the acceptability of the risk.

3. The implementation of the risk based inspection and maintenance at the Hungarian Oil and Gas Company

The *risk based inspection and maintenance* methodology is capable to answer the challenges described above. The principles of risk based inspection and maintenance has been known and used for several years, but its implementation is going on recently even in the Western-European countries. Its practical application started in the USA. The first standard was published in 2000 by API (American Petroleum Industry) for petrochemical industry.[1] In Europe there is a recent European project (RIMAP) which aims to develop a unified methodology of risk based inspection and maintenance that can be used in different industrial sectors (like power generation, chemical industry, steel production, etc.).

The BAY-LOGI institute is just currently implementing this methodology at MOL Refinery in Hungary as main contractor. Within this project a *complex system* has been developed using appropriate IT means (software, hardware), implementing the RCM (Reliability Centred Maintenance) and the RBI (Risk Based Inspection) methodology. With this system the following objectives can be achieved in short and in long term:

- To reduce the number and period of the non planned shut downs
- To increase the ratio of the planned and unplanned maintenance work
- To extend the intervals between shut downs
- To provide fast and accurate access to system data for development, inspection and maintenance tasks
- To optimise (minimise) the maintenance costs taking into account the expected level of system reliability

3.1. BASIC STRUCTURE OF THE CONDITION MONITORING SYSTEM

In order to obtain the above-mentioned objectives, a COMPLEX SYTEM should be developed. As an example, the structure of the system currently installed at MOL Refinery is shown in Figure 10. However this system could be adopted for other industries as well (e.g. for chemical plants, power plants, etc.) implementing the relevant risk based inspection methodology (e.g. developed in the framework of the RIMAP project).

The base of the whole system is a common database which is uploaded with all the relevant data of the equipment (drawings, inspection history, inspection data, process data, fluid data, etc.) that have to be collected from different sources. The CADMATIC software is for the graphical visualisation (in 2D or 3D) of the system and for storing the geometrical data correlated to the intelligent objects. The integration tool for the RCM and RBI software modules, as well as for the other additional expert system modules is the Expert System Shell. The system is connected to other existing systems like SAP through specific interfaces.

The whole software system is web based, thus can be used through the Internet. The whole system – including the database – has been designed in a way that could be easily adopted (modified or extended) to the need of another customer, or the specific need of the industrial sector.

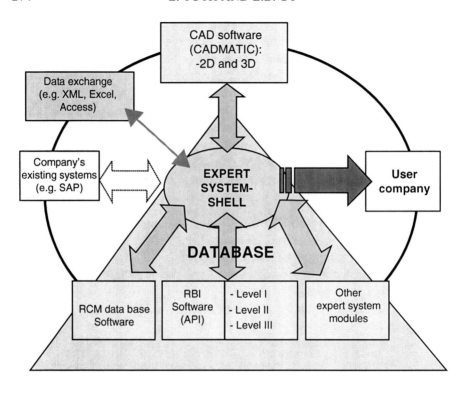

Figure 10. The structure of the complex condition monitoring system (installed at MOL Refinery).

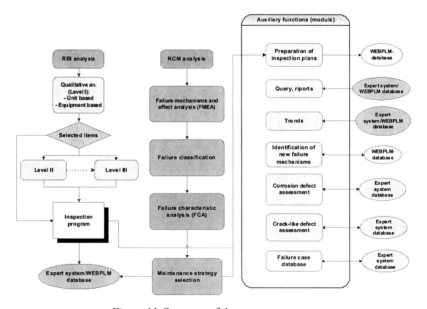

Figure 11. Structure of the expert system.

THE BASIC FUNCTIONS OF THE EXPERT SYSTEM

The system combines two methodologies:

- Inspection planning on the basis of RBI (Risk Based Inspection) strategy. The RBI methodology is based on API (American Petroleum Institute) 581 standard.
- Maintenance program planning based on RCM (Reliability Centred Maintenance) strategy. More detailed see in the references.[2-13]

Figure 11 shows how the two methodologies are linked in the system, and what additional software modules support the work of the experts.

3.2. BASIC PRINCIPLES OF RISK BASED INSPECTION

The objective of RBI is to improve the effectiveness of the inspection and maintenance processes on the basis of the analysis of all the data available about the equipment of a unit. The application of the RBI methodology provides a good basis for implementing a rational and cost-effective decision making mechanism that assure a required level of safety at the same time, since:

- It is possible to identify the most and least risky systems or system elements.
- It makes possible to develop a strategy for decreasing the risk.
- It can be determined what, where, when and how to inspect.
- The basic requirements for the inspection methods could be defined.

The basic principle of the RBI methodology is that it takes into account the probability (PoF) and the consequence of failure (CoF) of each piece of equipment, and the risk is defined as the product of these two (risk = PoFx-CoF). The CoF includes the consequence of a possible failure on health, environment, safety and production, and could be expressed in money as well. Thus using the RBI methodology an optimal inspections and maintenance strategy can be developed which besides decreasing the risk below an acceptable level, also optimises the costs of the inspection and maintenance tasks.

For the analysis the operational conditions of the equipment and the characteristics of the possible failure mechanisms should be considered. This analysis has industrial specific features, i.e. could be different for petrochemical, chemical, pharmaceutical and power industry. One of the most well established methodology is described in the API581 (Risk-Based Inspection. Base Resource Document) for the petrochemical industry. The analysis can be performed at different levels, considering different details and amount of data.

The risk analysis in the system is done in two steps:
- A qualitative screening phase: including a large amount of equipment with the aim of identifying the most critical ones. This qualitative screening is based on a RBI level 1.
- A quantitative phase limited to previously selected items.

Qualitative RBI procedures have three functions:

Screening the units within the site to select the level of analysis needed and to ascertain the benefit of further analyses (quantitative RBI or some other techniques),

Rating the degree of risk within the units and assigning them to a position within a risk matrix,

Identifying areas of potential concern at the plant, which may merit enhanced inspection programs.

Quantitative RBI is equipment-level risk assessment approaches that permit to calculate the risk associated with each piece of operating equipment in a process unit. This method integrates the inspection process in the probability of failure definition through the notion of PoD (probability of detection). Thus, the likelihood of failure with the number and effectiveness of the performed inspection.

The quantitative RBI is based on a series of calculations to assess the likelihood and consequence of failure of the pressure boundary of each piece of analyzed equipment. The product of the likelihood and consequence numbers provides a measure of the risk associated with the corresponding equipment. Based on calculated risk, the prioritized equipment list can then be used to focus the inspection program.

In order to implement these general principles, the five following questions have to be addressed:

- What type of defect to look for?
- Where to look for this defect?
- Which is the best technique?
- When is the best moment to inspect?
- What type of defect to look for?

To answer these questions, the detailed analysis of general equipment data, previous inspections' data, material and process data is needed, as it is shown in Figure 12.

The result of the risk-based analysis is the risk category of the equipment or component, which can be presented in a risk matrix as it is shown in Figure 13. Then the optimisation of the inspection strategy is always an iterative process, which changes the adopted inspection interval and technique until all the criteria are met.

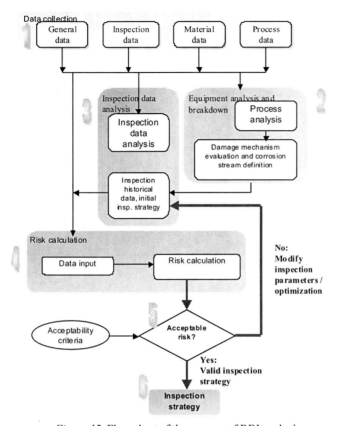

Figure 12. Flow chart of the process of RBI analysis.

On the basis of the result of a risk based analysis one can make more well-founded decision about the following measures: the number, methods and extent of the necessary inspection tasks, about the interim inspections and the related costs, and as a consequence about the modification of the risk category of a given equipment in the future. So this kind of approach has its potential to develop cost-effective inspection and maintenance strategies.

The risk based methodology gives also the possibility to compare the risk level of the different units of a plant, as it is shown in Figure 14. As can be seen, the risk level in Unit 1 is much lower in general, therefore more resources and efforts should be concentrated for Unit 2, and also the length of the inspection period of Unit 1 should be reviewed and reconsidered.

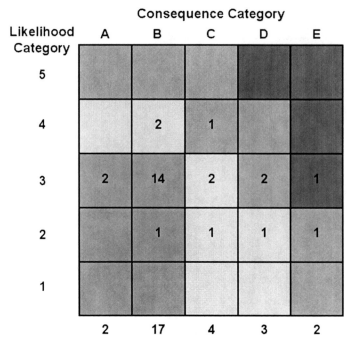

Figure 13. Presentation of the risk of the equipment in a unit in a risk matrix (from left to right – different coloured areas: low-medium-high risk level – and the numbers in each box mean the number of equipment.

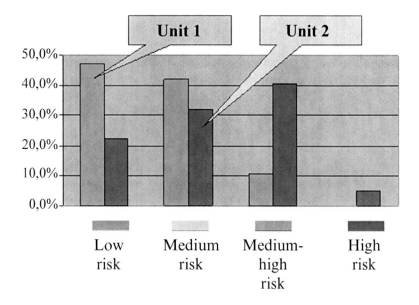

Figure 14. Comparison of the risk levels of the equipment of two units.

4. Conclusions

On the basis of this paper the following conclusions can be drawn:

1. The "basic words" of the technical–economical life is "safety, reliability, risk". From these the *safety* reflects a present state of a plant; the *reliability* includes all the means that are available at the present technical level; the *risk* can be expressed in term of "money", since it is the product of the probability of failure and its consequence.
2. An optimal, "cost effective" strategy for planning the inspection and maintenance tasks is the ***RBI*** – Risk Based Inspection and Maintenance methodology which is only the tool for a common interface between the economists and engineers.
3. A complex system has been developed for MOL Refinery in Hungary, which implements the RBI approach (first time in Hungary). The pilot phase of the project including two units just has been finished, but the first results concerning cost effectiveness can already be foreseen. An additional outcome of the project is the collection and systematization of a large amount of data and documents.

References

1. Risk Based Inspection, Base Resource Document. API Publication 581, first edition May 2000, American Petroleum Institute.
2. D J Smith: Reliability Maintainability and Risk. Practical Methods for Engineers including Reliability Centred Maintenance Safety-related Systems. Butterworth-Heinemann. Oxford. 2000.
3. T Wireman: World Class Maintenance Management. Industrial Press. New York. 1990.
4. T Wireman: Maintenance Management and Regulatory Compliance Strategies. Industrial Press. New York. 2003.
5. J D Andrews, T R Moss: Reliability and Risk Assessment. Professional Engineering Publishing. London. 2002.
6. A M Smith, G R Hinchliffe: RCM – Gateway to World Class Maintenance. Butterworth-Heinemann. Oxford (Copyright: Elsevier), 2004.
7. J Levitt: Managing Maintenance Shutdowns and Outages. Industrial Press. New York. 2004.
8. R B Jones: Risk-Based Management: A Reliability-Centred Approach. Practical, Cost-Effective Methods for Managing and Reducing Risk. Gulf Professional Publishing. Houston. TX. 1995.
9. R E Megill: An Introduction to Risk Analysis. PenWell Company. Tulsa, Oklahoma. 1992.
10. Y Y Haimes: Risk Modelling, Assessment, and Management. Wiley. New York. 1998.
11. GY Vajda: Risk and Safety (in Hungarian) Akadémia Kiadó. Hungary. 1998.
12. R C Hansen: Overall Equipment Effectiveness. Industrial Press. New York. 2001.
13. V Narayan: Effective Maintenance Management. Risk and Reliability Strategies for Optimizing Performance. Industrial Press. New York. 2004.

ASSESSING THE DEVELOPMENT OF FATIGUE CRACKS: FROM GRIFFITH FUNDAMENTALS TO THE LATEST APPLICATIONS IN FRACTURE MECHANICS

DONKA ANGELOVA
University of Chemical Technology and Metallurgy,
Kl. Ohridsky 8, 1756 Sofia, Bulgaria

Abstract Engineering analysis of fatigue crack growth depend on different loading conditions, wherever those conditions have been studied, in laboratory or real-world practice. Such analyses can be done on basis of different parameters: the stress intensity factor range, ΔK, introduced in linear elastic fracture mechanics; the J-integral range, ΔJ, employed in elastic-plastic material characterization; the square-root area parameter of Murakami, effective in the presence of small defects and non-metallic inclusions. An alternative presentation of fatigue data has been proposed that uses the crack growth rate against a newly introduced parameter, namely an energy fatigue-function ΔW based at different conditions on different parameters, ΔK or ΔJ or the square-root area parameter of Murakami. This alternative presentation shows fatigue data as forming an almost straight line, which may be termed the "natural fatigue tendency" of a material, and specified more precisely at a given stress range. Also the present study introduces a physical interpretation of the line presentation of fatigue data and some illustrations of the "natural fatigue tendency" for different materials under different conditions.

Keywords: Fatigue modeling, fatigue-crack growth rate, stress intensity factor range, J-integral range, square-root area parameter, energy fatigue-function ΔW, natural fatigue-tendency of materials

1. Introduction

Although major advances have been made in fatigue modelling, the application of fatigue concepts to different practical situations is highly individual and often involves empirical and semi-empirical approaches including a large number of specifying constants. In linear elastic fracture mechanics terms (LEFM) it is well-known that one of the most used presentations "Long fatigue crack growth rate, da_l/dN against Stress-intensity factor range ΔK_l" on log-log scales or $\log da_l/dN - \log \Delta K_l$ includes three regimes of crack growth, Suresh[1]: **I**, of threshold behaviour; **II**, of Paris linear presentation, $\log(da_l/dN) = C(\log \Delta K)^m$, where C and m are scaling constants; and **III**, of rapid increasing of da_l/dN leading to final failure. On another hand, the approach of short fatigue crack propagation represents plots "Fatigue crack growth rate da_{sh}/dN against Crack length a_{sh}" on log-log scales or $\log da_{sh}/dN - \log a_{sh}$ for two regimes corresponding to so called small and long crack stages revealed by Brown and Hobson,[2] and in more precise terms – for three regimes introduced by Angelova and Akid[3]: **I**, of short crack growth, mode II; **II**, of physically small crack growth, mode I; and **III**, of long crack growth. The first regime of Brown–Hobson's model and the first two regimes of Angelova–Akid's model are described by parabolas, and the final regime in both models by lines. Sometimes short fatigue crack data can be represented in the same way as is used for long fatigue crack data, namely: "Short fatigue crack growth rate, da_{sh}/dN against Short fatigue crack equivalent to ΔK_l, $\Delta K_{sh} = k\sigma\sqrt{a_{sh}}$" or $\log da_{sh}/dN - \log \Delta K_{sh}$, Gangloff.[4] At elastic-plastic fracture mechanics conditions (EPFM), a presentation "Short fatigue crack growth rate da_{sh}/dN against J-integral range, ΔJ" or $\log da_{sh}/dN - \log \Delta J$ takes place for some steels and non-ferrous alloys as it is shown for example in Hoshide,[5,6] Dowling,[7] Suresh.[1] A new mechanism of fatigue failure at inclusions presence in ultralong life regime, when $N > 10^7$ cycles, has been investigated, described and supplied with an efficient model for practical uses by Murakami.[8] This model is based on the parameter of a specific optically dark area (ODA) introduced in 2000 by Murakami,[9,10] having to play a critical role for the ultralong fatigue conditions of materials containing small defects and inclusions.

In the present study, a method of fatigue data presentation is proposed different from those just described, comprising a more precise fatigue testing of engineering materials and a specific presentation of fatigue-crack growth data, and based on the energy fatigue-function ΔW introduced by Angelova[11,12] in its four versions – (a) $\Delta W_K = (da/dN)\Delta K$ (or $kW_K = k(da/dN)\Delta K$); (b) $\Delta W_J = (da/dN)\Delta J$; (c) $\Delta W_a = \Delta\sigma(da/dN)a^{1/2}$; and

(d) $W^*_{ODA} = f(ODA, N)$. A generalization of the fatigue-function ΔW is made employing material Vickers hardness HV and its physical sense revealed.

2. Method of fatigue data analysis

The presentations $\log da_l / dN - \log \Delta K_l$ and $\log da_{sh} / dN - \log \Delta K_{sh}$ include the parameter ΔK with a dimension of $MPa\sqrt{m}$ and the presentation $\log da_{sh} / dN - \log \Delta J$ – the parameter ΔJ with a dimension of $[N/m]$. Now, we will discuss these dimensions in more details.

2.1. GRIFFITH'S FRACTURE CONCEPT

Taking into consideration Griffith's and Inglis' fracture concepts[13] a crack with length $2a$ at the interior of a plate and an edge crack with length a can produce the same effect on the fracture behaviour; such an edge-crack is shown in Figure 1a.

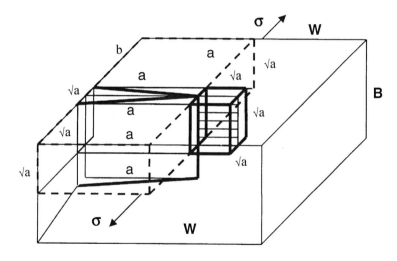

Figure 1a. Plate with notch-crack or notch-edge.

The total change in potential energy ΔU of this crack resulting from the crack creation includes elastic strain U_E and surface U_S energies, Eq. (1) residing in a cylindrical volume V_{cyl} around the crack, which has radius a and height B, Figure 1a:

$$\Delta U(a) = U_E + U_S, V_{cyl} = \pi a^2 B. \tag{1}$$

According to Griffith's criterion, the crack will propagate under a constant applied stress σ if an incremental increase in crack length produces no change in the total energy of the system ($d(\Delta U(a))/da = 0$) for either plane stress or plane strain. Irwin[13] has modified Griffith's criterion by replacing the plastic work for extending the crack wall, which is hard to measure, with the strain-energy release rate. The latter one is related to K^2 in a simple way.

2.2. STRAIN ENERGY IN THE PRISM VOLUME $V_{B,prism}$

The total strain energy in a rectangular-prism volume $V_{B,prism}$ (instead in that of the above mentioned cylinder) which has a square base a^2 and height B and is located around the edge-crack or edge-notch with size a is shown in Eq. (2):

$$U_{B,E} = k\sigma a^2 B, \quad V_{B,prism} = a^2 B, \tag{2}$$

while the total strain energy in a unity-prism obtained from the same rectangular-prism by its dividing into B prisms, each with unity height is given in Eq. (3); k is a constant depending on the nature of the metal and loading conditions ($k = f(\sigma/E)$, E is the Young modulus, and at the same time a characteristic concern with the theoretical strength of a given material):

$$U_{1,E} = k\sigma a^2, \quad V_{1,prism} = a^2. \tag{3}$$

Now the total strain energy in an elementary rectangular-prism volume $V_{e,prism}$ with sizes a and $\sqrt{a}$, $\sqrt{a}$ (shown in Figure 1a) around the edge-crack or edge-notch with size a is represented in Eq. (4):

$$U_{e,E} = k\sigma a^2, \quad V_{e,prism} = (\sqrt{a})^2 a = a^2. \tag{4}$$

But at the same time, the energy from Eq. (4) is equal to the energy in the unity-prism volume $V_{1,prism}$ from Eq. (3), so we may accept that $V_{1,prism}$ is fractured when the elementary volume $V_{e,prism}$ is fractured. Then we can accept that the fracture of $V_{e,prism}$ leads as well to

1. The fracture of a rectangular-prism volume $V_{\sqrt{a},prism}$ with sizes a, a and height $\sqrt{a}$ (and consequently to that of the dash-line prism volume with sizes a, b and $\sqrt{a}$, shown in Figure 1a)
2. The reach of the total energy in the volume $V_{\sqrt{a},prism}$ (sized a, a, $\sqrt{a}$) as given in Eq. (5):

$$U_{\sqrt{a},E} = k\sigma a^2 \sqrt{a}, \quad V_{\sqrt{a},prism} = a^2 \sqrt{a} \qquad (5)$$

The energy $U_{\sqrt{a},E}$ can be treated as an intensified energy (by a coefficient $\sqrt{a}$) if we compare it to the energy in the volume of the unity prism $V_{1,prism}$, or as an energy corresponding to an intensified stress (by a coefficient $\sqrt{a}$) - $\sqrt{a}\sigma$ – if we compare it to the applied stress σ.

2.3. STRAIN ENERGY IN THE ELEMENTARY VOLUME $V_{e,cube}$

The total energy in the volume of an elementary *square-root sided* cube $V_{e,cube}$ with size $\sqrt{a}$, located at the end-point of the edge-notch a and shown in thick line in Figure 1a is given in Eq. (6):

$$U^E_{e,\sqrt{a}^3} = k\sigma a\sqrt{a}, \quad V_{e,cube} = a\sqrt{a}. \qquad (6)$$

The total energy $U^E_{e,\sqrt{a}^3}$ in $V_{e,cube}$ per unity of the new *square-root surface* $\sqrt{a}\sqrt{a}$ (that can develop in this volume and correspond to an edge-notch growth by $\sqrt{a}$) is represented in the expression (7-i) of Eq. 7; the new *square-root surface* $\sqrt{a}\sqrt{a}$ is shown as a stripey area in the thick-line cube in Figure 1a:

$$(7\text{-i}) \Rightarrow \frac{U^E_{e,\sqrt{a}^3}}{\sqrt{a}\sqrt{a}} = \frac{k\sigma a\sqrt{a}}{a} = k\sigma\sqrt{a}$$

$$(7\text{-ii}) \Rightarrow K = \frac{Y}{k}\frac{U^E_{e,a^{3/2}}}{\sqrt{a}\sqrt{a}} = \frac{Y}{k}\frac{k\sigma a\sqrt{a}}{a} = \frac{Y}{k}(k\sigma\sqrt{a}) \qquad (7)$$

The expression (7-i) differs from the stress intensity factor K in (7-ii), both from Eq. 7, only by the constant Y/k as it is shown in Eq. (7) (where

Y is the finite size correction factor for the plate under consideration). So we may accept K as the total volume energy in $V_{e,cube}$ per unity new *square-root surface* $\sqrt{a}\sqrt{a}$, that can develop in $V_{e,cube}$. Under cycling loading it is well known[1] that K and σ should be replaced by ΔK and $\Delta \sigma$, Eq. (8):

$$\Delta K = (Y/k)k\Delta\sigma\sqrt{a}. \qquad (8)$$

2.4. GRIFFITH'S CRITICAL CRACK

According to Griffith when a from Eq. 1 fulfils the condition $a = 2E\gamma_0/\pi\sigma^2 = c_k$ it becomes a critical crack, c_k. At the same time the analysis of the function $U(a)$

$$U(a) = U_E + U_S,$$

$$U(a) \neq const,$$

when a grows at $\sigma = const$, shows $U(a)$ as a parabola with extremum (maximum) at

$$a = c_k = 2E\gamma_0/\pi\sigma^2,$$

Figure 1b. From the point of c_k onwards the decrease in $U(a)$ leads to decrease of K.

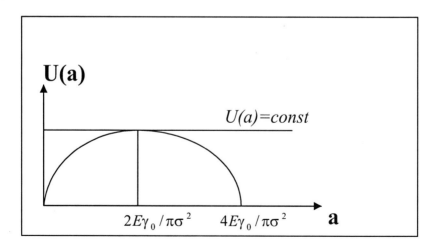

Figure 1b. Graphical presentation of energy $U(a) \neq const$ associated with the growing crack a.

2.5. DIFFERENT SURFACE AND VOLUME ENERGIES

The expression (9) is obtained as a multiplication between ΔK from Eq. (8) and crack growth rate $\dfrac{da}{dN} \underset{0 \leftarrow \Delta}{\leftarrow} \dfrac{\Delta a}{\Delta N}$

$$\Delta K \frac{\Delta a}{\Delta N} = \frac{Y}{k} k \Delta \sigma \sqrt{a} \frac{\Delta a}{\Delta N} = Y \Delta \sigma \sqrt{a \Delta a_N} = Y \Delta \sigma \left(\sqrt{\Delta a_N}\right) \sqrt{a}$$

$$\text{at dimension} \left[\frac{J}{m^3} \frac{m}{cycle} \sqrt{m} \right], \qquad (9)$$

and that dimension can be transformed into one shown in Eq. (10)

$$\left[\frac{J}{m^3} \frac{m}{cycle} \sqrt{m} \right] = \left[\frac{J}{m^2} \frac{1}{cycle} \frac{\sqrt{m}\sqrt{m}}{\sqrt{m}} \right] = \left[\frac{J}{m^2} \frac{1}{cycle} \frac{\left(\sqrt{m}\right)^2}{\sqrt{m}} \right]. \qquad (10)$$

From Eqs. (9) and (10) we can obtain Eq. (11):

$$\Delta K \frac{\Delta a}{\Delta N} = Y \Delta \sigma \left(\sqrt{\Delta a_N}\right) \sqrt{a} \left[\frac{J}{m^2} \frac{1}{cycle} \frac{\left(\sqrt{m}\right)^2}{\sqrt{m}} \right] \qquad (11)$$

where $\Delta a_N = \Delta a / \Delta N$ is the increment of the edge-notch per cycle. If we have to be more precise and assume that the edge-notch begins its growth (or its length has been last recorded) at size a, the next recording will for example show an increment Δa for ΔN cycles, meaning that the absolute size of the edge-notch is grown from a to $a + \Delta a$. So at the end-point $a + \Delta a$ of the edge-notch we can imagine a cube with size $\sqrt{a + \Delta a}$ similar to the thick-line cube characterized by size $\sqrt{a}$, and between them there can be located N-1 cubes with sizes $\sqrt{a + i \Delta a_N}$, $i = 1, 2, \ldots$ (N − 1). Comparing the sizes of the thick-line cube ($\sqrt{a}$) and the next one ($\sqrt{a + \Delta a_N}$) corresponding to the edge-notch growth by Δa_N for the first cycle, at a distance between the two cubes Δa_N and in terms of Eq. 11, we can note the following:

1. The difference between sizes $\sqrt{a}$ and $\sqrt{a + \Delta a_N}$ is small and as each successive pair of cubes is formed this difference is slowly decreased, while all the cubes are located at equal distances Δa_N.

2. In the thick-line cube with size $\sqrt{a}$ the energy is that of ΔK (*total volume energy* per unity *square-root surface* $\sqrt{a}\sqrt{a}$ or per unity edge-notch size $a-$ see Eq. (7)). Then the edge-notch growth rate $\Delta a/\Delta N$ transforms ΔK (accordingly to Eq. (11)) so in the next cube with size $\sqrt{a+\Delta a_N}$, the dimension of $\Delta K \dfrac{\Delta a}{\Delta N}$ is already that of the *total surface energy* $\left[\dfrac{J}{m^2}\dfrac{\sqrt{m}\sqrt{m}}{1}\right]$ of the *specific square-root cube surface* $\sqrt{a+\Delta a_N}\sqrt{a+\Delta a_N}$ which is of the same kind of the stripy *square-root surface* $\sqrt{a}\sqrt{a}$ in the thick-line cube, and which is necessary for eventual development of a unity of the *specific square-root edge-notch size* $\sqrt{a+\Delta a_N}$ per cycle. That kind of surface energy obtained after the transformation of ΔK into $\Delta K \dfrac{\Delta a}{\Delta N}$ (which is a transformation of *volume energy* of $\Delta\sigma \to [J/m^3]$ into *specific surface energy* of $\Delta\sigma(\Delta a_N) \to [(J/m^3)(m)]$) is *effective surface energy* as well considering the parameter Δa_N obtained at each calculated edge-notch growth rate, which differs from one rate to another.

2.6. THE J-INTEGRAL RANGE

The *J*-integral range, ΔJ has a dimension of *surface energy*, $[N/m] = [J/m^2]$ and the crack (or CTOD $\Rightarrow \delta$) growth rate transforms ΔJ, similarly to ΔK from Eq. (11), into Eq. (12) which means a dimensional transformation into *effective specific surface energy*

$$\Delta J \frac{\Delta a}{\Delta N} = \Delta J(\Delta a_N) \left[\frac{J}{m^2}\frac{m}{cycle} = \frac{J}{m^2}\frac{\sqrt{m}\sqrt{m}}{cycle}\right] \qquad (12)$$

Here that surface energy, on the one hand is *effective surface energy* depending on the edge-notch growth rate as explained in Section 2.5, and on the other hand is *specific surface energy* per unity surface from the kind of the *square-root surface* $(\sqrt{a+\Delta a_N})^2 [\sqrt{m}\sqrt{m}]$ that eventually can be created for 1 cycle.

2.7. THE SQUARE OF EFFECTIVE SURFACE ENERGY

The expression in Eq. (13) represents the square of *effective surface energy* necessary for an increment Δa_N of the edge-notch per cycle; this effective surface energy depends on the concrete edge-notch growth rate in a given cycle range:

$$E\Delta J \frac{\Delta a}{\Delta N} = E\Delta J \Delta a_N, \left[\frac{J}{m^3} \frac{J}{m^2} \frac{m}{cycle} \right] \text{ or}$$

$$E\Delta J \frac{\Delta a}{\Delta N} = (\Delta K)^2 \Delta a_N \left[\left(\frac{J}{m^2}\right)^2 \frac{1}{cycle} \right], \left(E\Delta J = (\Delta K)^2 \right) \quad (13)$$

where E is Young's modulus.

3. Materials, specimens and experimental procedures

The fatigue data used in this study come from papers and books published by myself and others:

- (i) *Low-carbon roller-quenched tempered steel, RQT501, own results* The steel, suitable for offshore applications, was subjected to tension-tension loading at a stress ratio $R = 0.1$ using a servo-electric fatigue rig with a load capacity of 100 kN. Tests were performed[3] in load control at stress levels of 396, 470 and 516 MPa in environment of 0.6 M *NaCl*. A sinusoidal waveform was used at frequencies 0.2, 0.5 and 1 Hz. The chemical composition and microstructural and mechanical properties of the *RQT501* steel[3] are: in wt % C 0.12 Si 0.30, Mn 1.45, Cr 0.02, Ni 0.02, P 0.011, Mo 0.01, S 0.003, Cu 0.02, Al 0.045, V 0.01,Ti 0.004, Nb 0.003; average grain size 8.6 μm; 0.2% proof stress (523 MPa); tensile strength (608 MPa); hardness (737 HV).
- (ii) *High-strength spring steel, existing results* The experimental work of Akid and Murtaza employed a high-strength spring steel under fully reversed torsion fatigue conditions of different $\Delta \tau$ in air and 0.6 M *NaCl*. For this steel, pit and short-crack initiation and growth have been monitored by the surface replication of the standard hour-glass specimens followed by direct microscopical observation. The chemical composition and microstructural and mechanical properties of the spring steel[14] are: in wt % C 0.56, Mn 0.81, Si 1.85, Cr 0.21, Ni 0.15, P 0.026, Mo 0.025, S 0.024; average grain size 30 μm; 0.2% proof stress (1,440 MPa); ultimate tensile strength (1,610 MPa); % elongation (9.3); hardness (480 HV).

- (iii) *β-Ti-6Al-4V alloy, existing results* The fully reversed fatigue tests of Hoshide were done under axial and combined axial-torsional loading of solid cylindrical specimens (with a circumferential blunt notch) made of Ti alloy, water-quenched after a heat-treatment of 30 min in β-region; the behaviour of small fatigue cracks was observed by the replication technique.[5] The chemical composition and microstructural and mechanical properties of the alloy are: in wt % Al 6.52, V 4.00, Fe 0.16, O 0.182; average grain size 400 μm; 0.2% proof stress (849 MPa); ultimate tensile strength (1,016 MPa); % elongation (6.8); Young's modulus (127 GPa).
- (iv) *Al–Mg Alloy and S35C Steel, existing results* All fatigue tests of Hoshide were carried out under fully reversed push-pull loading of solid cylindrical specimens with a thinner flat cylindrical part, annealed at different conditions for the different materials; the behavior of small fatigue cracks were observed by the replication technique.[6] The chemical composition and microstructural and mechanical properties of the alloys are: (a) for Al–Mg alloy in wt % Si 0.12, Fe 0.28, Cu 0.04, Mn 0.06, Mg 2.6, Cr 0.25, Zn 0.02, Ti 0.01; average grain size 47 μm; yield strength (134 MPa); tensile strength (239 MPa); % elongation (30.6); Young's modulus (76 GPa); (b) for S35C steel in wt % *C 0.37, Mn 0.77, Si 0.24, Cr 0.04, Ni 0.02, P 0.019, S 0.023, Cu 0.01*; average grain size 9.7 μm; yield strength (382 MPa); tensile strength (668 MPa); % elongation (37); Young's modulus (206 GPa).
- (v) *SCM435 Steel, existing results.* The fatigue experiments of Murakami employed Cr-Mo steel under tension-compression symmetric loading of solid cylindrical specimens quenched and tempered at different conditions; all fracture origins are at the internal inclusions and parameters of those inclusions and of the surrounding ODA areas are measured under an optical microscope.[8-10] The chemical composition and microstructural and mechanical properties of *SCM435* steel are: in wt % *C 0.36, Mn 0.77, Si 0.19, P 0.014, S 0.006, Cr 1.0, Mo 0.15, Cu 0.13 and 8 ppm O_2*; hardness of the specimens treated at different conditions (500–560 *HV*).

4. Different data presentations

The well known data presentations

1. "Long fatigue-crack growth rate, da_l/dN against Stress-intensity factor range ΔK_I" (or on log-log scales $\log da_l/dN - \log \Delta K_I$),

2. "Short fatigue-crack growth rate, da_{sh}/dN against Short fatigue-crack equivalent to ΔK_I, $\Delta K_{sh} = k\sigma\sqrt{a_{sh}}$" (or in log-log $\log da_{sh}/dN - \log \Delta K_{sh}$), for most engineering alloys exhibit sigmoidal curves with three distinct regimes of crack growth: (a) threshold behaviour, (b) the dominant Paris line and (c) catastrophic failure. In many cases, the dominant Paris linear regime – $\log(da_I/dN) = C(\log \Delta K)^m$ – is applied to all data, as it can be found in Murtaza[14] and shown as **M**(K) in Figure 2a–c. At the same time the presentation "Short fatigue-crack growth rate, da_{sh}/dN against Short fatigue crack length, a_{sh}" (or $\log da_{sh}/dN - \log a_{sh}$) can employ a mathematical parabolic-linear description based on the Brown–Hobson model[2] or statistical approach.[15] In elastic-plastic conditions modelling is based on $\log da/dN = C_J(\log \Delta J)^n$ (where C_J, n are scaling constants) and is presented by a straight line as in Hoshide.[5,6] In these publications there are other presentations of the kind "Long or short fatigue crack growth rate, da_{sh}/dN or da_I/dN against the J-integral range, ΔJ" (or on log-log scales $\log da_{sh}/dN - \log \Delta J$ and $\log da_I/dN - \log \Delta J$) which result not in a straight line but in curves.

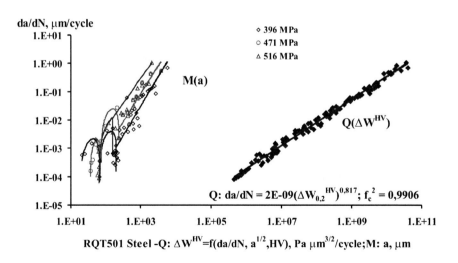

RQT501 Steel -Q: $\Delta W^{HV} = f(da/dN, a^{1/2}, HV)$, Pa $\mu m^{3/2}$/cycle; M: a, μm

Figure 2a. Proposed alternative (**Q**) and conventional (**M**) short-crack data presentations of fatigue at frequency 0.2 Hz.

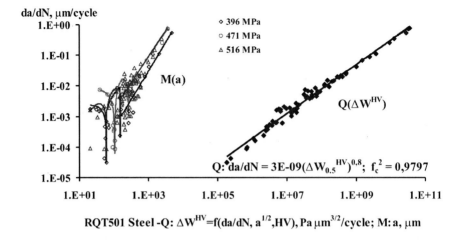

Figure 2b. Proposed alternative (**Q**) and conventional (**M**) short-crack data presentations of fatigue at frequency 0.5 Hz.

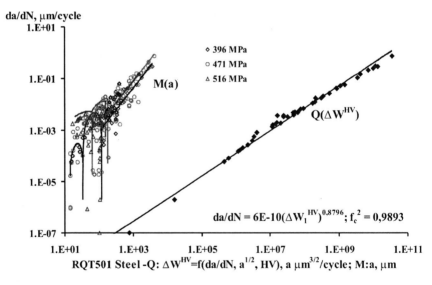

Figure 2c. Proposed alternative (**Q**) and conventional (**M**) short-crack data presentations of fatigue at frequency 1 Hz.

An alternative method is proposed comprising: (a) fatigue testing of engineering materials; (b) measuring long-crack or main short-crack lengths a and the corresponding number of cycles N; (c) calculating the crack rates da/dN and the ranges ΔK or ΔJ of the stress-intensity factor K or the J-integral, and (d) calculating a newly-introduced energy fatigue-function in its four versions:

$$W = (da/dN)\Delta K \text{ (or } kW = k(da/dN)\Delta K),$$
$$W = (da/dN)\Delta J,$$
$$W = (da/dN)a^{1/2}HV \text{ (or } W = (da/dN)a^{1/2}),$$
$$W^* = f(ODA, N),$$

where k is a normalizing constant of the type $k = N_f/(a_f(K_{max}(a_f)))$, a_f and N_f are respectively the final length of fatigue crack and the number of cycles at failure, $K_{max}(a_f)$ is the stress intensity factor at σ_{max} from the applied stress range $\Delta\sigma = \sigma_{max} - \sigma_{min}$, HV is the Vickers hardness and kW is a non-dimensional expression.

The presentation $\log da/dN - \log W$ is a straight line $Q(K)$, $Q(HV)$, $Q(J)$ or $Q^*(\sqrt{A'})$, shown as the thickest line in *Figures 2–7*, which may be termed the *"natural fatigue tendency"* of material ($Q(HV)$, $Q^*(\sqrt{A'})$) or of material at a given stress range($Q(K)$, $Q(J)$); this is mentioned for the first time and only for ΔK in Angelova.[11,12,16,17]

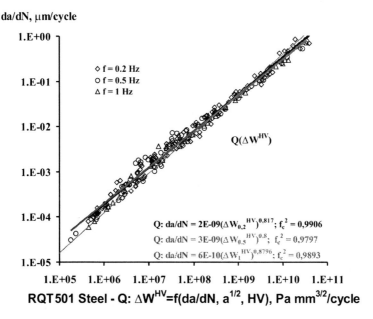

Figure 2d. Proposed alternative short-crack data presentations (**Q**) of fatigue at frequencies 0.2, 0.5 and 1 Hz.

The presentation $\log da_{sh}/dN - \log W$ which can be rewritten as $\log da_{sh}/dN - \log[(da/dN)\Delta K_{sh}]$ is applied to the fatigue data obtained by

Murtaza (accordingly, Section 2.5) and expresses a straight line, $Q(K)$ in Figure 3a, while the presentations $\log da_{sh}/dN - \log[(da_{sh}/dN)\Delta J]$ and $\log da_l/dN - \log[(da_l/dN)\Delta J]$ using the fatigue data for a β-Ti-6Al-4V alloy, Al–Mg alloy and S35C steel after Hoshide (considering Sections 2.6. and 2.7) show straight lines $Q(J)$ or $Q(JE(da/dN))$ in *Figures 4–6*. If we use a generalization of the fatigue-function ΔW based on employing material Vickers hardness HV and replace applied stress ranges by the corresponding HV for a given material, new $Q(HV)$ straight-line presentations can be seen in *Figures 2d* and *3b* (the original model in Figure 2 is of Yordanova[18]). Such a presentation reveals the *natural fatigue tendency* of a given material and may be used as one of its general fatigue characteristics.

Under conditions of superlong fatigue and in the presence of non-metallic inclusions Murakami expresses the relationship between the sizes of an inclusion area A_0 and surrounding ODA area A_1, which takes part in a ratio $\sqrt{A_0 + A_1}/\sqrt{A_0} = \sqrt{A'}/\sqrt{A_0}$ plotted against the cycles to failure N_f, multitude $M(\sqrt{A'})$ in Figure 7. To apply our alternative approach to Murakami's data, we calculate (i) the average crack growth rate $da/dN = (A'-A_0)^{1/2}/(N_f - N_0)$ for each specimen fractured at a specific inclusion and the ODA around it and, (ii) the energy function $W^* = (da/dN)\Delta\sigma(A'/A_0)^{1/4}$. On log-log scales the presentation $Q^*(\sqrt{A'})$ – $\log da/dN - \log W^*$ – is a straight line, Figure 7.

Note that all straight lines in *Figures 2–6* express **effective surface energy** depending on change of crack growth.

In terms of ΔK, the straight lines in *Figures 2, 3* and *7* express the **total effective surface energy** of a specific *square-root sized surface*, necessary for the creation of a unity *square-root edge-notch size* per cycle. In terms of ΔJ, the straight lines in *Figures 4–6* express the **effective surface energy** necessary for the creation of a unity of *specific square-root sized surface* per cycle.

A comparative analysis between **M** and **Q** presentations shows that at the same number of crack-size measurements, the precision of the proposed method is significantly higher, expressed quantitatively by the corresponding correlation coefficients f_c shown in *Figures 2–7*. Using this approach, therefore, fewer fatigue measurements may be needed than have conventionally been required. Confirmation is given in Angelova (2003),[16,17] where descriptions of the fatigue behaviour of 15 different materials are derived from an economical use of data.

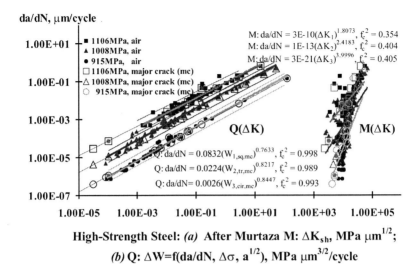

High-Strength Steel: *(a)* After Murtaza M: ΔK_{sh}, MPa $\mu m^{1/2}$;
(b) Q: $\Delta W = f(da/dN, \Delta\sigma, a^{1/2})$, MPa $\mu m^{3/2}$/cycle

Figure 3a. Proposed alternative (**Q**) and conventional (**M**) short-crack data presentations of fatigue for different stress levels in air medium.

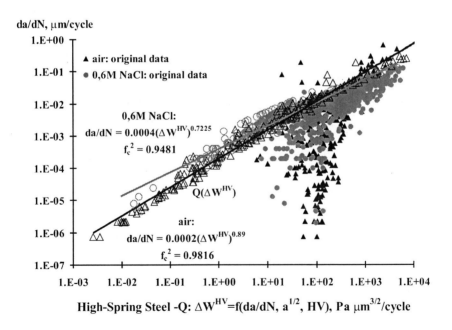

High-Spring Steel -Q: $\Delta W^{HV} = f(da/dN, a^{1/2}, HV)$, Pa $\mu m^{3/2}$/cycle

Figure 3b. Original short-crack data presentation of fatigue and the newly-proposed one (**Q**) based on the Vickers hardness for different media.

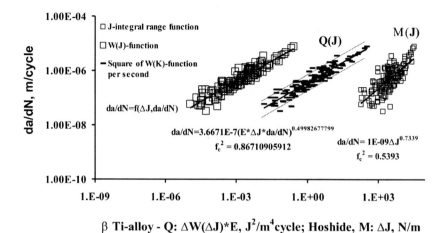

Figure 4. New (**Q**) and original (**M**) short fatigue-crack data presentations.

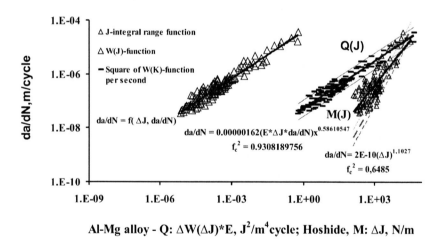

Figure 5. New (**Q**) and original (**M**) short fatigue-crack data presentations; the dash line shows long crack data.

DEVELOPMENT OF FATIGUE CRACKS 297

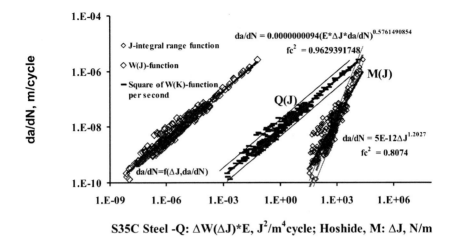

Figure 6. New (**Q**) and original (**M**) short fatigue-crack data presentations; the dash line shows long crack data.

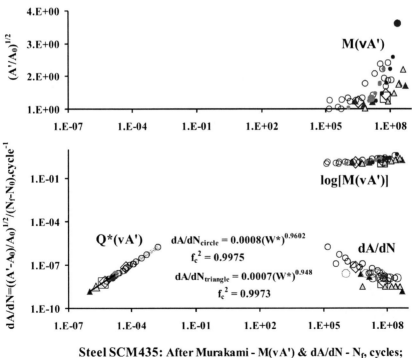

Figure 7. Alternative (**Q**) and original ODA (**M**) presentations of fatigue.

5. Final notes

An alternative approach to conventional fatigue data presentations is developed clarifying the physical sense of the parameters which it offers. This approach transforms the presentations of crack-growth rate against ΔK and ΔJ, and of the ODA parameter of Murakami against cycle to failure, into a linear presentation of crack-growth rate against a *specific effective surface energy* function ΔW. The line obtained may be termed the *natural fatigue tendency* of material under a given stress range, and because of its simplicity may be helpful for fatigue testing and real-world practice in terms of precision and the reduction of the number of fatigue characterizing tests, given by our technical standards. A generalization may be used when Vickers Hardness replaces applied stress ranges; then the *natural fatigue tendency* of a given material would depend only on its basic strength (hardness) characteristics.

ACKNOWLEDGEMENTS The author thanks Professor Yukitaka Murakami, Kyushu University, Japan for very useful discussions about the applicability of the square root parameter method, and of the superlong fatigue at small defects and non-metallic inclusions, described by ODA model. The author thanks Professor Toshihiko Hoshide, Kyoto University, Japan for the fatigue data of ΔJ investigations 5,6 and for useful discussions about the specific experiments and the quantitative evaluation of short fatigue crack growth based on EPFM analysis.

References

1. S. Suresh, *Fatigue of Materials* (Cambridge University Press, Cambridge, UK, 1998).
2. P. D. Hobson, M. Brown and de E.R. los Rios, in: *ECF Publication 1. Mechanical Engineering Publications*, (London, UK, 1986) pp. 441–459.
3. D. Angelova and R. Akid, A note on modelling short fatigue crack behaviour, *Fatigue Fract. Eng Mater. Struct.* **21**, 771–779 (1998).
4. R. P. Gangloff, The criticality of crack size in aqueous corrosion fatigue *Res Mech Lett.* **1**, 299–306 (1981).
5. T. Hoshide, T. Hirota and T. Inoue, Fatigue behaviour in notch component of titanium alloys under combined axial-torsional loading *Materials Sci. Research Int.* **1**(3), 169–174 (1995).
6. T. Hoshide, T. Yamada, S. Fujimura and T. Hayashi, Short crack growth and life prediction in low-cycle fatigue of smooth specimens *Eng. Fract. Mech.* Vol. **21**(1), 85–101(1985).
7. N. Dowling, *Mechanical Behaviour of Materials* (Prentice-Hall, NJ, 1999).

8. Y. Murakami, *Metal Fatigue: Effects of Small Defects and Nonmetallic Inclusions* (Elsevier, Oxford, UK, 2002).
9. Y. Murakami, On the mechanism of fatigue failure in the superlong life regime ($N>10^7$ cycles). Part I: Influence of hydrogen trapped by inclusions, *Fatigue Fract. Eng Mater. Struct.* **23**, 893–902 (2000).
10. Y. Murakami, On the mechanism of fatigue failure in the superlong life regime ($N>10^7$ cycles). Part II: A fractographic investigation, *Fatigue Fract. Eng Mater. Struct.* **23**, 903–910 (2000).
11. D. Angelova, A new normalized characterizing fatigue function, In *Proceedings of ECF 13 "Fracture Mechanics: Applications and Challenges"* (San Sebastian, Spain, 2000).
12. D. Angelova, A new method for fatigue testing and data presentation: An illustration on some metallic ceramic and polymer materials, in *Proceedings of ECF 14* (Krakow, Poland, 2002).
13. G. Dieter, *Mechanical Metallurgy* (McGraw-Hill Book Company, UK 1988).
14. G. Murtaza, Ph.D. Thesis (University of Sheffield, UK, 1992).
15. D. Angelova, Modelling of short crack growth in a low-carbon steel subjected to rotation-bending fatigue, in *Proceedings of the ECF16,* (Alexandroupolis, Greece, 2006).
16. D. Angelova, Basic fatigue conceptions and new approaches to fatigue failure, in *Proceedings of IFMASS 8*, (Belgrade, Serbia, 2003).
17. D. Angelova, Method for Testing Structural Materials Fatigue, *Patent Cooperation Treaty International Request for all Designated States*, (PTC/BG02/00032, 19 December 2002).
18. R. Yordanova, Ph.D. Thesis (University of Chemical Technology and Metallurgy, Sofia, Bulgaria, 2005).

EVALUATION OF SERVICE SECURITY OF STEEL STRUCTURES

EDWARD PETZEK[1], RADU BĂNCILA[2]
[1] *"Politehnica" University of Timisoara, Romania & SSF-RO Ltd. T. Vladimirescu Str. No. 12, 300.195 Timişoara, Romania, E-mail: epetzek@ssf.ro*
[2] *"Politehnica" University of Timisoara, Romania, Ion Curea Str., No.1, 300.224 Timişoara, Romania, E-mail: radu.bancila@ct.upt.ro*

Abstract The verification of existing steel structures especially steel bridges is in present one of the main problems of the structural engineers. The majority of existing railway steel bridges that have been built at the turn of the last century are riveted structures. Today many of these structures have already achieved a considerably age; therefore the establishment of the remaining fatigue safety of these structures is one of the most important tasks of contemporary society. Many of these bridges are still in operation after damages, several phases of repair and strengthening. The problem of these structures is the assessment of the present safety for modern traffic loads and the remaining service life. Along with the classical method of damage accumulation, a new approach based on the fracture mechanics principles is proposed. The paper presents the Romanian Methodology in this field with some case studies.

Keywords: Existing steel bridges, verification, safety, fracture mechanics

1. Introduction

Rehabilitation and maintenance of existing steel constructions, especially steel bridges is one of the most important actual problems.[1-3]

The infrastructure in Romania and in other East-European countries has an average age of about 70 to 90 years (Figure 1). Many of these structures, particularly railway bridges, have already achieved an age of 90, 100 or even more years and are still in operation after damages, several phases of repair and strengthening. To maintain these structures is one of the most important tasks of our society. Replacement with new structures raises financial, technical and political problems.

During service, bridges are subject to wear. In the last decades the initial volume of traffic has increased. Therefore many bridges require a detailed investigation and control (Figure 2). The examination should consider the age of the bridge and all repairs, the extent and location of any defects etc.

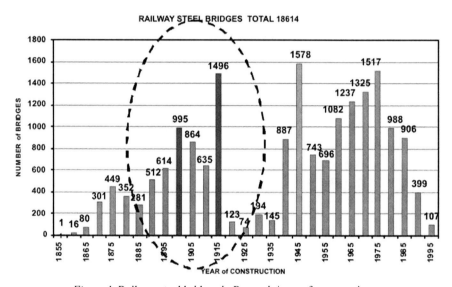

Figure 1. Railway steel bridges in Romania/year of construction.

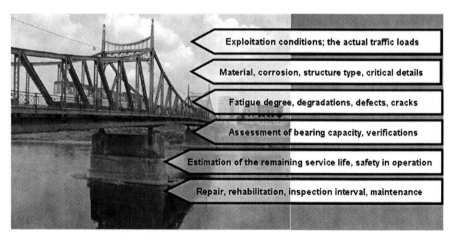

Figure 2. Assessment and control of existing steel bridges.

2. Technical condition of existing bridges

Carefully inspection of the structure is the most important aspect in evaluating the safety of the bridge. On the accuracy of the in situ inspection depends the level of evaluation.

The check of existing structures should be based on the complete bridge documentation (drawings with accuracy details, dimensions and cross sections of all structural elements, information about structural steel, stress history. However, in many cases these documentations are incomplete or missing. But these information can be recovered due to the carefully investigations and inspections of the structures, experimental determination of the material characteristics and stresses in structural elements, full scale in situ tests (static and dynamic), calibration of structure and spatial static analysis.

Bridge life is generally given by fatigue; difficult is the estimation of the loading history. For bridges where the stress history is known the fatigue life may be calculated using the Miners's rule and an appropriate S–N curve; also the assumption of the same spectrum for bridge life (or for certain periods) must be made. During the process of assessment the fatigue life of old riveted bridges is important to establish the proportion of the whole fatigue life that has been already got through. For stringers and cross girders of existing railway bridges the number of 10^7 cycles is exceeded. It might be affirmed that for such bridges if no fatigue cracks can be detected, no fatigue damage has occurred! Subsequently, if the loading spectrum remains the same in the future as in the past, fatigue cracking might not take place! In almost all the cases the loadings increased; in this situation survival for 100 years (or more) without cracking would not justify the assumption that no damage has yet occurred! Minor cracking is difficult to detect during usual inspections.

From the overall examination of a large number of bridges many defects can be pointed out. The defects are widespread, having a heterogeneous character from the point of view of location, development and development tendency; their amplification was also due to the climate and polluting factors that caused the reduction of the cross section due to corrosion. Statistically, in 283 from among 1,088 welded bridges, and in 356 from among 3,201 steel riveted bridges cracks were detected and repaired. It is not allowed to weld cracks. Old bridges can have welds executed in the early years; a special attention must be paid to these parts. Generally the riveted connections have a good behavior in time due to the initial prestressing force which can reach 70–80 N/mm².

3. Characteristics of materials

The following facts show that a material analysis for old riveted bridges is very useful:

- Old bridges are in many cases erected using material with very poor welding qualities and basing on railway administration data and specialised literature it is known that cast iron was used to build bridges.
- The specialised literature doesn't offer enough information about this structural steel.
- The structural material (Table 1) comes from several producers (for South Eastern Europe mostly from Reschitz – Romania and Györ – Hungary).

TABLE 1. The bridges on which the material study basis.

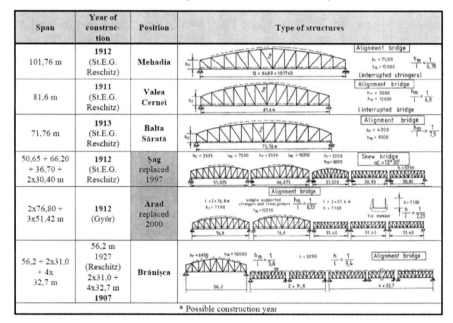

* Possible construction year

The study's results can be extended to Middle and South Eastern Europe when the history of communication ways and the state of old railway and highway steel bridges in this region is regarded.

In this context we mention the following event: on 1 January 1855 the "Kaiserliche und Königliche Privilegierte Österreichische Staatseisenbahngesellschaft" (St.E.G.) took over all steel producers in Banat. The investments in Reschitz turn the steel mill into an important bridges' factory.

EVALUATION OF SERVICE SECURITY OF BRIDGES   305

The production of steel bridges reached 3,960 t in 1910, whilst bridges made by St.E.G. Reschitz are still in use in Romania, Austria and Hungary. Between 1911–1913, 1,620 t of bridge structures made of cast iron were replaced in the western part of Romania (Banat), namely on the railway segment Timişoara–Orşova.

In this sense the material study took into account bridges from this region, built around 1911.

Following material analysis were performed in order to determine the characteristics of the material: chemical analysis, metallographic analysis, tensile tests, Brinell tests, Charpy "V Notch" tests.

The samples were taken from secondary elements, but also in some cases (Bridges in Arad and Şag which were replaced) from main elements: stringers, cross girders, main girders.[4]

The results of the chemical analysis are presented in Table 2.

TABLE 2. Chemical analysis results.

No.	Bridge / Specimen number & position	Chemical composition						
		C %	S %	Mn %	P %	Si %	Ni %	N %
1.	Valea Cernei Bridge / S1-VCB / Secondary elements	0,12	0,034	0,39	0,021	0,023	-	-
2.	Valea Cernei Bridge / S2-VCB / Sec. Secondary elements	0,11	0,020	0,32	0,024	0,010	-	-
3.	Mehadia Bridge / S1-MB / Secondary elements	0,11	0,014	0,50	0,028	0,050	-	-
4.	Balta Sărată Bridge / S1-BSB / Secondary elements	0,07	0,072	0,38	0,016	0,010	-	-
5.	Şag Timiş Bridge / S1-STB / Secondary elements – Span I	0,16	0,058	0,46	0,035	0,035	-	-
6.	Şag Timiş Bridge / S2-STB / Secondary elements – Span II	0,14	0,066	0,63	0,030	0,061	-	-
7.	Şag Timiş Bridge / S3-STB / Secondary elements – Span III	0,13	0,060	0,50	0,051	0,112	-	-
8.	Şag Timiş Bridge / S4-STB / Secondary elements – Span IV	0,18	0,054	0,48	0,066	0,056	-	-
9.	Şag Timiş Bridge / S5-STB / Secondary elements – Span V	0,18	0,017	0,44	0,059	0,010	-	-
10.	Brănişca Bridge / S1-BB / Secondary elements	0,12	0,070	0,39	0,013	0,060	-	-
11.	Brănişca Bridge / S2-BB / Secondary elements	0,11	0,055	0,38	0,014	0,060	-	-
12.	Arad Bridge / S1-AB / Secondary elements – Span I	0,14	0,061	0,53	0,038	0,069	-	-
13.	Arad Bridge / S2-AB / Secondary elements – Span II	0,19	0,033	0,64	0,051	0,018	-	-
14.	Arad Bridge / S3-AB / Main elements – Stringers Span III	0,089	0,032	0,531	0,009	0,018	0,067	-
15.	Arad Bridge / S4-AB / Main elements – Cross girders Span III	0,058	0,059	0,485	0,017	0,018	0,037	-
16.	Arad Bridge / S5-AB / Main elements – Main girder Span III	0,056	0,032	0,493	0,001	-	0,031	-
17.	Arad Bridge / S6-AB / Secondary elements – Span IV	0,18	0,035	0,46	0,063	0,078	-	-
18.	Arad Bridge / S7-AB / Secondary elements – Span V	0,1	0,047	0,43	0,020	0,030	-	-
19.	St 37 (STAS 500/2-80)	0,25	0,065	0,85	0,065	0,07	0,30	0,015
20.	St 34 (STAS 500/2-80)	0,17	0,055	0,60	0,055	-	-	-
21.	INCERTRANS Bucharest Research for cast iron	0,04…0,11	0,014…0,043	0,15…0,48	0,121…0,32	0,07…0,31	-	-
22.	German Research for cast iron*	0,16	0,056	0,100	0,470	0,100	0,007	-

* German research – Stahlbau 05.1985 (Prof.Dr.Ing. D. Kosteas, Ing. W. Stier, Ing. W. Grap)

The statistical interpretation of the tensile tests results shows a minimal value for the yield stress of 230 N/mm².

The impact tests on Charpy V Notch specimens lead to conclusion that the transition temperature is situated in many cases in the range from −10°C to 0°C (Figure 3).

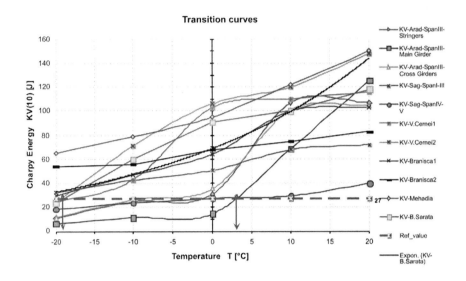

Figure 3. Transition curves for the analyzed bridge structures.

By analysing the laboratory results we can conclude that the steel is a mild one, that could be associated to the present steel types St 34 or St 37.1.

Also, on the two dismantled bridges – Arad and Şag –fracture mechanics tests were made[4] in order to establish the integral value J_c (according to ASTM E813-89), the CTOD and to determine the fatigue crack growth rate and the material constants C and m (according to ASTM E647-93). For these tests compact specimens CT (thickness 8 mm) as well as bending specimens have been used (see Chapter 6) (Figure 4).

They have been obtained from the stringers, cross girders and main girder – lower chord. The minimal value of material toughness in term of J-Integral for these old riveted steel bridges is J_{crit} = 10 N/mm for a temperature of −20°C.

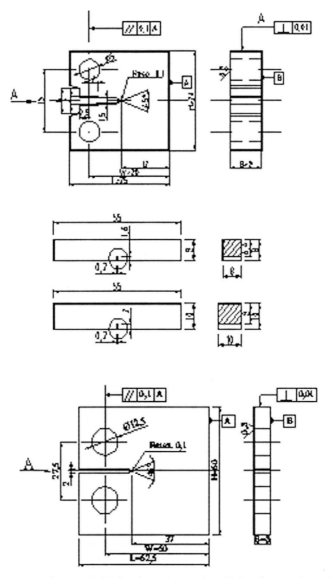

Figure 4. FM tests specimens. (a) CT-Specimen for J_c value. (b) Bending specimen for CTOD. (c) CT-Specimen for crack growth rate.

4. Present verification concept

During service, bridges are subject to wear. Therefore many bridges require an inspection. The examination should consider the age of the bridge and all repairs, the extent and location of any defects etc.[5] A continuous maintenance, which generally must increase in time, is important in order to assure the safety in operation of the existing structures. The present methodology includes the following stages (Figure 5):

STEP 1: estimation of the loading capacity of the structure based on a detailed inspection; analysis of drawings, inspection reports, repairs, reinforcements, analysis of the general behaviour of the bridge (displacements, vibrations, corrosion, cracks). In this phase the stresses in the structure can be calculated with the usual simplified hypothesis.

STEP 2: the accurate determination of the stresses in the structure and of the remaining safety of the elements. This phase includes: tests on materials, computer aided analysis of the space structure, remaining safety calculated on the base of the real time–stress history.

STEP 3: in situ static and dynamic tests.

This methodology adopted by the Romanian standard is illustrated in Figure 5.

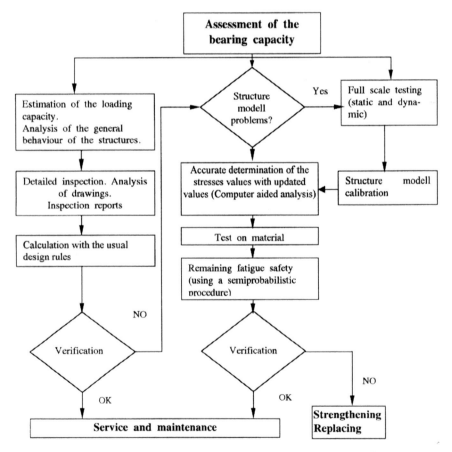

Figure 5. Methodology of the Romanian standard SR 1911–98.[3]

The calculation of remaining fatigue life is normally carried out by a damage accumulation calculation. The cumulative damage caused by stress cycles will be calculated; failure criteria will be reached.

$$D = \sum \frac{n_i}{N_i} \leq 1 \quad (1)$$

The classical fatigue concept is based on the assumption that a constructive element has no defects or cracks. However, discontinuities and cracks in the components of structures are unavoidable, basically because of the material fabrication and the erection of structures. It is very clear that the kind of fatigue cracks, which are initiated by structural non-homogeneity (possible non-metallic inclusions or other impurities), surface defects (including corrosion) and the stress factor, are present in the old riveted structures.

The presence of cracks in structural elements modifies essentially their fracture behavior. Fracture, assimilated in this case as crack dimensions growth process under external loadings, will be strongly influenced by the deformation capacity of material. The FM approach has acceleration in damage increase; with increasing damage a smaller stress range contribute to the damage increase. The authors proposed a complementary method based on the fracture mechanics basic concept

$$K_I \leq K_{Ic} \quad (2)$$

in order to calculate the remaining fatigue life. The principle of the new approach is presented above; these steps can be described in Figure 6.[6]

In practice two situations can be distinguished:

- $D < 0.8$ the probability to detect cracks is very low. The inspection intervals (generally between 3–6 years) can be established on criteria independent of fatigue. Nevertheless, a special attention must be paid to critical details.
- $D \geq 0.8$ cracks are probable and possible. An in situ inspection and the analysis of critical details are strongly necessary. Also a fracture mechanics approach is recommended.

Generally, the establishing of the maintenance program, the determination of inspection intervals, the inspection priorities of structural elements and finally the calculation with high accuracy of the remaining service life of old riveted bridges takes into account the following main data:

- Type of structure and exploitation conditions (traffic events)
- Information about structural steel (mechanical properties – yield strength, tensile strength, hardness, transition curve ductile – brittle and transition temperatures, chemical composition, metallographic analysis)

- Determination of critical members and details
- Crack detection and inspection techniques for evaluation of the initial crack size – a_0 and crack configuration
- Recording of the stress spectrum for the critical members under the actual traffic loads
- Evaluation of the critical crack size – a_{crit} based on failure assessment diagrams
- Fracture mechanics parameter – K_{crit}, δ_{crit}, J_{crit} (fracture toughness)
- Simulation of the fatigue crack growth
- Temperature, environment conditions

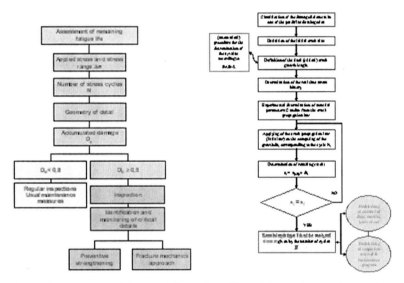

Figure 6. Assessment of the remaining fatigue life and the crack growth procedure.

The methodology is conceived as an advanced, complete analysis of structural elements containing fatigue defects, being founded on fracture mechanics principles and containing two steps; namely one of determination of defects' acceptability with the help of Failure Assessment Diagrams (level 2)[7] and of determination of final acceptable values of defect dimensions; this is followed by a second step which in fact represents a fatigue evaluation of the analyzed structural elements basing on the present stress history recorded on the structure, on the initial and final defect dimensions and the FM parameters, namely the material characteristics C and m from the Paris relation (crack growth under real traffic stress) and further on the exact determination of the number of cycles N needed in order that a fracture take place, respectively the determination of the remaining service life of the structural elements (years, months, days).

5. Fatigue crack propagation and crack propagation laws

The method of fatigue assessment for structural elements with defects was developed basing on the possibility of modelling, on the propagation rate of crack dimensions under fatigue loads and with the help of known laws. The method is founded on the recommendations of the BS 7910:1999.

In the present state of knowledge it is generally accepted that the fatigue failure of materials is a process containing three distinct steps: (1) initiation of defect (crack), (2) crack propagation in material, (3) separation through complete failure of the material in two or more pieces. Practically, the safety service life of an element under fatigue conditions can be expressed as follows (Figure 7):

$$N_f = N_i + N_p \qquad (3)$$

N_i = number of cycles necessary for the initiation of the defect (crack)
N_p = number of cycles necessary for the propagation of the defect until the occurrence of failure

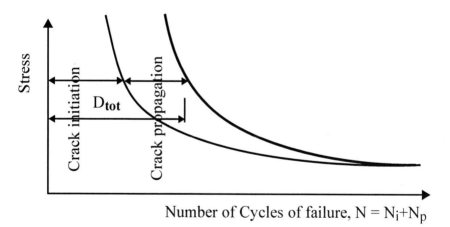

Figure 7. Fatigue life of structural elements.

The evaluation of crack propagation conditions can be accomplished with the help of characteristically values, which are founded on fracture mechanics concepts: material toughness express by the stress intensity factor K or J integral value and the crack growth rate da/dN (crack growth for each load cycle). A relation of the following type can express the crack growth rate (Figure 8):

$$\frac{da}{dN} = f(\Delta K, R, H) \qquad (4)$$

da/dN – crack extension for one load cycle

ΔK – stress intensity range, established basing on the stress range $\Delta\sigma$;

R – stress ratio $R = \dfrac{\sigma_{min}}{\sigma_{max}}$; H – indicates the stress history dependence.

The crack growth rate da/dN, defined as a crack extension da obtained through a load cycle dN (it can also be defined as da/dt, in which case the crack extension is related to a time interval), represents a value characteristic of the initiation phases respectively the stable crack propagation. It has been experimentally observed that the connection between the crack growth rate and stress intensity factor variation represents a suitable solution for the description of the behaviour of a metallic material containing a crack, as in the case of steel. In a logarithmic graphical representation of the crack growth rate da/dN versus the stress intensity range ΔK a curve as the one Figure 8 is obtained.

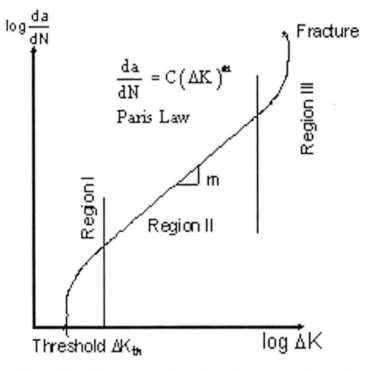

Figure 8. Logarithmic representation of the fatigue crack growth in steel.

In the technical literature a large number (over 60) of crack propagation rate laws can be found. These crack propagation equations can be divided into three groups by taking into account the parameters they contain:

a) $da/dN = C_1 f_1(a)$; b) $da/dN = C_2 f_2(\Delta\sigma)$; c) $da/dN = C_3 f_3(\Delta K)$ (5)

a – crack length;
N – number of cycles;
$\Delta\sigma$ – stress range;
ΔK – stress intensity factor range;
C_i cu i = 1...3 – material parameters;
f_i cu i = 1...3 – functions.

Obviously, the existing formulas have different degrees of complexity including more or less parameters. Beyond doubt, the equations belonging to group (c) are the most valuable, as the use of fracture mechanics parameters offers a series of advantages. These advantages base on the fact that within fracture mechanics a well defined relation between the material parameters – stress – dimension and geometry of the defect has been established.

The most important equation of group (c) is the Paris and Erdogan law:

$$da_1 = C \cdot \Delta K^m \qquad (6)$$

The calculation of the structural elements remaining service life can be done basing on the Paris law, more precisely by integration of this law:

$$N = \int_0^N dN = \int_{a_0}^{a_{crit}} \frac{da}{C \cdot \Delta K^m} \qquad (7)$$

N – number of stress cycles necessary in order that the crack extends from its initial dimension a_0 to the critical value a_{crit}, where failure occurs;

a – crack length
C, m – material constants from the crack propagation law
ΔK – stress intensity factor range

This integral can be numerically calculated by taking into account a critical detail knowing the crack values (initial and critical), basing on the following relation:

$$N = \int_{a_0}^{a_{crit}} \frac{da}{C \cdot \Delta\sigma^m \cdot Y^m \cdot (\pi a)^{-m/2}} \qquad (8)$$

The number of cycles N_i obtained with the help of relation (8) represents the remaining service life of the detail, by regarding the initial length a_0 up to the critical length a_{crit}, by admitting a stable crack propagation (Figure 9).

The critical crack value can be calculated basing on the K criterion respectively on the J or CTOD criterions or with the help of the failure assessment diagram.

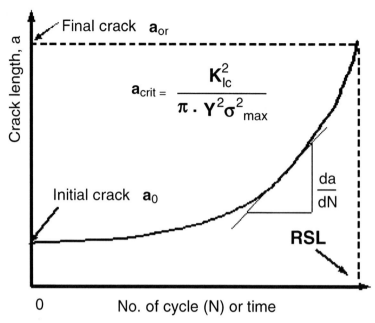

Figure 9. Principles for determining the remaining service life.

6. Measurement of fatigue crack growth rates

The C and m material constants from the Paris law are experimentally determined by fracture mechanical tests. In this sense in most cases compact specimens C(T), three point bended specimens SEN(B) and middle central panels M(T) are used. Such a standard which describes the test methods for the determination of the crack growth rate is the American Standard ASTM E 647 (*Standard Test Method for Measurement of Fatigue Crack Rates*).

The procedure for the determination of crack growth rate in metallic materials bases on the use of specimens containing a fatigue defect (crack). This precrack has a well established length and is placed at the top of the machined notch. The precracked specimens are then tested at a fatigue cycle; during the test the crack length extension is measured according to the number of cycles which correspond to these extensions. Basing on the recordings the curve crack length extension versus applied number of cycles is drawn. With the help of these curves the crack propagation rate da/dN is determined by using one of the standard methods: secant method or incremental polynomial method.

For tests, which were performed in accordance with ASTM E647-93, compact specimens C(T) (thickness 8 mm) have been used. They have been sampled from the stringers, cross girders and main girder – lower chord of the bridge (Figure 10).

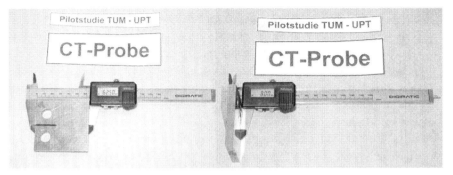

Figure 10. CT specimens.

It should be mentioned here that the test have been performed in the laboratory of the Technical University of Munich.

Basing on the determined values da/dN and ΔK the program also automatically determines the C and m material constants by the Paris relation:

$$\ln(\frac{da}{dN}) = \ln C + m \cdot \ln \Delta K \qquad (9)$$

The experimental tests on 26 CT specimens from two old bridges have shown that for the oldest mild steels the values of the material constants from the Paris relation are in the following intervals:

m = 2.05 ... 5.65
C = 2.2 × 10^{-11} ... 10^{-18}

Relatively large value of m corresponding to very small values of C, for example for m > 4 → C ≅ 10^{-15} ... 10^{-18}.

7. Case studies

7.1. HIGHWAY BRIDGE IN ARAD

The bridge was built between 1910–1913 being a reference work for that period and a symbol for the city Arad. The steel structure was erected at the Bridge Factory Reschitz (Reşiţa) between 1910–1912 – Kaiserliche und Königliche Privilegierte Österreichische Staatseisenbahngesellschaft" – **St.E.G.** Nowadays it is under technical monument protection.[8]

The bridge is located on the national highway DN 69, at km 49 + 621, crossing the Mureş River (not yet navigable) and connecting the suburb Aradul Nou to the centre of the city Arad. After the Worlds Wars and different events, the bridge is still in operation with some restrictions.

The bridge is a cantilever trough truss girder with three spans, L = 50.05 + 85.30 + 50.05 = 185.40 m (Figure 11); the width of the carriageway is 8.05 m (thus assuring a dual carriage, one for each direction). The bridge's width between the main truss girder axis is 9.6 m and the footpath on the lateral cantilever is 1.5 m. Due to the elegant curved forms, these structures give the impression of suspension bridges but they in reality being classical cantilever structures.

From the beginning the bridge had one tramway line. After the Second World War when the structure was damaged and repaired the tramway line was doubled.

The bridge was verified many times. Based on the laboratory tests, the base material is a mild steel comparable to the present St 34–37.n (STAS 500/2-80). The conclusions of the last report, basing also on the results of the in situ tests on the structure (with 30 t trucks) are rather unfavorable. Some restrictions were introduced: trucks circulation is prohibited, only light cars are allowed to cross the bridge; tramway has only two wagons (maximum speed 10 km/h), only one tramway being allowed on the bridge (Figure 12).

After the classical verification of the bridge stresses (Figure 13) the fatigue assessment of the structure was performed.

Also, the complementary method of fracture mechanics was applied. For the material characteristics followings values were considered: the material is a mild steel similar to the present St 34–37. n (STAS 500/2 – 80); yield stress is σ_y = 230 N/mm^2; tensile stress σ_{ult} = 360 N/mm^2.

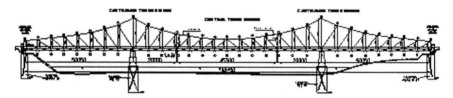

Figure 11. General view of the bridge.

Figure 12. The bridge in past (1912) and present (2007).

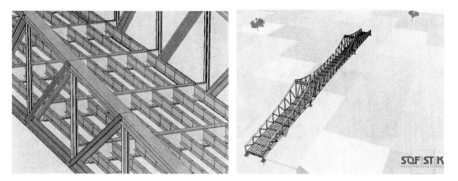

Figure 13. The static model of the structure.

For the material toughness in terms of J_{crit} a minimal value of 15 Nmm at a temperature of −20°C was chosen. For the life prediction procedure in the case of the material constants following values have been chosen: m = 3 and C = 3 × 10^{-12} (see also Ref.[9]).

A stress history was established for double line tram traffic, 15 min. tact, in the time interval 5:00 a.m.–12:00 p.m. ⇒ 76 tram pairs/day in tandem on the bridge (Figure 14).

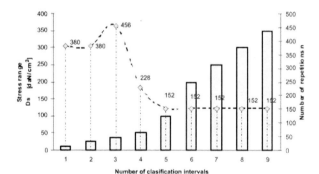

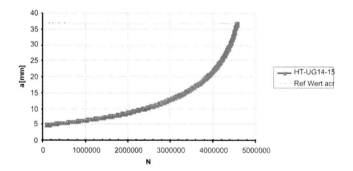

Figure 14. Stress history and crack growth curve.

In the case of the main girder lower chord (Figure 15) (independent central truss girder) for an initial crack value of $a_0 = 5$ mm it has been obtained N = 4,559,000 cycles up to the critical crack value → RFL = 6.95 years. In the same case for a value of $a_0 = 2$ mm (undetectable!) → RFL = 15 years.

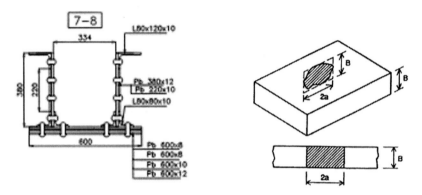

Figure 15. Cross section lower chord and analyzed crack model.

7.2. RAILWAY BRIDGE IN ARAD

Another case study is the old riveted railway steel bridge in Arad over the Mureş River (Figure 16). This structure has been dismantled in February 2000, after being 88 years in operation; as a consequence, a detailed analysis on main structural elements could be made. This structure has been chosen because specimens could be sampled from most stressed elements and also because many similar structures built in the same period of time (1900–1930) are still in operation.

The main girder is a truss girder with 5 spans L = 2 × 76.80 + 3 × 51.42 m (see Figure 9b). During the Second World War the bridge was damaged. In 1986 a systematic assessment of the bearing capacity of the bridge – including in situ tests with railway convoys – was fulfilled. The time–stress history of the bridge (on the base of an analysis of different periods) was carefully established.

After cutting the rivets a detailed visual inspection of the elements was possible. It is known that the cracks in existing riveted bridges most probably initiate at the riveted holes under the riveted heads and propagate through the plate thickness and widths (Figure 17).

The acceptability analyze (Figure 18) of existing and detected defects (Figure 19) and the determination of the critical cracks values was performed on the base of the failure assessment diagram FAD level 2.

EVALUATION OF SERVICE SECURITY OF BRIDGES 319

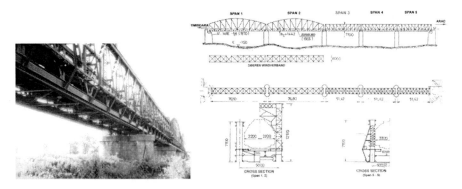

Figure 16. Arad Bridge/General view of Arad Bridge.

Figure 17. Cracks in elements.

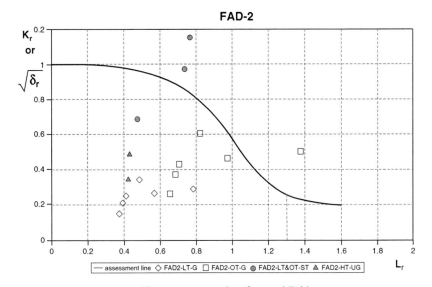

Figure 18. Assessment points for Arad Bridge.

By analyzing the assessment points from the fracture assessment diagram the following can be concluded:

- The structural defects present in the stringers and cross girders in the flange plate, analyzed with the help of models 1 and 2,[10] show admissible dimensions that don't endanger the structure's service safety (Figure 19).
- In the case of the chord of the stringers and the cross girder, the problem is more serious, because the majority of the assessment points are in the unacceptable area and even more, in an area of combined but manly fragile collapse. Here it must be told, that the analyzed theoretical defects are unacceptable, i.e. cracks that reach a sufficient length so that they can be discovered (practically, cracks that are minimum 2 mm longer than the edge of the connection angle and which in this case would have a length of 92 mm). But there haven't been discovered any such defects in the investigated samples (from the stringers and cross girders), at which the area covered by the connection angle has been analyzed by cutting the rivets.
- Regarding the lower chord of the main girder, the assessment shows the admissibility of the defects' dimensions, although it presented a minimal value of material toughness.

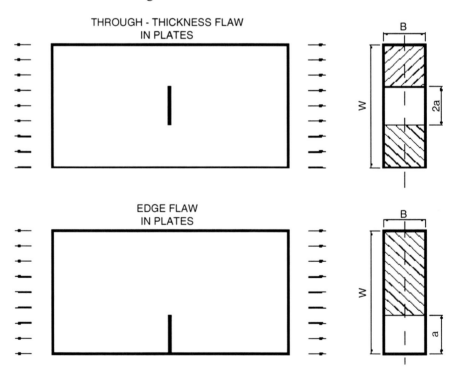

Figure 19. Model 1 – Middle crack, Model 2 – Edge crack.

The fatigue analysis have used the recordings] on the railway Arad–Timisoara in the section Arad bridge through "in situ" measurements with real trains in a time period of 14 days (Figure 20).

For each analyzed defect diagrams as in Figure 21 have been drawn while the results are synthesized in Table 3. They are presented in comparison with results obtained through a classical calculus based on the hypothesis of damage accumulation.

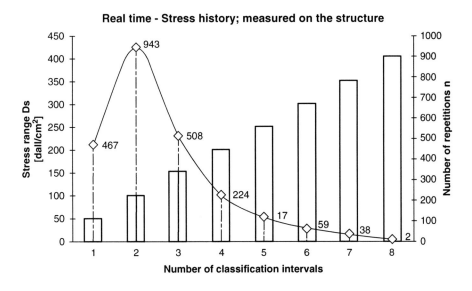

Figure 20. Stress history – example for stringers (Arad bridge/Span 3).

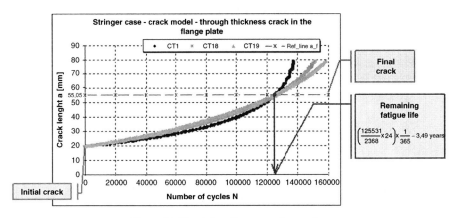

Figure 21. Crack length vs. number cycles.

TABLE 3. Results of the life prediction analysis.

Analysed Element	Remaining service life (years)	
	Classical Method PLM	Complementary Method FM
Cross girder	ca. 5.5	2.63
Final stringer	–0.9 !	3.49
Central stringer	–13.9 !	3.29

8. Conclusions

- The introduction of a complementary method based on fracture mechanics principles along with the classical method for the evaluation of the safety in operation of existing steel bridges is necessary.[11]
- A better knowledge of the fatigue resistance of riveted details and of the repair and strengthening of riveted bridge members damaged by fatigue, could extend the service life of a large number of bridges. In many cases there is a need to retain particular bridges as historical monuments.
- The methodology establishes clear rules for the dimension determination of initial defects, the definition type of design stress in the analysed elements and the material characteristics needed for the calculation of evaluation parameters. The determination procedure of the durability of structural elements is conceived as a logical succession of calculation steps.
- The procedure based on the fracture mechanics principles was simplified and can be applied along with the classical methods.
- For a reliable assessment of existing bridges a unified methodology is needed including damage accumulation method and fracture mechanics concepts.
- In the frame of the present rehabilitation of principle railway lines in Romania fracture mechanics concepts are used for the assessment of remaining safety in operation of existing steel bridges.
- The conclusions can be extended to other countries from Middle and Southeast Europe, where the situation of the existing bridges is similar.

References

1. Code UIC 778-2R; Recommandations pour la détermination de la capacité portante des structures métalliques existantes; Union Internationale des Chemins de fer, Paris, 1986.
2. DS 805 Bestehende Eisenbahnbrücken. Bewertung der Tragsicherheit und konstruktive Hinweise, August 2002.
3. SR 1911-98, Poduri metalice de cale ferată. Prescripții de proiectare", Institutul Român de Standardizare, Bucuresti, 1998.

4. Petzek, E., "Safety in Operation and Rehabilitation of Steel Bridges", Doctoral Thesis, UP Timişoara, 2004.
5. I-AM 08/2002. Richtlinie für die Beurteilung von genieteten Eisenbahnbrücken, SBB CFF FFS.
6. Petzek, E., Kosteas, D., Bancila, R., 2005. "Sicherheitsbestimmung bestehender Stahlbrücken in Rumänien", Stahlbau Nr. 8, 9, ISSN 0038-9145, Ernst und Sohn.
7. BS 7910:1999, "Guide on the Methods for Assessing the Acceptability of Flaws in Metallic Structures", British Standards Institution, London, 1999.
8. Petzek, E., Băncilă, R., Schmitt, V., General Rehabilitation Concept of a Historical Representative Cantilever Truss Girder Bridge in Arad, Proceedings of Eurosteel Conference Graz, Sept. 2008.
9. Eriksson, K., Toughness requirements for old structural steel, IABSE Report Congress, 2000.
10. ESDEP – European Steel Design Education Programme, "Structural Systems – Refurbishment", Lecture 16.5, Vol. 28 Londra, 1995.
11. Băncilă, R., Petzek, E., "Experience in Management of Ageing of Bridges in Eastern Europe ", IIW Conference, Bucharest, 2003.

FATIGUE DESIGN OF NOTCHED COMPONENTS BY A MULTISCALE APPROACH BASED ON SHAKEDOWN

K. DANG VAN[1], H.M. MAITOURNAN[1], J.F. FLAVENOT[2]
[1]*LMS (CNRS UMR 7649), Department of Mechanics,*
Ecole Polytechnique – 91128 Palaiseau Cedex, France,
E-mails: habibou@lms.polytechnique.fr,
dangvan@lms.polytechnique.fr
[2]*CETIM 52 Avenue Félix Louat – 60304 Senlis Cedex, France,*
E-mail: jean-francois.flavenot@cetim.fr

Abstract This paper shows that fatigue limit of notched specimens under cyclic loading can be simply and accurately estimated by using elastic-plastic computations and averaging stress over a critical volume obtained by an optimisation process minimizing the dispersion between experiments and simulations. The fatigue limit criterion considered is the Dang Van one. Fatigue tests (tension-compression, bending and torsion) carried out by the Cetim, are used to calibrate the critical volume.

Keywords: High cycle fatigue design, notched components, multiaxial cyclic loading

1. Introduction

The high-cycle fatigue design of industrial structures is still a not resolved problem especially in the presence of stress concentrations areas (holes, notches, indentations etc.). Taking account of these phenomena is particularly important for predicting the fatigue behaviour of such structures. In the current industry approaches, it is carried out by semi-empirical correction requiring a large number of tests and which domain of validity must be carefully delimited.

The most common approach is based on the concept of stress concentration factor K_t, defined as the ratio of the maximum stress at the notch evaluated under the assumption of a purely elastic behaviour and the nominal stress. However, the fatigue limit of a notched specimen is different from that of the smooth specimen divided by K_t, which led engineers to introduce the concept of fatigue strength reduction factor K_f, which is the ratio of the fatigue limit of the smooth specimen and the fatigue limit of the notched specimen calculated with the assumption of a purely elastic behavior. This coefficient is, however, not general. In the presence of residual stresses, the correction factor previously determined for a particular load a particular material will be no longer valid for another load nor another material. Various proposals empirical corrections exist; they require tests database which are used to carefully delimit the range of validity of the proposed factors. Another type of approach was proposed by G. Pluvinage,[6] based on an adaptation of the stress intensity factor concept to notch. This modelling, like fracture mechanic, works well for simple uniaxial loading.

Another way to take account of the effects of stress concentration is the use of theories of critical distances based on elastic stresses. Depending on the specific form of these theories, the relevant stresses for fatigue analysis are estimated at a certain distance from the notch (point method) or averaged along a line (line method) or along an area (area method).[2,3]

The aim of this paper is to study the suitability of the Dang Van criterion[1,4,5] associated with an adequate representative volume element (or critical volume) for the prediction of the fatigue limit of notched specimens. This criterion, quite widely used in mechanical industry, is based on a multiscale approach[5] and on the shakedown assumption. In order to check this point, elastic-plastic simulations are performed and calibrated with tests conducted by the CETIM on smooth and notched specimens subjected to:

- Tension-compression with and without mean value
- Rotating bending
- Alternated torsion

The critical volume is determined as the one which permit to obtain the best correlation between calculated fatigue limits and experimental ones for all the tests.

2. The proposed methodology

2.1. DESCRIPTION OF THE MODEL

The methodology used is based on the following features:

1. Cyclic elastic-plastic finite element simulations to calculate the stabilized mechanical state
2. Use of the Dang Van fatigue limit criterion associated with a material representative volume (critical volume)
3. Use of an optimization process coupling simulations and experiments to determine the critical volume

First, we recall the basic assumptions of the Dang Van fatigue approach. This model considers two scales, mesoscopic grains scale on one hand and a macroscopic scale for engineers attached to the notion of representative volume element (RVE) widely used in the polycrystalline theory. The RVE must contain a sufficient number of grains to be representative of average macroscopic mechanical properties. The model assumes that if the material can elastically shakedown at the two scales (mesoscopic and macroscopic), there will be no fatigue. The fatigue limit corresponds to the limit of the ability of the material to shakedown elastically.

The relation between the macroscopic stress tensor $\Sigma(M, t)$ at point M (which is a Representative Elementary Volume, RVE) and the mesoscopic stress tensor $\sigma(M, t, m)$ at point m of the RVE (M), is given by the following expression:

$$\sigma(M,t,m) = A(M,m) : \Sigma(M,t) + \rho(M,t,m) \quad (1)$$

ρ is the mesoscopic residual stress tensor, A is the localization tensor, which is taken here for simplicity, as equal to the identity tensor and ρ is the mesoscopic residual stress tensor. We assume that the material is in an elastic shakedown state at the macroscopic level; this means that there is a field of time independent macroscopic residual stress $\mathbf{R}$ such as

$$\Sigma(M,t) = \Sigma^{el}(M,t) + \mathbf{R}(M)$$

$\Sigma(M,t)$ does not violate the macroscopic criterion of plasticity. $\Sigma^{el}(M,t)$ is the macroscopic purely elastic stress calculated with the same loading under the assumption that the material has a purely elastic behavior. The mesoscopic stress is then:

$$\sigma(t) = \Sigma^{el}(t) + \mathbf{R} + \rho.$$

In the previous equation, for sake of simplicity, only time dependencies are explicitly written.

The first step of our approach, in the presence of a notch, is therefore the evaluation of the macroscopic stress state composed of the purely (variable) elastic stress $\Sigma^{el}(t)$ and the constant residual stress $\mathbf{R}$. This consists in a cyclic elastoplastic computation of the structure, under the fatigue loading, until a stabilized state (shakedown in this case) is obtained.

The second step is the determination of the mesoscopic residual stress. They are obtained by constructing the center of the smallest hypersphere circumscribing the deviatoric purely elastic stress $S^{el}(t)$.

And then, the Dang Van criterion is applied. Different RVE are used in order the find the one giving the best correlation between simulations and experiments.

2.2. MATERIAL AND TEST SPECIMENS

The fatigue tests under constant amplitude loading were carried out on 42CrMo4 steel. The material properties are the following: Re = 928 Mpa, Rm = 1024 Mpa.

Cylindrical specimens are used. These tests consist of tension-compression fatigue tests (Figure 1 through 3), pure bending, pure torsion and tension–torsion.

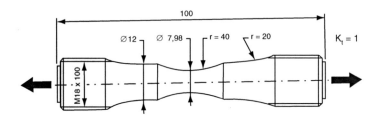

Figure 1. Tension-compression test specimen with $k_t = 1$.

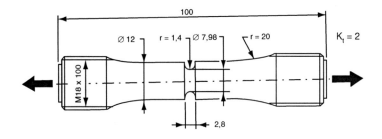

Figure 2. Tension-compression test specimen with $k_t = 2$.

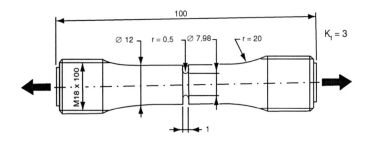

Figure 3. Tension-compression test specimen with $k_t = 3$.

The first step of the methodology is the identification of the material elastoplastic behavior from the experimental results (thin curve in Figure 4). In this case, the material is assumed as von Mises elastoplastic with linear kinematic hardening (thick curve in Figure 4).

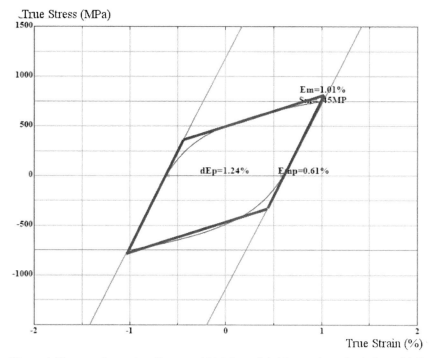

Figure 4. The experimental cyclic curve (thin) is modeled by an elastoplastic law with linear kinematic hardening (thick).

The material cyclic behavior is modeled as von Mises elastoplastic with linear kinematic hardening and the following parameters: Young modulus E = 209,000 MPa, tension yield stress σ_y = 650 MPa and hardening modulus H = 35,294 MPa. This constitutive law is used for the determination of the residual stress in the macroscopic specimen. The calculation is performed by the finite element method. We will present, in what follows, the results of elastoplastic calculations for the case of tensile load amplitude of 165 MPa around an average value of 500 MPa.

2.3. EXAMPLE: CASE OF THE TENSION CYCLIC LOADING 500 ± 165 MPa

The test specimen with k_t = 3 (Figure 3) subjected to a cyclic tensile loading of 500 ± 165 MPa. Figure 5 to 7 show the different results obtained. The mesh used and the refinement at the notch are shown in Figure 5. Examples of meshes used for other specimens are shown on Figure 6 and 7.

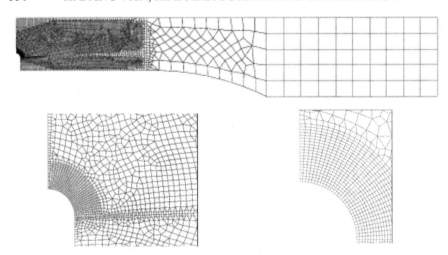

Figure 5. Mesh of the specimen ($k_t = 3$) with different zooms.

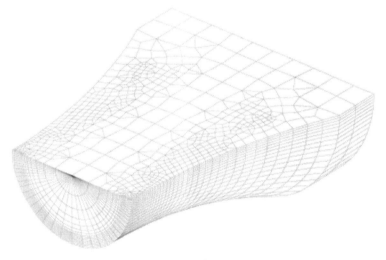

Figure 6. Mesh of the central part of bending specimen ($k_t = 2$).

Figure 8 shows the evolutions of the axial stress σ_{zz} along with a radius of the center section of the specimen, and calculated:

1. In linear elasticity (square); we can see that the stress concentration factor is 3.13.
2. In elastoplasticity (diamond).

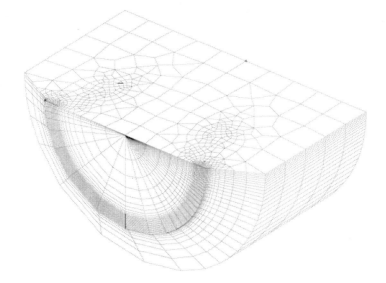

Figure 7. Mesh of the central part of bending specimen ($k_t = 3$).

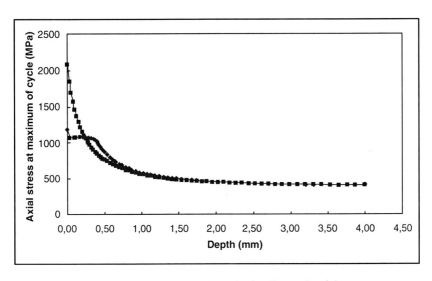

Figure 8. Elastic (square) and elastoplastic (diamond) axial stresses.

A cyclic elastoplastic simulation is then conducted to determine the stabilized mechanical cycle. This state is obtained after one loading cycle. The axial stresses obtained are shown on Figure 9 and the stabilized axial plastic deformation on Figure 10. The axial residual stresses are shown on Figure 11.

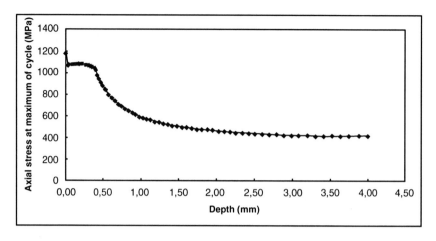

Figure 9. Axial elastoplastic stresses along a radius.

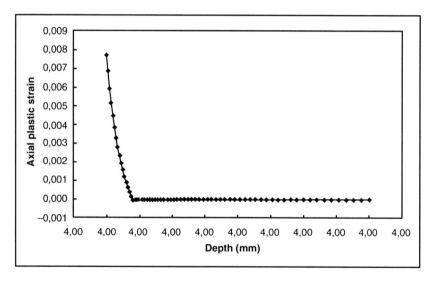

Figure 10. Stabilized axial plastic deformations along a radius.

After determining the macroscopic stabilized mechanical state (sum of the purely elastic response and the residual one), Dang Van criterion, which practical application is recalled in the next section, is used to study the fatigue limit.

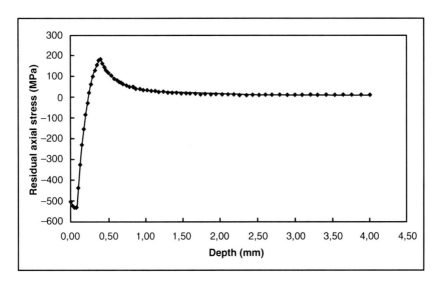

Figure 11. Axial residual stresses after total unloading.

2.4. FATIGUE ANALYSIS

The Dang Van criterion assumes that the material sustain a very high number (theoretically infinite) of loads cycles without being broken if only elastic shakedown occurs at the macroscopic and mesoscopic scales. The basic idea is to write that the mesoscopic must satisfy the plasticity criterion. The approach can be summarized as follows:

- From the macroscopic loading cycle (i.e. the evolution of the macroscopic stresses $\underline{\underline{\Sigma}}(t)$), calculate the mesoscopic stabilized stress cycle $\underline{\underline{\hat{\sigma}}}(t)$.
- The mesoscopic stabilized stress cycle $\underline{\underline{\hat{\sigma}}}(t)$ must satisfy the plastic criterion of the grain.
- Rewrite the obtained limitation in terms of the macroscopic stresses.

As a first step, Dang Van proposes to evaluate the stabilized mesoscopic stresses by the following equation:

$$\underline{\underline{\hat{\sigma}}}(t) = \underline{\underline{\Sigma}}(t) + \underline{\underline{\rho}}^*$$

where $\underline{\underline{\rho}}^*$ is a time independent stress field. Moreover, the mesoscopic stress field verifies the plastic criterion:

$$\forall t \quad f\left(\underline{\underline{\hat{\sigma}}}(t)\right) \leq 0$$

To take into account the experimental observations on the influence of the hydrostatic pressure on the fatigue limit, Dang Van postulates in 1973 a criterion for grain plasticity in the following form:

$$\forall n \forall t \quad \hat{\tau}(\underline{n},t) + a\,\hat{p}(t) \leq b$$

where $\hat{p}(t)$ and $\hat{\tau}(\underline{n},t)$ are the mesoscopic hydrostatic stress and the shear on a plane with normal vector $\underline{n}$. This criterion can be rewritten as:

$$\max_{\underline{n}} \left\{ \max_{t} \left[\hat{\tau}(\underline{n},t) + a\,\hat{p}(t) \right] \right\} \leq b$$

The problem now is to assess $\hat{p}(t)$ and $\hat{\tau}(\underline{n},t)$ from the macroscopic stresses $\underline{\underline{\Sigma}}(t)$. It is very easily done for the hydrostatic stress since:

$$\hat{p}(t) = P = \frac{1}{3} tr\left(\underline{\underline{\Sigma}}(t)\right)$$

The mesoscopic shear is the maximum shear stress due to the mesoscopic stresses.

2.5. RESULTS

Figure 12 represents the loading paths obtained at the fatigue limit for the following fatigue tests:

1. Fully reversed tension-compression (R = −1) on the specimen with $k_t = 1$ (Figure 1), with a stress amplitude of 508 MPa
2. Push-pull on the specimen with $k_t = 1$ (Figure 1), with a stress amplitude of 467 MPa around a mean stress of 400 MPa
3. Fully reversed tension-compression (R = −1) on the specimen with $k_t = 3$ (Figure 3), with a stress amplitude of 220 MPa
4. Push-pull on the specimen with $k_t = 3$ (Figure 3), with a stress amplitude of 165 MPa around a mean stress of 500 MPa
5. Fully reversed torsion (R = −1) with a stress amplitude of 320 MPa
6. Rotating bending (R = −1) on the specimen with $k_t = 1$, with a stress amplitude of 540 MPa
7. Rotating bending (R = −1) on the specimen with $k_t = 2$, with a stress amplitude of 267 MPa
8. Rotating bending (R = −1) on the specimen with $k_t = 3$, with a stress amplitude of 180 MPa
9. Push-Pull $k_t = 2$, 252 MPa around σm = 500 Mpa

The results show that, except for the cases $k_t = 2$ and 3 in bending, a critical layer of 100 μm is suitable (Figure 13).

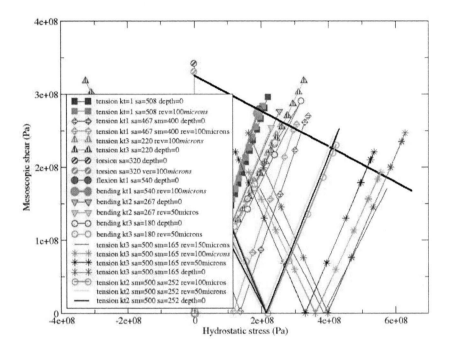

Figure 12. Loading paths in Dang Van diagram (hydrostatic stress (Pa) – mesoscopic shear (Pa)).

3. Conclusion

In Figure 13 are represented all the loading paths resulting from elastic-plastic computations, which allows us to take into account the residual stresses when they are not null. The fatigue limits are very well predicted if one calculates the macroscopic stress Σ as an average over a depth of 100 μm, excluding results for alternating bending on notched specimens with $k_t = 2$ and 3, where these values are too low. But a depth of 50 μm is more suitable for these cases. However, the experimental results for these two cases seem to be problematic when compared to those of similar but notched specimens subjected to cyclic tension around an average value of high tension. The stress amplitudes in alternate bending are comparable to the stress amplitudes in tension around very high mean values. In addition, it is well known that in general the fatigue limits are higher in bending than in tension. This leads us to believe that there is a problem with these tests bending. Apart from this observation, we can see that taking into account the residual macroscopic associated with a REV of the order of 100 μm allows interpreting simply and in a consistent manner, the overall results for smooth and notched. This method can be easily and unambiguously extended to other geometry accidents and defects of any shapes.

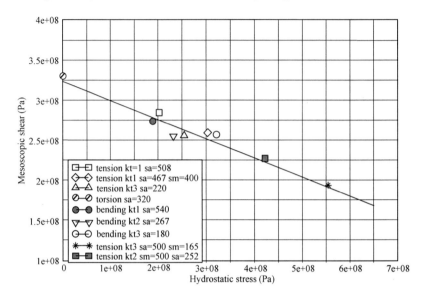

Figure 13. Limits points for the different loading path in the (p,τ) diagram.

References

1. K. Dang Van, *High-cycle metal fatigue in the context of mechanical design*, K. Dang Van, I.V. Papadopoulos, Eds, CISM Courses and Lectures no 392, Springer-Verlag, Wien-New-York, 1999, 57–88.
2. J.F. Flavenot, N. Skalli, A critical depth criterion for the evaluation of long life fatigue strength under multiaxial loading and a stress gradient. *Conference on life assessment of dynamically loaded materials and structures*, vol. I, Lisbon, Portugal, 17–21 Sept. 1984, 1985, 335–344.
3. D. Taylor, Analyse of fatigue failures in components using the theory of critical distances, *Eng. Failure Anal*, 12, 2004, 906–914.
4. P. Ballard, K. Dang Van, A. Deperrois, I.V. Papadopoulos, *Fatigue Fract. Eng. Mater. Struct.*, 18 (3), 1995, 397–411.
5. A. Constantinescu, K. Dang Van, M.H. Maitournam, *Fatigue Fract. Eng. Mater. Struct.*, 26 (6), 2003, 561–568.
6. G. Pluvinage, Fatigue and fracture emanating from notch; the use of notch stress intensity factor, *Nucl Eng. Des.* 185, 1998, 173–184.

MEASUREMENTS FOR MECHANICAL RELIABILITY OF THIN FILMS[*]

DAVID T. READ
Materials Reliability Division, National Institute of Standards and Technology, Boulder, Colorado

ALEX A. VOLINSKY
Department of Mechanical Engineering, University of South Florida, Tampa, Florida

Abstract This paper reviews techniques for measurement of basic mechanical properties of thin films. Emphasis is placed on the adaptations needed to prepare, handle, and characterize thin films, and on adaptations of fracture mechanics for adhesion strength. The paper also describes a recent development, the use of electrical current as a controlled means of applying thermomechanical stresses to electrical conductors to characterize their fatigue behavior.

Keywords: Delamination, grain size, strain, strength, substrate, testing, tensile, yield strength, Young's modulus

1. Introduction

From the time of Galileo to the late twentieth century, mechanical testing evolved at the macro scale, with specimen dimensions of the order of centimeters and even meters in some cases. This was a natural match to the structures being analyzed, which included bridges, pipelines, pressure vessels, aircraft and their engines, rockets, and so on. The appreciation that larger structures produce mechanical constraints that promote brittle fracture became widespread only after many tragic failures, which are well documented. Wide plate testing was developed to examine the conditions under which a small crack or inhomogeneity such as a weld can be tolerated by a large

---

[*] Contribution of the U.S. National Institute of Standards and Technology. Not subject to copyright in the U.S.

structural element. The need to control catastrophic brittle failure led to the first serious attempt to understand size effects in mechanical behavior, specifically, the development of structural fracture mechanics in the mid-twentieth century. This understanding was applied both to the structures themselves and to the test protocols, so that specimens small compared to the structures of interest could be used to explore and verify material behavior. These specimens were and still are macroscale: they can be manufactured with lathes and milling machines; they can be mounted by hand with no risk of damage to the specimen; and they can be tested without the need for microscopy.

The next push toward smaller scale mechanical testing came with the rise of thin film technology for microelectronics, and the related micro-electro-mechanical systems (MEMS) technology, where photolithography is used to create structures that provide mechanical functionality with critical dimensions on the scale of micrometers. Analytical tools from the macro world were adapted to design against the surprisingly high stresses that arise in integrated circuits; the source of these stresses is the difference in thermal expansion rate among the different materials that are bonded together in thin layers to produce integrated circuits, combined with the severe temperature excursions seen in these structures in both production and use. But numerical analysis alone was not enough; the results of a numerical analysis depend on the material properties data used, and thin film materials have properties much different from those of the same materials in bulk form. These property differences are a natural consequence of the much different microstructures between thin film and bulk materials. The microstructures are a consequence of the novel production methods used for the thin film materials, for example, physical vapor deposition for thin films as opposed to rolling and annealing for bulk materials. While measurements of basic mechanical properties, such as elastic modulus and yield and ultimate strength, of materials with dimensions around 1 μm are well established and widely practiced, fracture mechanics has recently found a mode of application in the microscale that differs in emphasis from the practice in macroscale structures.

Delamination of thin films from rigid substrates, of which delamination of an aluminum film from a silicon wafer would be a simplified example, is a critical issue for integrated circuits and other thin film structures. This failure mode is of relatively low importance in macroscale structures, which rarely utilize bonds between large flat sheets as critical structural elements. The testing of the adhesion between film and substrate has been attempted by a multitude of approaches, but it is now recognized that the strength of an interface can most accurately and usefully be described in terms from fracture mechanics, in particular, energy per unit area of the bond between film and substrate and magnitude of stress singularities at critical locations. Inspection for delaminations, for example by ultrasonic means, has been applied to larger scale features of integrated circuits. Acceptance criteria are couched more in terms of the quality of the bond than in critical crack sizes; such "quality" criteria are reminiscent of earlier practices in welding in macroscale structures. Finer-scale inspection and quality control techniques, down to the use of atomic force microscopy, are being developed for application to the understanding and detection of delamination. However, 100% inspection of every interface will never be applied to structures as complex as modern integrated circuits; the structures are simply too complex, too numerous, and too cheap, to allow such an effort.

This paper reviews techniques for measurement of basic mechanical properties of thin films, including adaptations of fracture mechanics for adhesion strength. For films with thicknesses on the order of micrometers, these measurements are well developed. The materials are generally well understood, and their behavior can be interpreted using concepts such as grain size and dislocation-mediated plastic strain, which are familiar from macroscale materials. Progress in extending these methods to films with nanoscale thicknesses will be noted. The paper will also describe a recent development, the use of electrical current as a controlled means of applying thermomechanical stresses to electrical conductors.

2. Mechanical properties measurements at the micrometer scale

This section draws heavily on the book chapter "*Thin Films for Microelectronics and Photonics: Physics, Mechanics, Characterization, and Reliability,*" by D. T. Read and A. A. Volinsky.[1] The main methods in current use for mechanical characterization of thin films include microtensile testing and instrumented indentation, also referred to as nanoindentation (NI). Other methods in wide use include wafer curvature, the pressurized bulge test, and a variety of tests of the adhesion of a film to its substrate.

2.1. MICROTENSILE TESTING

Tensile testing is the standard means of obtaining basic mechanical properties of structural metals. Because the stress field is uniform throughout the gage section, the Young's modulus, yield strength, and ultimate tensile strength can be obtained from an accurate force-displacement record. So it was natural to apply this time-tested method to thin films. Early attempts to pull thin films in conventional testing machines used specimens lifted from their substrate. This operation depended on special separation layers beneath the specimen film, such as water soluble sodium chloride. Excessive wrinkling often occurred during placement of the specimen on the grips. Despite the obstacles, meaningful data were gradually obtained. Early tests of metal films revealed the main phenomena still seen today: high strength, and low elongation to failure.[2] There is at present no standard test method for microtensile testing of thin films; individual investigators adapt the standard methods for bulk metal specimens to fit their specific specimen geometry. Standardization is hindered by the multitude of specimen sizes and designs that are in use, which has resulted from the difficulty of fabricating microtensile specimens.

The problems with the early methods led to improved procedures. It became evident that since films in actual devices are always produced on substrates, the use of the substrate to support the thin film specimen is appropriate. But the substrate is always much more massive than the film, so it must be removed at least from beneath the gage section of the specimen. Ding et al.[3] reported the use of a silicon frame design for testing doped silicon. The first realization of this scheme for metal films was the silicon frame tensile specimen.[4] Bulk micromachining of MEMS devices had been developed by this time, demonstrating the concept of etching away a selected portion of the substrate to form a useful device. To produce the silicon frame tensile specimen, photolithographic patterning was used to form a straight and relatively narrow gage section with larger grip sections on a silicon frame. The substrate beneath the gage section was removed by a suitable etchant. The silicon frame, carrying its tensile specimen of a thin film, was mounted on a purpose-built test device capable of supplying force and displacement.[4] The silicon frame was cut, while leaving the specimen undamaged. This step has been accomplished manually with a dental drill, using a temporary clamp to hold the specimen in place, and by the use of a cutting wheel mounted on a moveable stage.[5]

All the tensile testing techniques include measurements of force and displacement. The force is measured using a load cell, either commercial or custom-built. For specimens of thin films with cross sectional areas of the order of 200 μm^2, the force might amount to 0.1 N. Commercial load cells with this range are available. Displacement has been measured by interferometric techniques such as electron speckle pattern interferometry (ESPI), for example as in,[6] or by diffraction from markers placed on the specimen surface.[7] Even with measurement of displacement directly on the specimen gauge section, modulus measurements are difficult. Successful attempts to use grip or crosshead displacement for accurate strain measurements are unknown to this author.

The specimen fabrication challenge with these techniques was the chemical selectivity required to etch through hundreds of micrometers of silicon without damaging the metal specimen. Aqueous hydrazine has been used, but this material is hazardous. Another disadvantage is the large width of the gage section, 100 µm or more, by comparison with the line widths used in interconnect and also with typical film thicknesses of the order of 1 µm.

A new generation of smaller-scale specimens, and complementary test techniques, has been developed. In this version, the specimen width is around 10 µm and the gage length is around 200 µm, while the thickness remains near 1 µm, Figure 1.[8] The surface micromachining concept is used; the substrate is removed to a depth of around 100 µm beneath the specimen by use of xenon difluoride. This etchant is less hazardous than hydrazine, and is very selective for silicon masked by SiO_2, aluminum, copper, etc. Young's modulus can usually be measured in these specimens, but Poisson's ratio has been measured only by special techniques on relatively large specimens,[5,9] because the transverse displacements are so small on a few-micrometer wide specimens. In an early version of this test, the specimen was loaded by engaging a tungsten probe tip, 50 µm in diameter, to a hole in the loading tab, Figure 2. A recent variant of the surface micromachining approach is the membrane deflection tensile test, applied to a series of face-centered-cubic (FCC) metals by Espinosa et al.,[10] Figure 3.

A new advance is the co-fabrication of a specimen and a protective frame that includes a force sensor, Figure 4.[11] This specimen is suitable for use inside a transmission electron microscope (TEM).

A recent round robin showed reasonable agreement among several laboratories in the strength of polySi (polycrystalline silicon), although most labs required their own unique specimen geometry. The different geometries were produced on the same MEMS chip.[12] The strength values obtained for polySi were impressively high, of the order of 1/30 of the polycrystalline Young's modulus, which is the usual estimate of the theoretical strength of a solid.

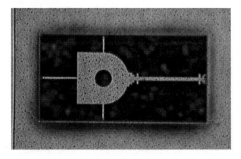

Figure 1. Microtensile specimen of aluminum, fabricated through the MOSIS process. The loading tab, with its 50 µm diameter hole, is to the left. The gauge section, with "ears" for use in digital image correlation for displacement measurement, is to the right. The silicon substrate has been etched away to a depth of 60 µm or more. The three slender aluminum lines connecting the field to the loading tab are tethers that are manually cut just before testing.

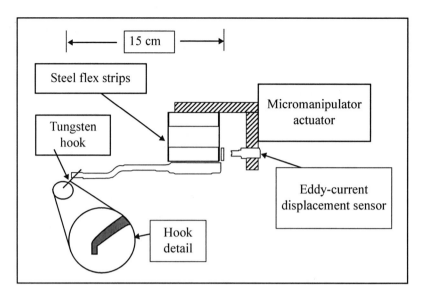

Figure 2. Tungsten "hook" carried by instrumented micromanipulator to load tensile specimen and measure the force.

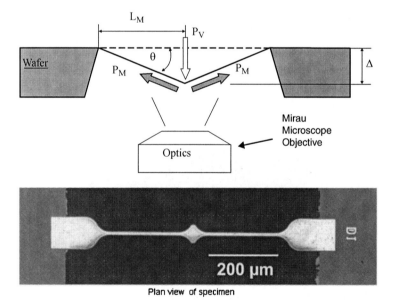

Figure 3. Setup for the membrane deflection tensile test.[10] (Figure courtesy of H. Espinosa).

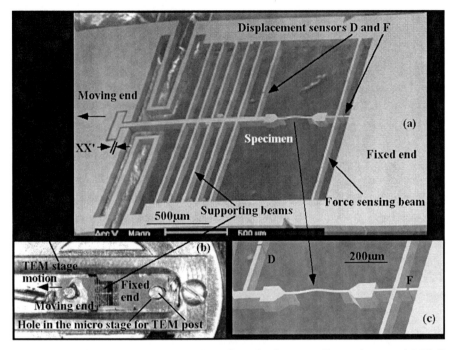

Figure 4. Tensile specimen assembly including aluminum tensile specimen and MEMS support assembly and force gage for use in the TEM.[11] (Figure courtesy of T. Saif.)

2.2. INSTRUMENTED INDENTATION

The nanoindentation test is similar to the conventional hardness test, but is performed on a much smaller scale using specialized equipment – a nanoindenter.[13] The force required to press a sharp diamond indenter into tested material is continuously recorded as a function of the indentation depth, as indicated schematically in Figure 5. The actuation mechanism can be based either on electromagnetic or electrostatic application of force. Since the depth resolution is on the order of angstroms, it is possible to usefully indent even very thin (~100 nm) films. The nanoindentation load–displacement curve, similar to one shown in Figure 6, provides a "mechanical fingerprint" of the material's response to contact deformation. Elastic modulus and hardness are the two parameters that can be readily extracted from the nanoindentation load-displacement curve. Elastic property measurements by nanoindentation were originally proposed by Loubet et al.[14] Later, Doerner and Nix[15] suggested that a linear fit to the upper 1/3 of the unloading portion of the indentation curve could be used to determine film stiffness, $S = dP/dh$, from which the reduced elastic modulus, E_r, could be found as

$$E_r = S\frac{\sqrt{\pi}}{2\sqrt{A}} \quad (1)$$

Here A is the contact area and the reduced modulus is a combined elastic property of the film and indenter material. Since the indenter material itself has finite elastic constants, its deformation contributes to the measured displacement. The reduced modulus E_r is

$$\frac{1}{E_r} = \frac{1-v_i^2}{E_i} + \frac{1-v_f^2}{E_f} \quad (2)$$

In this equation E is the elastic modulus, v is the Poisson's ratio, and the subscripts f and i refer to the film and the indenter materials respectively. A more elaborate power law fit to the unloading portion of the load-displacement curve was suggested by Oliver and Pharr,[16] and is widely known as the Oliver and Pharr method.

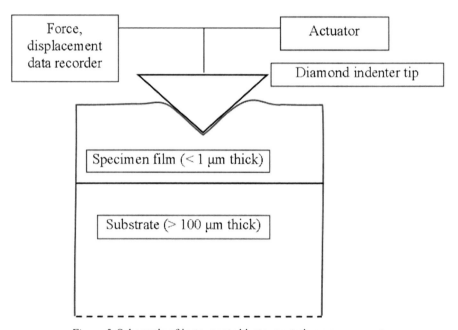

Figure 5. Schematic of instrumented instrumentation measurement.

Hardness H, a material's resistance to plastic deformation, is defined as

$$H = \frac{P_{max}}{A} \quad (3)$$

where A is the projected area of contact (a function of the indentation depth) at the maximum load P_{max}.

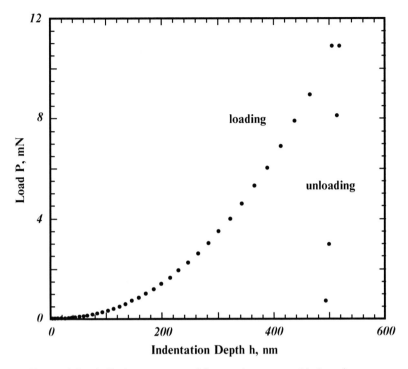

Figure 6. Load–displacement record from an instrumented indentation test.

The expressions for both elastic modulus and hardness contain the contact area, which is correlated to the indentation depth both theoretically, through the known geometry of the indenter, and experimentally, by indenting a material with known elastic modulus. This tip calibration procedure consists of indenting a standard material (often fused quartz or single crystal Al) to various maximum indentation depths. Since the contact area is determined from tip calibration, various tip geometries can be used, with the most common being the Berkovich three-sided pyramid geometry. From the manufacturing standpoint, a three-sided pyramid always ends as a point, and the tip radius can be as sharp as 10–50 nm. Other geometries are also used, and include Vickers (a standardized square pyramid), cube corner, conical and wedge indenters. The unloading slope, dP/dh is related to the tip geometry as

$$\frac{dP}{dh} = 2\beta \sqrt{\frac{A}{\pi}} E_r \qquad (4)$$

where h is the indentation depth, and β is a constant, near unity, for a given tip geometry. King et al. calculated β values for different tip geometries using finite element analysis.[17] One should note that the tip calibration does not account for either plastic pile-up or sink-in of both the standard and the specimen materials, which causes inaccuracies in indentation depth and contact area determination. In addition, the total test compliance, i.e., the inverse of stiffness, is affected by the indentation contact. One should also account for the test frame compliance, C_f, as it offsets the total test compliance:

$$C_{total} = C_f + \frac{\sqrt{\pi}}{2\sqrt{AE}} \qquad (5)$$

In order to avoid substrate effects on the measured mechanical properties, a film should be indented only up to a certain percentage of its thickness (up to 10–20%). There is also an influence of the residual stress and substrate effects that are hard to account for in the analysis.[18,19] Indentation curve analysis has been extended in the past few years with new finite-element-based models being developed.[20,21]

A comprehensive review of the method applied for magnetic storage and MEMS materials was reported by Li and Bhushan.[22]

2.3. OTHER TECHNIQUES

2.3.1. Wafer curvature

The basic principle of the wafer curvature technique is that differential thermal expansion between a specimen film and a silicon substrate produce measurable curvature of the substrate (the wafer); the curvature is related directly to the product of stress and thickness in the film through the Stoney equation.[23,24] This phenomenon is used in evaluating and adjusting film deposition procedures, to measure residual stress in the deposited films. High values of residual stress, especially tension, may make a film less resistant to delamination from the substrate.

Wafer curvature measurement was adapted for characterization of mechanical behavior by Nix.[25] The substrate with its film is placed in a furnace equipped for measurement of the substrate curvature. The temperature is cycled, while the curvature is recorded. Given the film thickness, the film stress can be plotted against temperature. The accessible range of temperature is limited only by the eventual breakdown of the specimen film by melting or chemical reaction. The stress depends in turn on the difference in thermal expansion between the specimen film and the substrate, and the elastic constants of the specimen film. Deviations from linear behavior with temperature imply plastic deformation of the specimen film; the nature of this deformation is confirmed by the hysteresis loop observed at least on the first temperature cycle. The advantages of the wafer curvature technique include the simplicity (in principle) of both the experimental technique and the specimen, which is a film on the same substrate used in actual manufactured products, without the necessity of selectively removing the substrate beneath the film. Analysis of the results using Eq. (1) does not require knowledge of the elastic properties of the deposited film, only those of the substrate. The disadvantage is that the ultimate tensile strength and elongation to failure cannot be measured, and that only certain combinations of Young's modulus, flow stress, and temperature are accessible. This technique has been very successful in providing insight and data on deformation mechanisms, particularly in aluminum films.[25]

2.3.2. *Pressurized bulge testing*

The name of the bulge test is descriptive: by etching away the substrate beneath a region of the specimen film, the film can be exposed to stress by a pressurized fluid introduced beneath the substrate. The mechanics of a pressurized membrane can be used to analyze the observed behavior. The shape of the pressurized region is chosen based on the purpose of the test; circular, square, and rectangular shapes have been explored. The out-of-plane deformation of the membrane can be measured by interferometry or related optical techniques. This technique has been used to explore the elasticity of thin films; care must be taken to properly characterize the initial state of the film, including the possibility of residual stress,[26,27] Figure 7. It has also been used to measure the adhesion between the film and the substrate.[28]

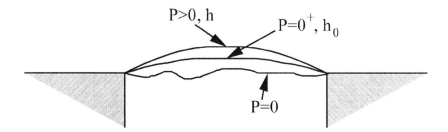

Figure 7. Schematic diagram of the bulge test specimen, showing stages in the loading: slack (zero pressure), infinitesimal, and finite pressure.

2.3.3. *Deformed and Resonant Cantilever*

Micromachined cantilevers have been used as specimens in thin film properties measurements.[29,30] Photolithography can be used to define the cantilever geometry. Cantilevers can be deformed by loading with, for example, an instrumented indenter, or can be excited to resonance, to measure film elastic properties. The relationship between the mechanical stiffness or the resonant frequency and the elastic constant of the film depends sensitively on the dimensions of the cantilever.[31] The ideas of the bulge test and resonance can be combined in the resonant membrane test, which can be used to determine the product of film elastic modulus and mass per unit area. If the thickness and mass density of the film are known, the elastic modulus can be measured.

3. Adhesion tests based on fracture mechanics

Adhesion between layers of different materials is a critical issue in microelectronic packages, and also within the chips themselves. While the time-honored "scotch tape" adhesion test is still in use, quantitative tests, developed in recent years based on the concepts of fracture mechanics, provide material characteristics that can be compared to calculable stress- and strain-based driving forces, and are therefore suitable for use in lifetime predictions.[32] Reviews by Volinsky et al.[33] and by Lane[34] provide useful summaries. The basic idea, as in macroscale fracture mechanics, is that it is useful to quantify the conditions under which an existing crack may advance. The crack, in this case, is assumed to be a small delamination of the film from the substrate. The driving force for crack propagation is taken as the strain energy release rate, which depends on the geometry and the stress state.

3.1. FRACTURE MECHANICS FOR DELAMINATION

Both tensile and compressive stresses in thin films promote adhesion failures; a thin film in compression buckles, delaminates and spalls from the substrate when its strain energy release rate exceeds a critical value that is characteristic of the adhesion between film and substrate.[35] A general, simplified form of the strain energy release rate, G, in a stressed film, regardless of the algebraic sign of the stress is

$$G = Z \frac{\sigma_f^2 h}{E_f}, \qquad (6)$$

where σ_f is the stress in the film, h is the film thickness, E_f is the modulus of elasticity, and Z is a dimensionless cracking parameter. More accurately, the energy release rate averaged over the front of advancing isolated crack is

$$G = g(\alpha,\beta)\frac{\pi(1-v^2)\sigma_f^2 h}{2E_f} \qquad (7)$$

where $g(\alpha,\beta)$ is a function of the Dundurs parameters α and β, and can be found in.[36,37] This strain energy release rate is the driving force for fracture. Film fracture or delamination is observed when the strain energy release rate exceeds the toughness of the film, G_f, or the interfacial toughness, Γ_I respectively ($G > G_f$, or $G > \Gamma_I$). One can avoid these types of failures by either reducing the film thickness, or the stress, or by increasing adhesion. Practically, the film thickness is easier to control. For a given stress level, there is a certain critical film thickness at which failures are observed. As an example, Figure 8 shows through-thickness cracks in a low-k dielectric film 2 μm thick. Thinner films showed no signs of failure. If a film has fractured, and if its residual stress and thickness are known, Eqs. (6) and (7) can be used as upper bound estimates for adhesion.

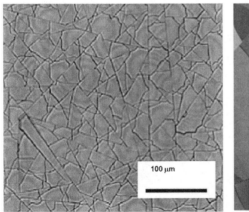
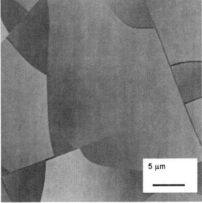

Figure 8. Optical and AFM images of cracks in low-k dielectric thin film.

In the case of compressed films, telephone cord delamination is commonly observed (Figure 9). The geometry of the buckles can be used to asses thin film adhesion. The following analysis is based on Hutchinson's and Suo's developments for buckling-driven delamination of thin films.[35] Upon buckling, the stress in the film, σ_B, is estimated as

$$\sigma_B = \frac{\pi^2}{12} \frac{E}{(1-v^2)} \left(\frac{h}{b}\right)^2 \tag{8}$$

where h is the film thickness, b is the blister half-width, and E and v are Young's modulus and Poisson's ratio, respectively. The buckling stress acts in the vertical direction. The compressive residual stress, σ_r, responsible for producing buckling delamination is

$$\sigma_r = \frac{3}{4}\sigma_B \left(\frac{\delta^2}{h^2}+1\right) \tag{9}$$

where δ is the blister height. The film steady state interfacial toughness in the direction of blister propagation (Figure 10a) can be estimated as

$$\Gamma_{SS} = \frac{(1-v^2)h\sigma_r^2}{2E}\left(1-\frac{\sigma_B}{\sigma_r}\right)^2 \tag{10}$$

Mode-dependent interfacial toughness in the buckling direction, perpendicular to blister propagation is:

$$\Gamma(\Psi) = \frac{(1-v^2)h}{2E}(\sigma_r - \sigma_B)(\sigma_r + 3\sigma_B) \tag{11}$$

Figure 9. Telephone cord delamination in a 1 μm tungsten film.

3.2. SUPERLAYER TEST WITH INDENTATION

The superlayer indentation test provides information on local film adhesion at the microscale. A superlayer film, selected for high stress, high strength, and high adhesion, is deposited on top of the film to be tested. Indentation is used to initiate delamination. The highly stressed hard superlayer provides additional driving force for interfacial crack propagation, and prevents plastic deformation of the tested film around the indenter. As the indenter tip is pressed against the superlayer film stack, it supplies additional energy necessary for crack initiation and propagation. The blister radius is measured optically (Figure 10a). The indentation volume is obtained from the plastic depth of the load–displacement curve (Figure 10b) and the tip geometry. Both the blister radius and the indentation volume are then used to calculate the strain energy release rate (measure of the practical work of adhesion). Calculations for adhesion measurements were made by following the solution developed by Marshall and Evans[38] that was further expanded by Kriese and Gerberich for multilayer films.[39,40] Figure 11a shows a typical delamination blister seen from making indents with a conical tip at 300 mN maximum load and a corresponding load–displacement curve. From Figure 11b, the plastic indentation depth is obtained by using the power law fit of the top 65% of the unloading curve,[16] and used to calculate the indentation volume, based on the tip geometry. It is assumed that the volume is conserved, and plastic deformation around the indenter results in the elastic displacement at the crack tip, allowing calculation of the indentation stress, and ultimately the strain energy release rate, a measure of the practical work of adhesion. Adhesion results for several microelectronics-relevant film materials are summarized in.[41]

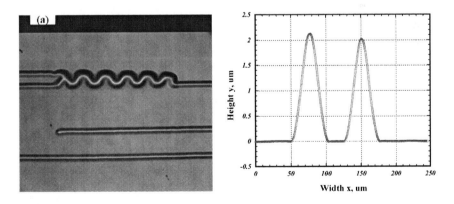

Figure 10. Analysis of the telephone-cord delamination of a tungsten film shown in the previous figure. (a) Telephone cord delamination in a 1 μm tungsten film on top of a 2 nm diamond-like carbon (DLC) film on Si. (b) Corresponding blister heights profile.

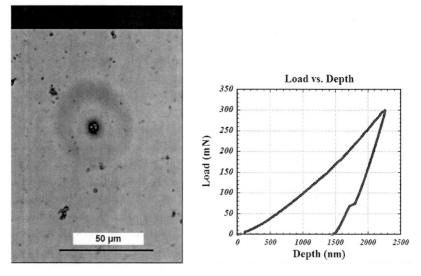

Figure 11. (a) Indentation-induced delamination blister in tungsten film; and (b) corresponding load–displacement curve.

3.3. FOUR-POINT-BEND TEST FOR THIN FILM ADHESION

Because the interfacial energies found in films are numerically much lower than those in bulk metals, for which fracture toughness testing was developed, the four-point-bend bar with a cracks propagating along its length from a central notch has been found useful.[32,42-44] Below, we briefly describe this technique, to show a specific application of fracture mechanics in thin film adhesion. The many reports of adhesion measurement methods in the literature testify to the importance of the problem, the difficulty of the measurement, and the ingenuity of the researchers, but a detailed review is beyond the scope of this article.

The delaminating beam test specimen, Figure 12, is a four-point-bend bar with an interface of interest built into the interior of the beam along the whole length. A "sandwich" beam made with the substrate on the top and bottom, and the surface layers bonded together in the center, is a typical geometry. The substrate layers are much thicker than the interface layer, and give the assembly sufficient stiffness to handle. In the bending beam, the outer fiber in tension is often located on the upper side, and is conventionally referred to as the top of the specimen. The bottom fiber is in compression. The top section is carefully cut without notching the bottom section. Cracks are intentionally nucleated to grow away from the notch along the interface layer being tested. While the crack length significantly exceeds the thickness of the cut layer, the energy release rate is constant until the cracks reach the inner loading points of the four-point-bend specimen.

The energy release rate is evaluated from the load and displacement, specimen geometry, and elastic properties of the support layers of the specimen. An advantage of this test is that the parameters needed to evaluate the adhesion do not include the residual stress on the film, which may be difficult to measure. Becker[44] points out that properties measured with this specimen may depend on the specific geometry, contrary to the case for standardized fracture toughness specimens. This is not considered to be a serious disadvantage for testing materials for chips and electronic packages, because actual-size specimens can be tested.

All the fracture toughness techniques highlight a critical problem in the design of electronic packages and chips: some commonly used interfaces, such as polymer-metal interfaces, have very low fracture toughness,[32] around 10 J/m^2.

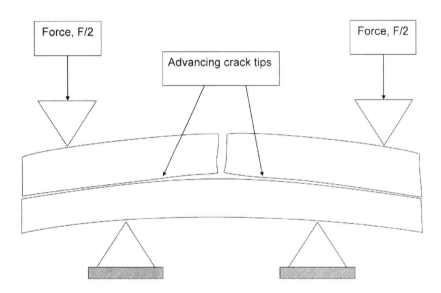

Figure 12. Delaminating beam specimen for measuring the energy required to separate an adhesive interface.

4. A new development: electrical testing for mechanical reliability

It has been proposed that the cyclic stresses produced by the combination of joule heating by AC (alternating current) and differential thermal expansion in conductive films on silicon substrates may be useful in evaluating the mechanical reliability of thin films. Interest has arisen in the use of this electrical test to extract information about the mechanical behavior of interconnect structures because of the problems associated with more conventional approaches to mechanical characterization of very small thin-film structures. Microtensile testing requires special test structures, which must become much more sophisticated as the linewidth of interest falls below 1 µm. Nanoindentation with conventional indenter shapes requires an area of at least a few square micrometers. The use of atomic force microscopy to extract mechanical properties is in its infancy. And all of these require that the film to be tested be exposed; none are applicable to buried lines. On the other hand, interconnect lines within the damascene structure are commonly tested electrically during development of advanced interconnect designs by industry. So a further development of electrical testing, to a point where it could produce mechanical information about narrow, buried lines or about other small, inaccessible structures, would be a significant advance.

The AC fatigue test technique[45] uses cyclic Joule heating to apply thermal cycles to metal lines and vias in damascene dielectric structures on silicon substrates. Cyclic stresses from differential thermal expansion produce elastic and possibly plastic deformation in the metal line and its surrounding dielectric. The use of high-amplitude, low-frequency alternating current in tests of thin-film copper lines was explored by Mönig et al.[45]; they reported surface topography changes that appeared to be mechanical in origin. Tests of aluminum lines under by AC fatigue produced topographic damage in the form of regular undulations or wrinkles.[46] Extensive TEM and SEM examination of these aluminum lines revealed that the AC stressing produced dislocations, grain growth, and grain rotation in various regions of the specimen.[47,48]

Barbosa et al. plotted the behavior of aluminum lines under AC stress as S-N curves, familiar from metal fatigue.[49] They showed that their data could be fit by the Basquin law for fatigue in the appropriate range of cyclic temperature, and that the values of the exponent in the fit were within the same range as those for mechanical fatigue of bulk metals. The stress prefactor in the Basquin law is an estimator of the ultimate tensile strength in metals; they proposed this same relationship for the AC fatigue test. They were able to deduce a value of this stress prefactor that agreed with the ultimate tensile strength for their thin film, as measured by the microtensile test.[49] AC fatigue data for copper and aluminum films on substrates appear consistent with conventional mechanical fatigue tests where the temperature reached in the AC fatigue test is not too high (R.R. Keller, 2008, Personal communication). Figure 13 shows mechanical fatigue data from the literature plotted for comparison with AC fatigue data. The stresses for the AC data points were calculated using the simple biaxial stress formula. For both the copper and the aluminum data in this figure, the AC data are offset vertically from the bulk data, indicating different values of the fatigue stress prefactors in the Basquin fits to the data sets. We ascribe these differences to the difference in grain sizes between the bulk and the thin film materials; other effects, such as crystallographic texture and effects of added mechanical constraint by the substrate, may also play a role. The grain sizes for the bulk materials are noted on the plot; the thin films have average grain diameters of less than 1 μm.

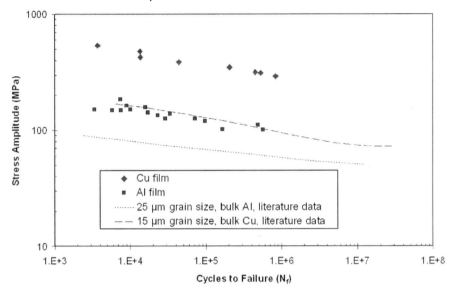

Figure 13. Fatigue stress vs lifetime data plotted as S-N curves. The plot includes literature data from mechanical fatigue tests and recent data obtained with the AC fatigue test, and shows similar behavior for all cases. The vertical offsets are a result of differences in the grain sizes and possibly other differences between bulk and film materials as noted in the text.

5. Conclusion

The state of the art of measurements of mechanical properties for thin films has advanced significantly in the past 20 years. Measurements of microscale films, with thicknesses of 0.5 μm and above, and in-plane dimensions of tens of micrometers, are made routinely by multiple techniques. The tensile properties of these micrometer-scale films can be understood by use of the Hall–Petch relation. Fracture of the films themselves has not proved to be a problem of general relevance, but fracture mechanics has been found to be the appropriate framework for quantitative treatment of layer-to-layer and film-to-substrate adhesion. Some of the tests now in use for interfacial adhesion have been described. Nanometer-scale materials now represent the latest new challenge in understanding and measuring the mechanical behavior of materials.

References

1. Read, D. T.; Volinsky, A. A. Thin Films for Microelectronics and Photonics: Physics, Mechanics, Characterization, and Reliability, in *Materials and Structures: Physics, Mechanics, Design, Reliability, Packaging: Volume 1 Materials Physics/Materials Mechanics.*, Suhir, E.; Lee, Y. C.; Wong, C. P., editors; Springer: New York, 2007; Chapter 4, pp. 135–180.
2. Brotzen, F. R. Mechanical Testing of Thin-Films, International Materials Reviews 39 (1), 24–45, 1994.
3. Ding, X. Y.; Ko, W. H.; Mansour, J. M. Residual-Stress and Mechanical-Properties of Boron-Doped P+-Silicon Films, Sensors and Actuators A-Physical 23 (1–3), 866–871, 1990.
4. Read, D. T.; Dally, J. W. A New Method for Measuring the Strength and Ductility of Thin-Films, Journal of Materials Research 8 (7), 1542–1549, 1993.
5. Sharpe, W. N.; Jackson, K. M.; Coles, G.; Eby, M. A.; Edwards, R. L. Tensile Tests of Various Thin Films, in *ASTM STP 1413: Mechanical Properties of Structural Films;* edited by Muhlstein, C.; Brown, S. B., editors; American Society for Testing and Materials: West Conshohoken, PA, 2001; pp. 229–247.
6. Read, D. T. Young's Modulus of Thin Films by Speckle Interferometry, Measurement Science and Technology 9 (4), 676–685, 1998.
7. Sharpe, W. N.; Yuan, B.; Edwards, R. L. A New Technique for Measuring the Mechanical Properties of Thin Films, Journal of Microelectromechanical Systems 6 (3), 193–199, 1997.
8. Read, D. T.; Cheng, Y. W.; Keller, R. R.; McColskey, J. D. Tensile Properties of Free-Standing Aluminum Thin Films, Scripta Materialia 45 (5), 583–589, 2001.
9. Ruud, J. A.; Josell, D.; Spaepen, F.; Greer, A. L. A New Method for Tensile Testing of Thin Films, Journal of Materials Research 8 (1), 112–117, 1993.
10. Espinosa, H. D.; Prorok, B. C.; Peng, B. Plasticity Size Effects in Free-Standing Submicron Polycrystalline FCC Films Subjected to Pure Tension, Journal of the Mechanics and Physics of Solids 52 (3), 667–689, 2004.
11. Haque, M. A.; Saif, M. T. A. Deformation Mechanisms in Free-Standing Nanoscale Thin Films: A Quantitative in situ Transmission Electron Microscope Study, Proceedings of the National Academy of Sciences of the United States of America 101 (17), 6335–6340, 2004.
12. LaVan, D. A.; Tsuchiya, T.; Coles, G.; Knauss, W. G.; Chasiotis, I.; Read, D. T. Cross Comparison of Direct Strength Testing Techniques on Polysilicon Films, in *ASTM STP 1413: Mechanical Properties of Structural Films*; edited by Muhlstein, C.; Brown, S. B., editors; American Society for Testing and Materials: West Conshohoken, PA, 2001; pp. 16–27.
13. VanLandingham, M. R. Review of Instrumented Indentation, Journal of Research of the National Institute of Standards and Technology 108 (4), 249–265, 2003.
14. Loubet, J. L.; Georges, J. M.; Marchesini, O.; Meille, G. Vickers Indentation Curves of Magnesium-Oxide (MgO), Journal of Tribology-Transactions of the ASME 106 (1), 43–48, 1984.
15. Doerner, M. F.; Nix, W. D. A Method for Interpreting the Data from Depth-Sensing Indentation Measurements, Journal of Materials Research 1 (4), 601–616, 1986.
16. Oliver, W. C.; Pharr, G. M. An Improved Technique for Determining Hardness and Elastic-Modulus Using Load and Displacement Sensing Indentation Experiments, Journal of Materials Research 7 (6), 1564–1583, 1992.

17. King, R. B.; Osullivan, T. C. Sliding Contact Stresses in A Two-Dimensional Layered Elastic Half-Space, International Journal of Solids and Structures 23 (5), 581–597, 1987.
18. Tsui, T. Y.; Oliver, W. C.; Pharr, G. M. Influences of Stress on the Measurement of Mechanical Properties Using Nanoindentation .1. Experimental Studies in an Aluminum Alloy, Journal of Materials Research 11 (3), 752–759, 1996.
19. Bolshakov, A.; Oliver, W. C.; Pharr, G. M. Influences of Stress on the Measurement of Mechanical Properties Using Nanoindentation .2. Finite Element Simulations, Journal of Materials Research 11 (3), 760–768, 1996.
20. Hainsworth, S. V.; Chandler, H. W.; Page, T. F. Analysis of Nanoindentation Load-Displacement Loading Curves, Journal of Materials Research 11 (8), 1987–1995, 1996.
21. Berriche, R. Vickers Hardness from Plastic Energy, Scripta Metallurgica et Materialia 32 (4), 617–620, 1995.
22. Li, X. D.; Bhushan, B. A Review of Nanoindentation Continuous Stiffness Measurement Technique and Its Applications, Materials Characterization 48 (1), 11–36, 2002.
23. Freund, L. B.; Suresh, S. *Thin Film Materials: Stress, Defect Formation and Surface Evolution*; Cambridge University Press: Cambridge, UK, 2003.
24. Ohring, M. *Materials Science of Thin Films, Deposition and Structure*; Academic Press: San Diego, CA, 2002.
25. Nix, W. D. Mechanical-Properties of Thin-Films, Metallurgical Transactions A-Physical Metallurgy and Materials Science 20 (11), 2217–2245, 1989.
26. Jankowski, A. F.; Tsakalakos, T. Effects of Deflection on Bulge Test Measurements of Enhanced Modulus in Multilayered Films, Thin Solid Films 291 243–247, 1996.
27. Small, M. K.; Nix, W. D. Analysis of the Accuracy of the Bulge Test in Determining the Mechanical-Properties of Thin-Films, Journal of Materials Research 7 (6), 1553–1563, 1992.
28. Liechti, K. M.; Shirani, A. Large-Scale Yielding in Blister Specimens, International Journal of Fracture 67 (1), 21–36, 1994.
29. Petersen, K. E.; Guarnieri, C. R. Youngs Modulus Measurements of Thin-Films Using Micromechanics, Journal of Applied Physics 50 (11), 6761–6766, 1979.
30. Osterberg, P. M.; Senturia, S. D. M-TEST: A Test Chip for MEMS Material Property Measurement Using Electrostatically Actuated Test Structures, Journal of Microelectromechanical Systems 6 (2), 107–118, 1997.
31. Weihs, T. P.; Hong, S.; Bravman, J. C.; Nix, W. D. Mechanical Deflection of Cantilever Microbeams – A New Technique for Testing the Mechanical-Properties of Thin-Films, Journal of Materials Research 3 (5), 931–942, 1988.
32. Dauskardt, R.; Lane, M.; Ma, Q.; Krishna, N. Adhesion and Debonding of Multi-Layer Thin Film Structures, Engineering Fracture Mechanics 61 (1), 141–162, 1998.
33. Volinsky, A. A.; Moody, N. R.; Gerberich, W. W. Interfacial Toughness Measurements for Thin Films on Substrates, Acta Materialia 50 (3), 441–466, 2002.
34. Lane, M. Interface Fracture, Annual Review of Materials Research 33 29–54, 2003.
35. Thouless, M. D. Cracking and Delamination of Coatings, Journal of Vacuum Science and Technology A-Vacuum Surfaces and Films 9 (4), 2510–2515, 1991.
36. Hutchinson, J. W.; Suo, Z. Mixed-Mode Cracking in Layered Materials, Advances in Applied Mechanics, 29, 63–191, 1992.
37. Beuth, J. L. Cracking of Thin Bonded Films in Residual Tension, International Journal of Solids and Structures 29 (13), 1657–1675, 1992.
38. Marshall, D. B.; Evans, A. G. Measurement of Adherence of Residually Stressed Thin-Films by Indentation .1. Mechanics of Interface Delamination, Journal of Applied Physics 56 (10), 2632–2638, 1984.

39. Kriese, M. D.; Gerberich, W. W.; Moody, N. R. Quantitative Adhesion Measures of Multilayer Films: Part I. Indentation Mechanics, Journal of Materials Research 14 (7), 3007–3018, 1999.
40. Kriese, M. D.; Gerberich, W. W.; Moody, N. R. Quantitative Adhesion Measures of Multilayer Films: Part II. Indentation of W/Cu, W/W, Cr/W, Journal of Materials Research 14 (7), 3019–3026, 1999.
41. Volinsky, A. A.; Moody, N. R.; Gerberich, W. W. Interfacial Toughness Measurements for Thin Films on Substrates, Acta Materialia 50 (3), 441–466, 2002.
42. Charalambides, M. Fracture Mechanics Specimen for Interface Toughness Measurement, Journal of Applied Mechanics 56 (0), 77–82, 1989.
43. Hofinger, I.; Oechsner, M.; Bahr, H. A.; Swain, M. V. Modified Four-Point Bending Specimen for Determining the Interface Fracture Energy for Thin, Brittle Layers, International Journal of Fracture 92 (3), 213–220, 1998.
44. Becker, T. L.; McNaney, J. M.; Cannon, R. M.; Ritchie, R. O. Limitations on the Use of the Mixed-Mode Delaminating Beam Test Specimen: Effects of the Size of the Region of K-Dominance, Mechanics of Materials 25 (4), 291–308, 1997.
45. Mönig, R.; Keller, R. R.; Volkert, C. A. Thermal Fatigue Testing of Thin Metal Films, Review of Scientific Instruments 75 (11), 4997–5004, 2004.
46. Keller, R. R.; Geiss, R. H.; Cheng, Y.-W.; Read, D. T. IMECE2004-61291: Microstructure Evolution During Alternating-Current-Induced Fatigue, in Proceedings of the International Mechanical Engineering Conference and Exposition 2004; American Society of Mechanical Engineers: 2004; pp. 107–112.
47. Geiss, R. H.; Read, D. T.; Keller, R. R. TEM Study of Dislocation Loops in Deformed Aluminium Films, Microscopy and Microanalysis 11 (S02), 1870–1871, 2005.
48. Keller, R. R.; Geiss, R. H.; Barbosa, N.; Slifka, A. J.; Read, D. T. Strain-Induced Grain Growth During Rapid Thermal Cycling of Aluminum Interconnects, Metallurgical and Materials Transactions A-Physical Metallurgy and Materials Science 38A (13), 2263–2272, 2007.
49. Barbosa, N.; Keller, R. R.; Read, D. T.; Geiss, R. H.; Vinci, R. P. Comparison of Electrical and Microtensile Evaluations of Mechanical Properties of an Aluminum Film, Metallurgical and Materials Transactions A-Physical Metallurgy and Materials Science 38A (13), 2160–2167, 2007.

FROM MACRO TO MESO AND NANO MATERIAL FAILURE. QUANTIZED COHESIVE MODEL FOR FRACTAL CRACKS

MICHAEL P. WNUK[*]
Department of Civil Engineering and Mechanics, College of Engineering and Applied Science, University of Wisconsin – Milwaukee, 3200 N. Cramer St Milwaukee, WI 53211, USA

Abstract A discretization procedure for the cohesive model of a fractal crack requires that all pertinent entities describing the influence of the cohesive stress that restrains opening of the crack, such as effective stress intensity factor, the modulus of cohesion, extent of the end zone and the opening displacement within the high-strain region adjacent to the crack tip are re-visited and replaced by certain averages over a finite length referred to as either "unit step growth" or "fracture quantum". Thus, two novel aspects of the model enter the theory: (1) degree of fractality related to the roughness of the newly created surface, and (2) discrete nature of the propagating crack. Both variables are shown to increase the equilibrium length of the cohesive zone. At the point of incipient fracture this length becomes the characteristic material length parameter L_c.

Novel properties of the present model provide a better insight and an effective tool to explain multiscale nature of fracture process and the associated transitions from nano- to micro- and macro-levels of material response to deformation and fracture. These multiscale features of any real material appear to be inherent defense mechanisms provided by nature.

As the degree of fractality increases, the characteristic material length is shown to rapidly grow to the levels around three orders of magnitude higher than those predicted for the classic case. Such effect is helpful in explaining an unusual size-sensitivity of fracture testing in materials with cementitious bonding such as concrete and certain types of ceramics, where fractal cracks are commonly observed.

In the limit of vanishing quantum fracture and/or reduced degree of fractality the quantized cohesive model of a fractal crack, as presented here, reduces to the well-known classic models of Dugdale–Barenblatt or to the LEFM or the QFM fracture theories.

---

[*] E-mail: mpw@uwm.edu

Keywords: Fracture, deformation, cohesion, fractals, discrete process, multiscale, nano, meso, macro

1. Introduction

Cohesive models of cracks cf. Figure 1 have been remarkably successful in explaining certain essential features of fracture process such as a finite stress at the crack tip, a non-zero crack opening displacement at the tip of the crack, and a certain equilibrium length of cohesive zone. This length increases with the applied load up to the point of incipient fracture and it serves as a measure of the material resistance to crack propagation.

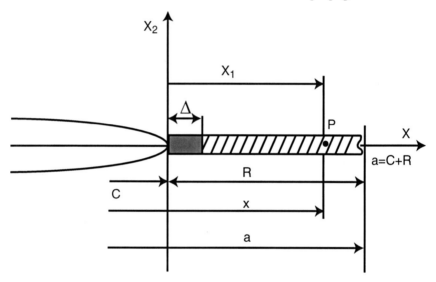

Figure 1. Cohesive crack model created by extending the crack to a new length, which incorporates the length of the physical crack and the length the equilibrium cohesive zone. For quantized fracture mechanics (QFM) the unit growth step or the fracture quantum Δ is embedded within the cohesive zone and located next to the crack tip. This quantity enters all equations pertinent to the QFM representation of fracture.

We shall begin with re-visiting this model and then go on to incorporate fractal and discrete nature of fracture occurring in real materials. Recent research on nano-cracks, cf. Isupov and Mikhailov (1998) and Ippolito et al. (2006), indicate that the classic failure criteria break down for very small cracks. Such examples show the need for novel non-local failure criteria for design of structures, in which multiscale fracture mechanisms are present. In order to refine available mathematical tools and to extend their validity for nano-scale and for fractal (rough) cracks we will merge here the basic concepts of the cohesive crack model with the fractal view of the decohesion process. At the same time we shall employ the non-local criteria, which remain valid at atomistic or nano-scale levels. The present model is used to predict the upper and the lower bounds for the cohesion modulus, which is used as a measure of material resistance to initiation and to propagation of fracture.

The concept of discrete nature of fracture has been investigated by many researchers, beginning with Neuber (1958), Novozhilov (1969), Wnuk (1974), Seweryn (1994) and Pugno and Ruoff (2004). Comparison of various fracture criteria used to describe discrete fracture process suggests that a certain finite length parameter, determined either by the microstructural or atomistic considerations, must be introduced into the basic equations underlying the theory of fracture. Even though various names and symbols have been used in describing this entity, such as "Neuber particle" by Williams (1965), "unit growth step" by Wnuk (1974) and "fracture quantum" by Novozhilov (1969) and Pugno and Ruoff (2006), the physical meaning of these length parameters is the same, and it can be accounted for mathematically by a discretization technique, cf. Wnuk and Yavari (2008) and Pugno and Ruoff (2006).

A discretization procedure for the cohesive model of a fractal crack requires that all pertinent entities describing the influence of the cohesive stress that restrains opening of the crack, such as effective stress intensity factor, the modulus of cohesion, extent of the end zone and the opening displacement within the high-strain region adjacent to the crack tip are re-visited and replaced by certain averages over a finite length referred to as either "unit growth step", cf. Wnuk (1974) or "fracture quantum", cf. Williams (1957, 1965), Novozhilov (1969) and Pugno et al. (2004). Thus, two novel aspects of the model enter the theory: (1) degree of fractality related to the roughness of the newly created surface, and (2) discrete nature of the propagating crack.

The basic concepts needed in the present analysis were discussed by Barenblatt (1959, 1962), Balankin (1997), Borodich (1997), Carpinteri (1994), Carpinteri and Chiaia (1996), Cherepanov et al. (1995), Mosolov (1991), Murray (1984) and Taylor (2008). The fractal crack model described here is based on a simplifying assumption, according to which the original problem is approximated by considerations of a smooth crack embedded in the stress field generated by a fractal crack, cf. Wnuk and Yavari (2003). The well-known concepts of the stress intensity factor and the Barenblatt cohesive modulus, which is a measure of material toughness, have been re-defined to accommodate the fractal view of fracture. Specifically, the cohesion modulus, in addition to its dependence on the distribution of the cohesion forces, is shown now to be a function of the "degree of fractality" reflected by the fractal dimension D, or by the Hausford fractal roughness parameter, H. It is also a function of the unit growth step, the so-called fracture quantum and depends on the form of the "decohesion law". As expected in the limit of the fractal dimension D approaching 1 and for the disappearing magnitude of the fracture quantum, one recovers the cohesive crack model known from the fracture mechanics of smooth cracks.

In what follows we shall solve the problem assuming that we are dealing with a fractal crack represented by a cohesive model and that propagation of fracture is not continuous but discrete. Naturally, when the problem is reduced to that of a smooth crack and when the discrete nature of decohesion act is neglected; the present model yields the results identical to those known in classical fracture mechanics. This is analogous to the "correspondence principle" used in Quantum Mechanics to recover the solutions pertinent to continuum.

2. Preliminaries – Classic cohesive crack model

When a cohesive model is designed, the length of the physical crack c is extended by an adding the cohesive zones at both ends of the crack, as shown in Figure 1. Within these zones the restraining stress S counteracts the separation process. If the length of each cohesive zone is $\tilde{R}$, then the half-length of the extended crack becomes a = c + $\tilde{R}$. In order to solve the pertinent mixed boundary value problem one needs to assume the following distribution of pressure applied along the surface of the extended crack

$$p(x) = \begin{cases} \sigma, 0 \leq |x| \leq c \\ \sigma - S, c \leq |x| \leq a \end{cases} \tag{1}$$

This is later superposed with the uniform tension $p(x) = -\sigma$ thus generating a stress-free crack with two cohesive zones, in which the S-stress is present. The second boundary condition is expressed in terms of the displacement component u_y, which is set equal zero along the symmetry axis outside the crack for $|x| \leq a$. The resulting solution is the familiar stress field, which in the vicinity of the crack tip contains the dominant term controlled by the stress intensity factor. This is the singular term, and it will be subject to annihilation. Such requirement of disappearance of the singular term is known as the "finiteness condition".

Before we can proceed to set up such condition for the crack pictured in Figure 1, we need to evaluate K-factors associated with stresses σ and S. Applying the well-known LEFM expression

$$K_I = 2\sqrt{\frac{a}{\pi}} \int_0^a \frac{p(x)}{\sqrt{a^2 - x^2}} dx \qquad (2)$$

we substitute Eq. (1) for the pressure p(x) to obtain the stress intensity factor describing a cohesive crack

$$K_{TOT} = 2\sqrt{\frac{a}{\pi}} \left\{ \int_0^c \frac{\sigma dx}{\sqrt{a^2 - x^2}} + \int_c^a \frac{(\sigma - S)dx}{\sqrt{a^2 - x^2}} \right\} =$$

$$= 2\sqrt{\frac{a}{\pi}} \int_0^a \frac{\sigma dx}{\sqrt{a^2 - x^2}} - 2\sqrt{\frac{a}{\pi}} \int_c^a \frac{S dx}{\sqrt{a^2 - x^2}} \qquad (3)$$

It is seen that both σ and S contribute to the total stress intensity factor. Using the notation

$$K_\sigma = 2\sqrt{\frac{a}{\pi}} \int_0^a \frac{\sigma dx}{\sqrt{a^2 - x^2}}$$

$$K_S = 2\sqrt{\frac{a}{\pi}} \int_c^a \frac{S dx}{\sqrt{a^2 - x^2}} \qquad (4)$$

we rewrite Eq. (3) as follows

$$K_{TOT} = K_\sigma - K_S \qquad (5)$$

With constant σ and S the integrals are easy to evaluate and the two K-factors become

$$K_\sigma = \sigma \sqrt{\pi(c + \tilde{R})}$$

$$K_S = 2S \sqrt{\frac{c + \tilde{R}}{\pi}} \cos^{-1}\left(\frac{c}{c + \tilde{R}}\right) = S\sqrt{\pi(c + \tilde{R})} \left[\frac{2}{\pi} \cos^{-1}\left(\frac{c}{c + \tilde{R}}\right)\right] \qquad (6)$$

These two stress intensity factors remain in equilibrium $K_\sigma = K_S$ during the process of loading up to the point of incipient fracture. The function K_σ^2 can be thought of as a "driving force", which forces the crack to open up, while K_S^2 represents material resistance. It is noted that K_S reduces to zero for a vanishing length of the cohesive zone. For a physically important case of $\tilde{R}$ being small compared to the crack length c the second equation in Eq. (6) acquires a simple form

$$[K_S]_{\tilde{R}\ll c} = \left[\sqrt{\frac{2}{\pi}}\int_0^{\tilde{R}}\frac{S(\lambda)d\lambda}{\sqrt{\lambda}}\right]_{S=const} = \sqrt{\frac{2}{\pi}}S\left(2\sqrt{\tilde{R}}\right) \quad (7)$$

This relation proves that the length of the cohesive zone $\tilde{R}$ can indeed be used as a measure of the fracture toughness. We will now proceed to show that this characteristic material length parameter assumes substantially different values depending upon the basic assumptions underlying the physical model of fracture. Four such models will be discussed here:

1. Griffith case as represented by the LEFM crack model
2. Dugdale–Barenblatt cohesive crack model for continuous fracture
3. Modification of the DB cohesive model for a discrete fracture
4. Fractal version of the cohesive crack propagating in a discrete manner

Variable λ in Eq. (7) represents the distance measured backwards from the outer edge of the cohesive zone. Restraining stress spatial distribution $S(\lambda)$ will of course influence the final form of the material cohesion characteristic K_S. In this limiting case the cohesion modulus does not depend on the crack size. It is expressed in terms of the magnitude of the cohesive stress and the length $\tilde{R}$ measured at the point of incipient fracture, when $\tilde{R} = \tilde{R}_c$. Equating expression (7) with fracture toughness K_c and inverting the equation produces the ubiquitous result

$$\tilde{R}_c = L_c = \frac{\pi}{8}\frac{K_c^2}{S^2} \quad (8)$$

This formula defines the characteristic material length parameter determined through the cohesive model of fracture. Three essential variables $\tilde{R}$, S and K_c are related via this equation (subject to the restriction $\tilde{R}_c \ll c$). When the restriction of R being much smaller than X is removed, the general expressions (6) must be used, and then the fourth variable enters all pertinent results. This variable is the fracture quantum a_0. Parallel with the characteristic length (8) we shall define an **equilibrium** cohesive zone length $R = \tilde{R}/a_0$. The term "equilibrium" used here means that for any given applied load there exists a single-valued cohesive zone length complying with the condition $K_\sigma = K_S$. At the point of incipient fracture this length becomes the characteristic material constant L_c. For arbitrary $\tilde{R}/c$ ratios the Eq. (6) should be applied along with the finiteness condition $K_\sigma = K_S$. Thus we arrive at

$$\frac{\sigma\pi}{2} - S\cos^{-1}\left(\frac{c}{c+\tilde{R}}\right) = 0 \tag{9}$$

With Q denoting non-dimensional loading parameter, $Q = \pi\sigma/2S$, one can readily solve Eq. (9) for the length of the cohesive zone, namely

$$\tilde{R}_D = c(\sec Q - 1), or$$
$$R_D = X(\sec Q - 1) \tag{10}$$

This is the celebrated Dugdale's formula valid for arbitrary ratios $\tilde{R}/c$ or R/X, and to emphasize this fact the subscript "D" has been added. For small Q the right hand side reduces to $(X/2)Q^2$, and Eq. (10) reduces then to the result analogous to the one given in Eq. (8). For $R(=\tilde{R}/a_0)$ small compared to the crack length $X(=c/a_0)$ the equilibrium length of the Dugdale cohesive zone reads

$$\tilde{R}_D = \frac{\pi}{8}\left(\frac{K_I}{S}\right)^2, or$$
$$R_D = \frac{\pi}{8a_0}\left(\frac{K_I}{S}\right)^2 \tag{11}$$

3. Discretization of the cohesive crack model

In order to account for the discrete nature of fracture process all K-factors discussed above need to be replaced by their averages. Applying the scheme

$$K_\sigma \to \langle K_\sigma \rangle_{c,c+a_0} = \left[\frac{1}{a_0}\int_c^{c+a_0} K_\sigma^2 dc\right]^{1/2}$$

$$K_S \to \langle K_S \rangle_{c,c+a_0} = \left[\frac{1}{a_0}\int_c^{c+a_0} K_S^2 dc\right]^{1/2} \tag{12}$$

we proceed to evaluate the averages defined in Eq. (12). Symbol a_0 denotes the fracture quantum, which is used as a normalization constant for the length-like variables X and R,

$$X = c/a_0$$
$$R = \tilde{R}/a_0 \tag{13}$$

The results are

$$\langle K_\sigma \rangle = \sigma\sqrt{\pi}\left[\frac{1}{a_0}\int_c^{c+a_0}(c+\tilde{R})dc\right]^{1/2} = \sigma\sqrt{\pi a_0}\left(c+\tilde{R}+\frac{a_0}{2}\right)^{1/2} = \sigma\sqrt{\pi a_0}\left(X+R+\frac{1}{2}\right)^{1/2}$$

$$\langle K_S \rangle = S\sqrt{\pi}\left(\frac{2}{\pi}\right)\left\{\frac{1}{a_0}\int_c^{c+a_0}(c+\tilde{R})\left[\cos^{-1}\left(\frac{c}{c+\tilde{R}}\right)\right]^2 dc\right\}^{1/2}$$

$$= S\sqrt{\pi a_0}\left(\frac{2}{\pi}\right)\left\{\int_X^{X+1}(X+R)\left[\cos^{-1}\left(\frac{X}{X+R}\right)\right]^2 dX\right\}^{1/2} = S\sqrt{\pi a_0}\left(\frac{2}{\pi}\right)I(X,R)^{1/2}$$

(14)

The auxiliary integral $I(X,R)$ is defined as follows

$$I(X,R) = \int_X^{X+1}(X+R)\left[\cos^{-1}\left(\frac{X}{X+R}\right)\right]^2 dX \qquad (15)$$

Substituting Eq. (14) into the equilibrium equation $\langle K_\sigma \rangle = \langle K_S \rangle$ yields

$$\frac{\sigma\pi}{2S}\left[X+R+\frac{1}{2}\right]^{1/2} = I(X,R)^{1/2} \qquad (16)$$

One can readily solve this equation for the loading parameter Q as a function of X and R. To distinguish this solution from the classic Dugdale solution (10) we shall use the subscript D for "Dugdale" and another subscript "d" for "discrete". From Eq. (16) it follows

$$Q_{Dd} = \frac{I(X,R)^{1/2}}{\sqrt{X+R+\frac{1}{2}}} \qquad (17)$$

An interesting simplification of Eq. (17) is obtained for the limiting case of small ratios R/X. For this case the integral (15) reduces to 2R, and with R in the denominator of Eq. (17) neglected vs. X, the Eq. (17) can be re-written as follows

$$[Q_{Dd}]_{R\ll X} = \sqrt{\frac{2R}{X+\frac{1}{2}}} \qquad (18)$$

Now the inverse relationship can be readily provided

$$[R_{Dd}]_{R \ll X} = \frac{\left(\frac{1}{2} + X\right)Q^2}{2}, \text{ or} \qquad (19)$$

$$[R_{Dd}]_{R \ll X} = [R_D]_{R \ll X}\left(1 + \frac{1}{2X}\right)$$

From the expression on the extreme right of Eq. (19) it is seen that the discretization of the cohesive crack alters the predicted characteristic length as compared to the Dugdale result R_D. The change, however, is significant only for the crack sizes comparable to the fracture quantum a_0 when $X \approx 1$. Otherwise, for large X, the effect disappears as then $R_{Dd} \geq R_D$. If arbitrary values of the R/X ratio are considered, a similar conclusion follows.

We conclude, therefore, that the length of the equilibrium cohesive zone is enhanced when the discrete nature of fracture is taken into account. Naturally, the differences are more pronounced for the nano scale levels when the initial crack length is comparable to the magnitude of the fracture quantum; examine graphs (a)–(c) in Figure 2.

An interesting limiting case is obtained for the crack length shrinking to zero, $X \geq 0$. From the first equation in Eq. (19) we obtain the relation between the applied load Q and the equilibrium length of the cohesive zone R, namely

$$[Q]_{X \to 0} = 2\sqrt{R} \qquad (20)$$

At the point of incipient fracture we set R equal to $R_c = (\pi/8)(K_c/S)^2$. Substituting this value of R in Eq. (20) and replacing Q by the dimensional load σ, we obtain the strength of a virgin material which contains no cracks, the so-called inherent strength

$$[\sigma_{crit}]_{X=0} = \sigma_0 = \sqrt{\frac{2}{\pi a_0}} K_c \qquad (21)$$

Not surprisingly this value of the critical stress is exactly the same as the inherent strength predicted by quantized fracture mechanics, cf. Pugno and Ruoff (2004). When this quantity is used as a normalization constant, then the stress at the onset of fracture due to a cohesive discrete crack subject to the condition R<<X can be expressed by a remarkably simple formula

$$\frac{\sigma_{crit}^{Dd}}{\sigma_0} = \frac{1}{\sqrt{2X+1}} \qquad (22)$$

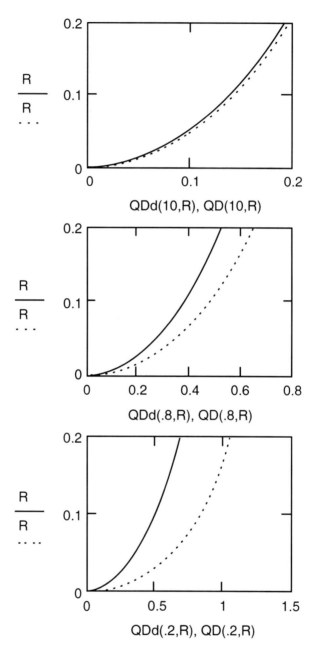

Figure 2. Equilibrium length of the cohesive zone R as a function of the nondimensional load for two models: cohesive, labeled by "D" and the discrete cohesive labeled by "Dd". It is seen that for cracks within the nano-scale range ($X \leq 1$) the differences between two models increase, but they disappear for $X \gg 1$, compare (a), (b) and (c).

This, again, is identical to the finite (or quantized) fracture mechanics result (Taylor et al., 2005; Pugno and Ruoff, 2004), and it should be compared to the Griffith (LEFM) result

$$\frac{\sigma_{Griffith}}{\sigma_0} = \frac{1}{\sqrt{2X}} \qquad (23)$$

Two curves resulting from Eqs. (22) and (23) are shown in Figure 3. The discrete model predicts a finite critical stress for a crack of zero length. The differences between the two results diminish as the crack length becomes much larger than the fracture quantum a_0, i.e., when $X \gg 1$.

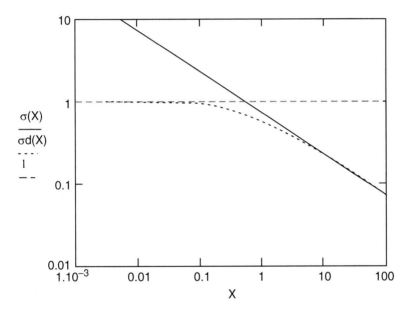

Figure 3. Log-log plot of the critical stress normalized by the inherent strength: straight line results from the LEFM and it approaches infinity for $X \geq 0$. The curve under this line resulted from the quantized cohesive zone model.

It is noteworthy to see that the equilibrium length of the cohesive zone is strongly affected by the discrete nature of the cohesive crack model. This effect is illustrated in Figure 4, which was drawn according to Eq. (17) valid for arbitrary R/X ratios.

Finally, let us take a look at the cohesion modulus of the discrete cohesive crack valid for a constant restraining stress S. From Eq. (14) we have

$$K_{coh}^d = \langle K_S \rangle = \frac{2}{\pi} S\sqrt{\pi a_0} I(X,R)^{1/2} \tag{24}$$

When the size of the cohesion zone is small compared to the crack length, the integral in the Eq. (24) reduces to $2\tilde{R}/a_0$. For this case we obtain a simplified expression for the cohesion modulus identical to the formula derived from a continuous cohesive crack model, see Eq. (7). It is noted that in this limiting case of R/X being small the fracture quantum drops out of the equation. For a general case, though, described by the Eq. (24) the fracture quantum remains as one of the independent variables.

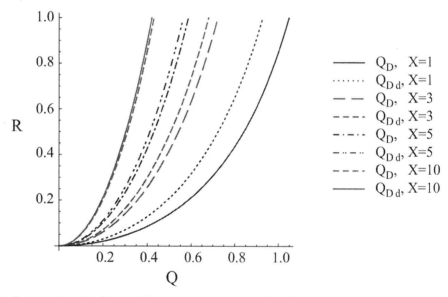

Figure 4. Length of the equilibrium cohesive zone predicted by the Dugdale equation and by the discrete cohesive crack model. The discrepancies become evident for shorter cracks and they tend to disappear for crack sizes large compared with fracture quantum. Discrete nature of fracture process results in an increased levels of the cohesive zone length for a fixed value of the applied load. When the crack length is substantially larger than the fracture quantum (X >> 1), these differences disappear.

The problem faced by the material engineers consists in finding a right proportion between S and $\tilde{R}$; there exists a conflicting trend between the magnitude of the cohesive strength (approximated by the inherent strength of the material) and length of the cohesive zone. Increasing S, by increasing strength without paying attention to cohesion toughness parameter, which describes the degree of ductility, lowers the length $\tilde{R}_c$ and thus it may lead to undesirable effects of increased brittleness. This is where the theoretical considerations involving the fracture quantum and the resulting discrete

MODEL FOR FRACTAL CRACKS

nature of fracture at nano levels may prove useful in designing new materials. The formula (24) shows that indeed there is an additional length involved in defining the cohesion modulus. This is the length a_0. As seen from Eq. (24) in addition to the inherent strength S there are two other factors that influence material toughness; these are the fracture quantum and the ratio of cohesive zone size to fracture quantum. This provides a material engineer with additional degrees of freedom to work with.

Yet another variable affecting the cohesion modulus and material toughness is the roughness of the surface created in the course of fracture process. This phenomenon will be discussed in the next section.

4. Discrete cohesive model for fractal cracks

In this section we shall incorporate two features typical of any fracture process:

1. Roughness of the newly created surface due to varying degree of fractality (as opposed to the assumption of perfectly smooth surface employed in the classical LEFM)
2. Discrete nature of the separation of two adjacent surfaces caused by decohesion (as opposed to continuous character of the propagation process commonly assumed in all local fracture criteria)

We shall consider a fractal crack equipped with the cohesive zone. Therefore, we shall now be using all equations derived in the previous sections and modify them to fit the fractal model of Wnuk and Yavari (2005). Although this model is only an approximation, but that is the only model available at the present time.

The cohesive model implies existence of two stress intensity factors, one associated with the applied stress, K_σ^f, and the other K_S^f assigned to the cohesive stress. Superscript "f" emphasizes the fact that we are dealing now with fractal cracks. When the discrete model is employed these two entities should be replaced by their averages taken over the interval $(c, c + a_0)$, where c denotes the half-length of the crack and a_0 is the fracture quantum. For each value of the applied load one can determine the corresponding equilibrium length of the cohesive zone. Let us proceed with the calculations invoking the finiteness condition

$$\langle K_\sigma^f \rangle - \langle K_S^f \rangle = 0 \tag{25}$$

One needs to evaluate both terms in Eq. (25). Let us use a procedure similar to the one used in previous section, in which we calculated the averages. Now we have the following scheme

$$K_\sigma \to \langle K_\sigma^f \rangle_{c,c+a_0} = \left\{ \frac{1}{a_0} \int_c^{c+a_0} \left[K_\sigma^f(c,\tilde{R},\alpha) \right]^2 dc \right\}^{1/2}$$

$$K_S \to \langle K_S^f \rangle_{c,c+a_0} = \left\{ \frac{1}{a_0} \int_c^{c+a_0} \left[K_S^f(c,\tilde{R},\alpha) \right]^2 dc \right\}^{1/2} \quad (26)$$

The resulting functions depend on the fractal exponent α and two length-like variables X and R. To calculate the K-factors for a fractal crack we employ the expression given by Wnuk and Yavari[1] (2003) and obtain

$$K_\sigma^f = \frac{a^{\alpha-1}}{\pi^{2\alpha-1/2}} \int_0^a \frac{(a+x)^{2\alpha} + (a-x)^{2\alpha}}{\sqrt{a^2 - x^2}} p(x) dx \quad (27)$$

In the first equation in Eq. (26) we substitute σ for $p(x)$ and $c+\tilde{R}$ for a. This leads to

$$K_\sigma^f(c,R,\alpha) = \chi(\alpha) \sigma \sqrt{\pi} (c+R)^\alpha \quad (28)$$

The scalar function appearing in this equation reads

$$\chi(\alpha) = \frac{1}{\pi^{2\alpha}} \int_0^1 \frac{(1+s)^{2\alpha} + (1-s)^{2\alpha}}{(1-s^2)^\alpha} ds \quad (29)$$

Next we evaluate the average

$$\langle K_\sigma^f \rangle = \chi(\alpha) \sigma \sqrt{\pi a_0^{2\alpha}} \left\{ \frac{1}{a_0} \int_c^{c+a_0} (c+R)^{2\alpha} dc \right\}^{1/2} = \chi(\alpha) \sigma \sqrt{\pi a_0^{2\alpha}} \left\{ \int_X^{X+1} (X'+R)^{2\alpha} dX' \right\}^{1/2}$$

$$(30)$$

The integral is elementary and we obtain

$$\langle K_\sigma^f \rangle = \chi(\alpha) \sigma \sqrt{\pi a_0^{2\alpha}} \frac{\left[(X+R+1)^{2\alpha+1} - (X+R)^{2\alpha+1} \right]^{1/2}}{\sqrt{2\alpha+1}} \quad (31)$$

It is readily observed that for the fractal dimension $D \to 1$ (or $\alpha \to 1/2$) expression (31) reduces to the equation valid for non-fractal discretized cohesive model

---

[1] The fractal crack model employed here is based on a simplifying assumption, according to which the original problem is approximated by considerations of a smooth crack embedded in the stress field generated by a fractal crack, cf. Wnuk and Yavari (2003).

MODEL FOR FRACTAL CRACKS

$$\langle K_\sigma \rangle = \sigma \sqrt{\pi a_0} \left(X + R + \frac{1}{2} \right)^{1/2} \tag{32}$$

Let us define the ratio of the last two K-factors

$$k_\sigma^f = \frac{\langle K_\sigma^f \rangle}{\langle K_\sigma \rangle} = a_0^{\alpha-1/2} \chi_\sigma(X, R, \alpha) \tag{33}$$

where the new scalar function is defined as follows

$$\chi_\sigma(X, R, \alpha) = \chi(\alpha) \left[\frac{(X+R+1)^{2\alpha+1} - (X+R)^{2\alpha+1}}{(2\alpha+1)(X+R+\frac{1}{2})} \right]^{1/2} \tag{34}$$

The plot of the function χ_σ vs. α is shown in Figure 5. Since χ_σ depends also on crack length and the cohesive zone size, X and R, these two length-like variables are used as parameters in plotting the graphs in Figure 5.

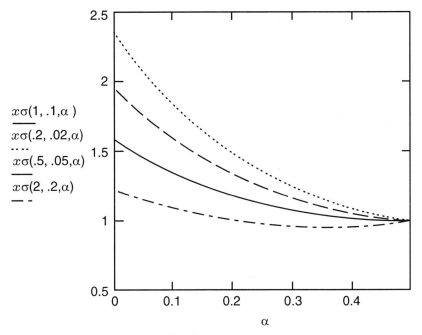

Figure 5. Nondimensional average $\langle K_\sigma \rangle$ plotted as a function of the fractal exponent α. Zero value of α corresponds to the fractal dimension D = 2, while ½ value of α coincides with D = 1; the case of a smooth (non-fractal) crack.

The discrepancies in the χ_σ function are particularly visible for the fractal dimension D approaching 2 (or α converging to zero). This effect is reflected by the larger spread of function values at the vertical axis, where α equals zero.

For the other limiting case of α approaching ½, when a fractal crack reduces to a smooth crack, all curves shown in Figure 5 converge to unity, which means that the effects of fractal nature of the crack disappear and the solution for the discrete cohesive model is recovered. It can also be seen that for the crack sizes comparable to fracture quantum the magnitude of the average $\langle K_\sigma^f \rangle$ is enhanced for any α, while for cracks much longer than a_0 the value of this average tends to drop below the value $\langle K_\sigma \rangle$ valid for a smooth cohesive crack. Therefore, one may conclude that both discrete nature of fracture and its fractal geometry alter the solutions obtained for the established cohesive crack models.

To further emphasize this point, we will evaluate and discuss the cohesion modulus as given by $\langle K_S^f \rangle$. This entity is a function of X, R and α, and its physical meaning is that of material resistance to initiation and propagation of cracks. Applying the second equation in Eq. (26) we have

$$\langle K_S^f \rangle = \frac{S}{\pi^{2\alpha-1/2}} \left\{ \frac{1}{a_0} \int_c^{c+a_0} (c+R)^{2\alpha} dc \left[\int_{\frac{c}{c+R}}^{1} \frac{(1+s)^{2\alpha}+(1-s)^{2\alpha}}{(1-s^2)^\alpha} ds \right]^2 \right\}^{1/2} \quad (35)$$

Introducing the nondimensional variables X and R we obtain

$$\langle K_S^f \rangle = S\sqrt{\pi a_0^{2\alpha}} \left\{ \int_X^{X+1} (X'+R)^{2\alpha} dX' \left[\int_{\frac{X'}{X'+R}}^{1} \Phi(s,\alpha) ds \right]^2 \right\}^{1/2} \quad (36)$$

With the inner integral denoted by H(X,R,α) and the auxiliary function Φ defined below

$$\Phi(s,\alpha) = \frac{1}{\pi^{2\alpha}} \frac{(1+s)^{2\alpha}+(1-s)^{2\alpha}}{(1-s^2)^\alpha}$$

$$H(X,R,\alpha) = \int_{\frac{X}{X+R}}^{1} \Phi(s,\alpha) ds \quad (37)$$

both integrals in Eq. (36) are evaluated numerically. With the notation (37) we have the final expression for the desired average

$$\langle K_S^f \rangle = S\sqrt{\pi a_0^{2\alpha}} \left\{ \int_X^{X+1} (X'+R)^{2\alpha} H^2(X', R, \alpha) dX' \right\}^{1/2} \quad (38)$$

Written in a somewhat shorter form expression (38) becomes

$$\langle K_S^f \rangle = S\sqrt{\pi a_0^{2\alpha}} \, G_{coh}^f(X, R, \alpha)$$

$$G_{coh}^f(X, R, \alpha) = \left\{ \int_X^{X+1} (X'+R)^{2\alpha} H^2(X', R, \alpha) dX' \right\}^{1/2} \quad (39)$$

To verify the correctness of these expressions we note that for $\alpha \to 1/2$ the function H reduces to

$$[H(X, R, \alpha)]_{\alpha=1/2} = \frac{2}{\pi} \cos^{-1}\left(\frac{c}{c+R}\right) \quad (40)$$

and the function G_{coh}^f becomes identical with the non-fractal result obtained previously for a discrete cohesive crack, namely

$$[G_{coh}^f]_{\alpha \to 1/2} = \left\{ \int_X^{X+1} (X'+R) \left(\frac{4}{\pi^2}\right) \left[\cos^{-1}\left(\frac{X'}{X'+R}\right)\right]^2 dX' \right\}^{1/2} =$$

$$\frac{2}{\pi} \left\{ \int_X^{X+1} (X'+R) \left[\cos^{-1}\left(\frac{X'}{X'+R}\right)\right]^2 dX' \right\}^{1/2} = \frac{2}{\pi} I^{1/2}(X, R) \quad (41)$$

Thus the cohesion modulus for $\alpha \geq \frac{1}{2}$ acquires the form

$$\langle K_S^f \rangle_{\alpha=1/2} = S\sqrt{\pi a_0} \left[\frac{2}{\pi} I^{1/2}(X, R)\right] \quad (42)$$

as expected.

To better understand the effects of fractality and discrete aspects of fracture on the cohesion modulus let us examine the ratio

$$\chi_S = \frac{\langle K_S^f \rangle}{\langle K_S \rangle} = \frac{G_{coh}^f(X, R, \alpha)}{\frac{2}{\pi} I^{1/2}(X, R)} \quad (43)$$

Figure 6 provides the diagrams illustrating the dependence of this function on the fractal exponent α and the two length-like variables X and R. Best way to visualize the effect of these variables is to think of the ratio R/X. As seen from the graph in Figure 6 the material toughness measured for fractal and discrete fracture, as given by expressions (39) and (43), can be either enhanced or reduced when the set of the independent variables (X,R,α) is manipulated. The range of $\chi_S > 1$ corresponds to an enhancement of material resistance to crack initiation due to variations in the degree of fractality of crack surface. The enhancement is seen to occur for the fractal dimension D approaching 2 and crack sizes small in comparison to the constant a_0. The effect is pertinent to nano-cracks.

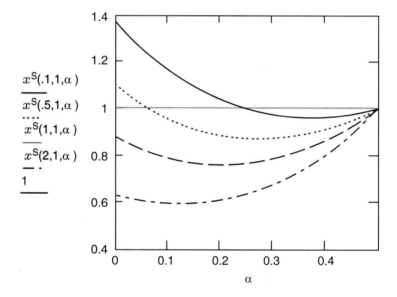

Figure 6. Effect of the measures of fractality on the modulus of cohesion for a discrete fractal crack. The extreme values of the measures of surface roughness are α = 0.5 or D = 1 and α = 0 or D = 2 as shown in the graph. The degree of fractality has a pronounced influence on the cohesion modulus. Note that the ratio of the cohesion zone length to the crack size, $\tilde{R}/c$, has been used as a parameter distinguishing the curves shown in the graph.

Now we have all the entities needed to set up the finiteness condition. Recalling the equality $\langle K_\sigma^f \rangle = \langle K_S^f \rangle$ and using Eq. (31) and (39) we arrive at

$$\chi(\alpha)\sigma\sqrt{\pi a_0^{2\alpha}}\left[\frac{(X+R+1)^{2\alpha+1}-(X+R)^{2\alpha+1}}{1+2\alpha}\right]^{1/2} = S\sqrt{\pi a_0^{2\alpha}}\,G_{coh}^f(X,R,\alpha)$$

(44)

MODEL FOR FRACTAL CRACKS 377

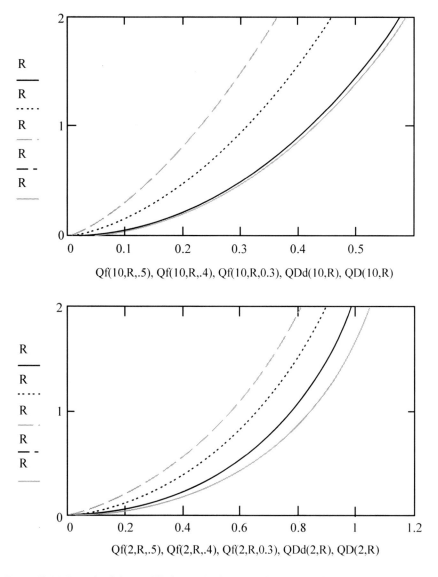

Figure 7. (a) Length of the equilibrium cohesive zone for discrete fracture represented by a fractal cohesive model plotted as a function of loading parameter Q. Top curve is drawn for the fractal exponent $\alpha = 0.3$, next down is valid for $\alpha = 0.4$ and the lowest is a superposition of two curves, one obtained from Eq. (45) at $\alpha = 0.5$ and the other described by Eq. (17) valid for a discrete cohesive crack. The lowest curve represents the Dugdale result. As the surface roughness increases the length of the cohesive zone size increases for a fixed external loading parameter. All curves are evaluated for the crack length $X = 10$. (b) The same functions plotted for the crack length $X = 2$.

This can be solved implicitly for the loading parameter $Q_f (= \pi\sigma_f/2S)$ as a function of X, R and α. The solution is

$$Q_f = \frac{\pi}{2\chi(\alpha)} G_{coh}^f(X,R,\alpha) \left[\frac{2\alpha+1}{(X+R+1)^{2\alpha+1} - (X+R)^{2\alpha+1}} \right]^{1/2} \quad (45)$$

What we would like to derive from this equation is the functional dependence of the length of cohesive zone R_f on the loading parameter Q_f. This can be done by inverting the function (45). A computer program is used to plot the inverse of the function defined in Eq. (45). The resulting graphs of R_f vs. Q_f are shown in Figure 7. When the exponent α varies within the interval [0,0.5] the curves in Figure 7 pertain to fractal and discrete fracture. For each case shown the length of the cohesive zone at fixed applied load is demonstrated to increase with increasing fractal dimension D. Such result proves that fractal crack geometry and/or the discrete nature of fracture tends to increase the material resistance to crack initiation and propagation. It also shows that a fractal crack is capable of relaxing the high stress near the crack tip by generating a larger cohesive zone.

5. Limiting case of cohesive zone length small compared to the crack length

Some of the expressions given in Section 4 can be considerably simplified (and inverted if necessary) when an additional assumption of R being much smaller than X is applied. Specifically we shall focus attention on (a) the cohesion modulus for a discrete fractal fracture and (b) the length of the equilibrium size of the cohesive zone associated with a fractal crack.

Let us begin with the expression for the cohesion modulus of a discrete fractal crack. From Eqs. (37) and (38) we have

$$\langle K_S^f \rangle = S\sqrt{\pi a_0^{2\alpha}} \left\{ \int_X^{X+1} (t+R)^{2\alpha} dt \int_{\frac{t}{t+R}}^{1} \frac{(1-s)^{2\alpha} + (1+s)^{2\alpha}}{\pi^{2\alpha}(1-s^2)^\alpha} ds \right\}^{1/2} \quad (46)$$

The physical meaning of the integration variable "t" is that of the nondimensional crack length X. For R being small compared to X the lower limit of the inner integral in Eq. (46) is close to one. Therefore, the variable "s" is also close to one. Expansion of the integrand for $s \geq 1$ into a power series gives the dominant term

$$[\Phi(s,\alpha)]_{s\to 1} \to \phi(s,\alpha) = \frac{2^{2\alpha}}{\pi^{2\alpha} 2^\alpha} \frac{1}{(1-s)^\alpha} + \ldots \quad (47)$$

Replacing $(1-s)$ by λ changes the inner integral H into

$$[H(t,R,\alpha)]_{s\to 1} = \frac{2^\alpha}{\pi^{2\alpha}} \int_m^1 \frac{ds}{(1-s)^\alpha} = \frac{2^\alpha}{\pi^{2\alpha}} \int_0^{R/t} \frac{d\lambda}{\lambda^\alpha} = \frac{2^\alpha}{\pi^{2\alpha}(1-\alpha)} \left(\frac{R}{t}\right)^{1-\alpha} \quad (48)$$

Here the symbol m denotes the ratio $t/(t+R)$, thus the upper limit in the integral in "λ" is found to be $R/(X+R) \approx R/X$. The expression (46) assumes now the form

$$\langle K_S^f \rangle_{R \ll X} = S\sqrt{\pi a_0^{2\alpha}} \left\{ \int_X^{X+1} t^{2\alpha} \left(\frac{2^\alpha}{\pi^{2\alpha}}\right)^2 \frac{(R/t)^{2-2\alpha}}{(1-\alpha)^2} dt \right\}^{1/2} \quad (49)$$

The integral can be solved in a closed form yielding the result

$$\langle K_S^f \rangle_{R \ll X} = S\sqrt{\pi a_0^{2\alpha}} \frac{2^\alpha R^{1-\alpha}}{\pi^{2\alpha}(1-\alpha)} F(X,\alpha) \quad (50)$$

in which the function F is defined as follows

$$F(X,\alpha) = \left\{ \frac{(X+1)^{1-2(1-2\alpha)} - X^{1-2(1-2\alpha)}}{1-2(1-2\alpha)} \right\}^{1/2} \quad (51)$$

for all "α" except 0.25. For $\alpha = \frac{1}{4}$ the auxiliary function reads

$$F(X) = \left\{ \ln\left(\frac{X+1}{X}\right) \right\}^{1/2} \quad (52)$$

Numerical verification of this approximate formula gives satisfactory results with exception of α being close to zero. To determine the equilibrium level of the cohesive length R_f (index "f" is added to distinguish so designated entity as "fractal") the expression (50) is set equal to the fractal stress intensity factor K_I^f, and then the resulting equation is solved for the characteristic material length associated with a fractal discrete fracture. The solution reads

$$\tilde{R}_f(X,\alpha) = \kappa(\alpha) a_0^{\frac{1-2\alpha}{1-\alpha}} \left(\frac{K_I^f}{S}\right)^{\frac{1}{1-\alpha}} F(X,\alpha)^{\frac{1}{1-\alpha}} \quad (53)$$

The coefficient κ is defined as follows

$$\kappa(\alpha) = \left[\frac{(1-\alpha)\pi^{2\alpha}}{2^\alpha \sqrt{\pi}}\right]^{\frac{1}{1-\alpha}} \quad (54)$$

Figure 8 shows the variations of the nondimensional equilibrium cohesive zone length

$$\Psi_f(X,\alpha) = \frac{8}{\pi}\kappa(\alpha)F(X,\alpha)^{\frac{1}{1-\alpha}} \qquad (55)$$

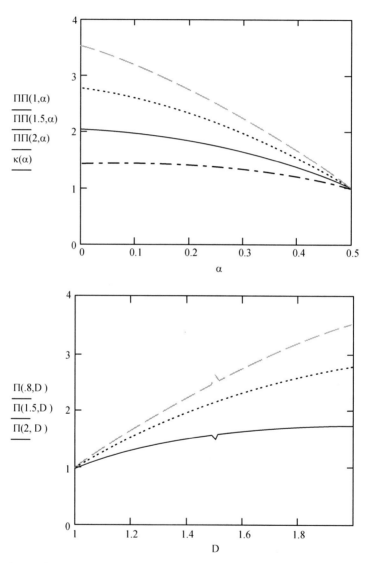

Figure 8. Nondimensional length of the equilibrium cohesive zone Ψ_f shown as a function of
   (a) Fractal exponent α and the crack length X
   (b) Fractal dimension D and the crack length X
In plotting the graphs shown Eq. (55) was used. Note that the "missing points" or the "bumps" on the curves correspond to $\alpha = \frac{1}{4}$ for which F assumes a logarithmic form defined by Eq. (52).

The fractal exponent α (or the fractal dimension D = 2(1 − α)) is used as an independent variable in plotting graphs shown in Figure 8, while the nondimensional length X is used as a parameter. It appears that for the fractal exponent α less than ¼ the proposed approximation exaggerates dependence of the cohesive zone on X. An alternative solution can be found provided that the discretization procedure is omitted. This leads to the cohesive toughness

$$K_S^f = S\sqrt{\pi a_0^{2\alpha}}\left(X+R\right)^\alpha \frac{2^\alpha}{\pi^{2\alpha}(1-\alpha)}\left(\frac{R}{X}\right)^{1-\alpha} \approx \frac{2^\alpha S\sqrt{\pi a_0^{2\alpha}}}{\pi^{2\alpha}(1-\alpha)}\frac{R^{1-\alpha}}{X^{1-2\alpha}} \quad (56)$$

When this expression is set equal the K_I^f, the following solution for the length R_f results

$$\tilde{R}_f(X,\alpha) = \kappa(\alpha) a_0^{\frac{1-2\alpha}{1-\alpha}} \left(\frac{K_I^f}{S}\right)^{\frac{1}{1-\alpha}} X^{\frac{1-2\alpha}{1-\alpha}} \quad (57)$$

Figure 9 illustrates the variations of the function $\tilde{R}_f(X,\alpha)$. In order to remove the dimensions from the entity plotted, the length $\tilde{R}_f$ has been first divided by $R_D = (\pi/8)(K_I/S)^2$ and then the following nondimensional ratio was defined

$$\Lambda_f(X,\alpha) = \frac{\tilde{R}_f}{\tilde{R}_D} \frac{\left(\frac{K_I}{S}\right)^2}{a_0^{\frac{1-2\alpha}{1-\alpha}}\left(\frac{K_I^f}{S}\right)^{\frac{1}{1-\alpha}}} = \frac{8}{\pi}\kappa(\alpha) X^{\frac{1-2\alpha}{1-\alpha}} \quad (58)$$

For the range of nano-scale cracks ($X \approx 1$) the nondimensional length of the cohesive zone associated with a fractal discrete fracture Λ_f varies between 1 for a smooth crack when D = 1, and $\left(8/\pi^{3/2}\right) \approx 1.44$ for a very rough crack, when the fractal dimension D = 2. This represents 44% increase in the equilibrium length of the cohesive zone due to an increase in the degree of fractality.

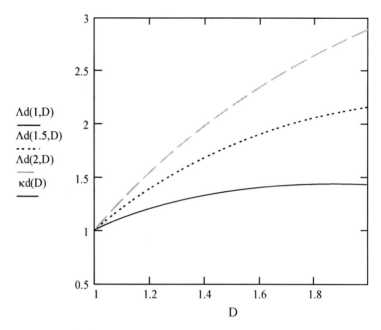

Figure 9. Characteristic material length parameter $\Lambda_f(X,\alpha)$ predicted by the discrete cohesive model of a fractal crack, cf. Eq. (58). Top part of the figure (a) shows the dependence on the fractal exponent α, and the lower portion of the figure (b) illustrates the dependence on the fractal dimension D. For the nano-size cracks ($X \approx 1$) the predicted increase in the characteristic material length parameter due to "roughness" of the crack is shown by the lowest curve. At D = 2 ($\alpha = 0$) the maximum increase compared to the L_c for a smooth crack is about 44%.

6. Conclusions

Relations between applied load and the equilibrium length of the cohesive zone have been established for three different mathematical representations of the crack, and these are

1. Cohesive crack model of Dugdale–Barenblatt for a smooth crack described by two K-factors K_σ and K_S corresponding to the applied stress and the cohesive stress, respectively
2. Discrete cohesive crack model described by the averages $\langle K_\sigma \rangle_{c+a_0}$ and $\langle K_S \rangle_{c+a_0}$
3. Discrete and fractal cohesive crack model involving the fractal equivalents of the averages used in second part, namely $\langle K_\sigma^f \rangle_{c+a_0}$ and $\langle K_S^f \rangle_{c+a_0}$

For a classic LEFM crack model there is no cohesive zone, and the crack itself provides a mechanism for relaxing the high stresses in the vicinity of a stress concentrator. Therefore, the classic representation may be thought of as a limiting case of a more general and refined mathematical formulation involving cohesive zone associated with a crack. For such a representation the equilibrium between the driving force K_σ^2 and material resistance K_S^2 defines a unique relation between the length of the equilibrium cohesion zone, say $\tilde{R}$, and the applied load, say Q. The equilibrium between R ($=\tilde{R}/a_0$) and Q is maintained during the loading process up to the point of incipient fracture.

Cohesive zone generated prior to fracture is related to the material fracture toughness. It also provides an important mechanism for relaxing stresses prior to fracture initiated in the immediate vicinity of a stress concentrator. Figure 10 shows schematically four sketches of a crack represented by four mathematical models; (a) the LEFM concept of a Griffith crack embedded in a linear elastic solid, (b) Dugdale–Barenblatt cohesive model, (c) discretetized cohesive model and, finally, by (d) a fractal cohesive and discrete model. The very first crack shown in the figure has no cohesive zone at all, and the stresses are singular at the tip of the crack. In this case fracture toughness must be measured by employing the ASTM standards that make no mention of cohesion. The second crack in the figure corresponds to a cohesive model suggested independently by Barenblatt (1959) and Dugdale (1960). With cohesion included we gain a better insight into material response to fracture by visualizing the effect of the cohesive zone. In the next two models considered we have generalized the cohesive crack model by incorporating material mesomechanical features related to discrete growth of the crack and with its fractal geometry taken into account. Therefore, in the most advanced model considered here two new variables enter the theory: (1) fracture quantum a_0, and (2) degree of fractality measured either by the fractal dimension D or by the fractal exponent α. Accounting for the discrete and fractal nature of fracture leads to a conclusion that the equilibrium length of the cohesive zone is indeed significantly influenced by both the quantum (discrete) and fractal aspects of the subcritical as well as the propagating crack.

At the micro- and nano-scale the size of the Neuber particle, or process zone in a more updated nomenclature, becomes important not only for mathematical treatment of the problem, but also for the physical interpretation of the decohesion phenomenon at the atomistic scale.

It is noteworthy that each of the successive models listed in Figure 10 predicts for a given fixed level of the applied load successively larger cohesive zone, namely

$$\tilde{R}^f_{Dd} \geq \tilde{R}_{Dd} \geq \tilde{R}_D \geq \tilde{R}_{LEFM}$$

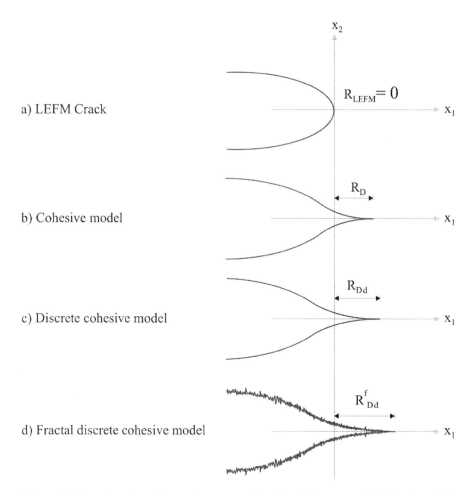

Figure 10. Near tip region of a crack represented by four different models. The size of the equilibrium cohesive zone associated with a crack serves as a measure of the degree of stress relaxation occurring in the crack vicinity. It appears that the two features of the model considered here, discreteness and roughness of the newly created surface are inherent mechanisms of stress relaxation provided by nature.

The interpretation of the subscripts is as follows: f – fractal, Dd – Dugdale discrete, D – Dugdale. Of course the LEFM value of $\tilde{R}$ is zero, but using the K_c value obtained from the tests specified by ASTM one could, in a hindsight, define an equivalent length $\tilde{R}$ that could be associated with a LEFM crack. It is noted that all equations given here reduce to the LEFM results when a fractal crack is replaced by a smooth crack and when the fracture quantum is allowed to vanish.

References

Balankin, A. S. (1997). Physics of fracture and mechanics of self-affine cracks. *Engineering Fracture Mechanics* **57**(2),135–203.

Barenblatt, G. I. (1959). Equilibrium Cracks in Brittle Solids, in Russian, PMM 3,4,5, Vol. 23.

Barenblatt, G. I. (1962). The mathematical theory of equilibrium of crack in brittle fracture. *Advances in Applied Mechanics* **7**,55–129.

Borodich, F. M. (1997). Some fractal models of fracture. *Journal of the Mechanics and Physics of Solids* **45**(2),239–259.

Carpinteri, A. (1994). Scaling laws and renormalization-groups for strength and toughness of disordered materials. *International Journal of Solids and Structures*, **31**,291–302.

Carpinteri, A. and Chiaia, B. (1996). Crack-resistance behavior as a consequence of self-similar fracture topologies. *International Journal of Fracture*, **76**,327–340, 1996.

Cherepanov, G. P., Balankin, A. S., and Ivanova, V. S. (1995). Fractal fracture mechanics. A review. *Engineering Fracture Mechanics* **51**(6),997–1033.

Dugdale, D. S. (1960) Yielding of steel sheets containing slits. *Journal of the Mechanics and Physics of Solids* **8**,100–104.

Goldshtein, R. V. and Mosolov, A. A. (1991a). Cracks with a fractal surface. *Soviet Physics Doklady* **38**(8),603–605.

Goldshtein, R. V. and Mosolov, A. A. (1991b). Fractal cracks. *Journal of Applied Mathematics and Mechanics* **56**(4),563–571.

Ippolito, M., Mattoni, A., Colombo, L., and Pugno, N. (2006). Role of lattice discreteness on brittle fracture: atomistic simulations versus analytical models. *Physical Review B* **73**,104111.

Isupov, L. P. and Mikhailov, S. E. (1998). A comparative analysis of several nonlocal fracture criteria. *Archive of Applied Mechanics* **68**,597–612.

Mosolov, A. A. (1991). Cracks with fractal surfaces. *Doklady Akademii Nauk SSSR* **319**(4),840–844.

Murray, J. D., (1984). *Asymptotic Analysis*, Springer-Verlag, New York.

Neuber, H. (1958). *Theory of Notch Stresses*, Springer-Verlag, Berlin.

Novozhilov, V. V. (1969). On a necessary and sufficient criterion for brittle strength. *Journal of Applied Mathematics and Mechanics-USSR* **33**,212–222.

Pugno, N. and Ruoff, R. S. (2004). Quantized fracture mechanics. *Philosophical Magazine* **84**(27),2829–2845.

Seweryn, A. (1994). Brittle-fracture criterion for structures with sharp notches. *Engineering Fracture Mechanics*, **47**,673–681.

Taylor, D. (2008). The theory of critical distances. *Engineering Fracture Mechanics*, **75**,1696–1705.

Taylor, D., Cornetti, P., and Pugno, N. (2005). The fracture mechanics of finite crack extension. *Engineering Fracture Mechanics*, **72**,1021–1038.

Williams, M. L. (1957). On the stress distribution at the base of stationary cracks. *Journal of Applied Mechanics* **24**,109–114.

Williams, M. L. (1965). Dealing with singularities in elastic mechanics of fracture. *11th Polish Symposium on Mechanics of Solids*, Krynica, Poland.

Wnuk, M. P. (1974). Quasi-static extension of a tensile crack contained in a viscoelastic-plastic solid. *Journal of Applied Mechanics* **41**,234–242.

Wnuk, M. P. and Yavari, A. (2003). On estimating stress intensity factors and modulus of cohesion for fractal cracks. *Engineering Fracture Mechanics* **70**,1659–1674.

Wnuk, M. P. and Yavari, A. (2005). A correspondence principle for fractal and classical cracks. *Engineering Fracture Mechanics* **72**,2744–2757.

Wnuk, M. P. and Yavari, A. (2008). Discrete fractal fracture mechanics. *Engineering Fracture Mechanics* **75**(5),1127–1142.

Xie, H. (1989). The fractal effect of irregularity of crack branching on the fracture toughness of brittle materials. *International Journal of Fracture* **41**,267–274.

Xie, H. and Sanderson, D. J. (1995). Fractal effects of crack propagation on dynamic stress intensity factors and crack velocities. *International Journal of Fracture* **74**,29–42.

Yavari, A. (2002). Generalization of Barenblatt's cohesive fracture theory for fractal cracks. *Fractals-Complex Geometry Patterns and Scaling in Nature and Society* **10**,189–198.

Yavari, A., Hockett, K. G., and Sarkani, S. (2000). The fourth mode of fracture in fractal fracture mechanics. *International Journal of Fracture* **101**(4),365–384.

Yavari, A., Sarkani, S., and Moyer, E. T. (2002). The mechanics of self-similar and self-affine fractal cracks. *International Journal of Fracture* **114**,1–27.

REPARABILITY OF DAMAGED FLUID TRANSPORT PIPES

PHILIPPE JODIN*
*University Paul Verlaine – Metz & ENIMetz, île du Saulcy,
F-57045 Metz cedex*

Abstract Most of energy fluids and also water are transported over very long distance by pipelines. There is a quite large spectrum of situations defined by: nature and pressure of the transported fluid, environmental and political situations, and economical constraints. If a pipeline is damaged and becomes unable to continue its service or presents some risks towards the local population, or is an economical challenge, the question of repairing appears. Therefore, the technical choices should be optimized with respect to the induced costs and to the social, economical and, perhaps, political costs. This paper will present various situations and the available repairing solutions, with respect to their costs in view to be a guide for the person in charge to make decision.

Keywords: Pipelines, safety, damage, repairing, economical costs, social costs, sustainable development, fracture mechanics, fatigue

1. Introduction

The total gas pipelines length was estimated in 2007 to 1 million km (Pluvinage and Elwany, 2007). For drinkable water, the total length is very difficult to establish, as networks exists for a time and corresponding data-bases are very diversely maintained in the different countries. However, the continuity of distribution of drinkable water is vital for peoples all over the world, as well as energetic fluids such as natural gas and oil. Moreover, the project of transporting hydrogen as an energetic vector in pipelines is a new challenge with respect to safety, due to the particular chemical reactivity of this gas. Batisse (2007) has presented an extensive review of repairing methods for pipes. He mainly focused on describing the different

---

* E-mail: jodin@univ-metz.fr

techniques that may be used. In this paper, we will be interested in detailing a method of repairing using a fiberglass-epoxy ribbon, which is wrapped around the damaged tube. This problem was already described for a pipe containing an external longitudinal notch (Jodin, 2006, 2007)

2. A methodology for cases analysis

For the analysis of a case of damage of a pipeline, it is necessary to establish a strict methodology, which ensures that all influent parameters that may correspond to a cost are taken into account. Thus, several steps are considered by order of occurrence in the process:

1. Situation analysis
2. Methods for establishing service again that are considered
3. Estimated residual life after repairing
4. Assessment of induced costs of the complete event, including economical, social and engineering costs
5. Economical and social costs that are considered
6. Final decision

These items are detailed and explained below.

2.1. SITUATION ANALYSIS

This first phase of the process will consider the extension of damage of the pipeline, with respect to the expected service of the pipe, say:

2.1.1. *State of the pipe after discovering damage*

First of all, is the pipe broken, and is there a leaking of fluid (gas or liquid)? Therefore, is it necessary to stop the service?

If the pipe is not broken, is there any leaking from the damaged place and is there any danger if the pipe service is not immediately stopped?

If the pipe is not broken, and if there is no leak, what is the estimated damage and what could be the estimated delay before possible break and leak?

2.1.2. *Consequences on environment*

In case of leaking, what are the consequences on environment? Fire and/or explosion danger, pollution of air or soils. Are there nearby people that are exposed to a danger or a contamination through air or drinking water?

Is the pipe easily accessible and what are the constraints for providing specialized technicians and materials and machines for repairing?

What are the economical consequences of stopping the service of the pipeline, at least for a time? Are there people that would suffer directly from the sopping of the pipe service? (loss of drinking water, loss of heating fuel, ...)

All these items have a cost:

- Human costs: loss of human lives, health consequences of a severe pollution
- Social costs: costs induced by the eventual rehabilitation of living space in a large area, with consequences on the quality of life in a given area
- Economical costs including:
  - Loss of production
  - Costs for rehabilitation of living space
  - Payments for financial compensation to human and social consequences

2.1.3. *Accessibility to the damage place*

The assessment of all these costs should take into account the evolution of the local situation if stopping the flow of the pipe is not instantaneously possible. Moreover, there could exist delays due to time to access the place to be repaired and, if any, delay due to local political situation, which could be dangerous for technicians that will operate on the pipe.

2.2. METHODS TO RESTART THE SERVICE

Methods to restart the service of the pipe are highly dependent on the analysis made before. On one hand, the safest but most expensive solution is to completely change a portion of the pipe for a new one, on the other hand the most hazardous one is to do nothing, which is also the cheapest. Of course, between these extreme solutions, there is a lot of intermediary ones that could be examined, especially in view of the economical costs and of the assumed service life after repairing. The optimal solution would probably be that which gives an acceptable probability of survival for the lowest cost. Here is a short review of repairing actions by increasing order of costs.

2.2.1. *Doing nothing*

This is the first extreme solution, which could be the applied when the detected damage is estimated to be without any influence on the integrity of the pipeline. Of course, this is also the cheapest.

Decision of application should be taken after examining the harmfulness of the damage with respect to maximum static loading and to fatigue crack propagation.

2.2.2. *Supervision of the damaged section*

In this situation too, no repairing action is undertaken, but a special supervision programme is launched if there is any doubt about the fatigue crack propagation phenomenon. Here also, fracture mechanics and fatigue crack propagation tools will be used.

2.2.3. *Repairing without service interruption*

There is no leak, but the damage is extended enough to present a danger with respect to fracture and crack propagation. In this case a repairing by different techniques such as welding a patch or wrapping and gluing a sleeve on or around the damaged zone can be used. Moreover, the welding technique allows metal refilling of external damaged zone, so that the integrity of the damaged zone is established again.

2.2.4. *Repairing with service interruption*

This could be the case when there is a leak, which can be dangerous with respect to repairing techniques. Moreover, the leak may represent a danger toward environment and a loss for the company which operates the pipeline. In this case, any repairing technique could be employed, including the installation of a bypass, which allows establishing again service during repairing of the damaged zone and delivering of products to final users. When repairing is completed, the normal service can be restored and the bypass suppressed.

In certain cases, it will be necessary to monitor the repaired portion to verify the efficiency of the repairing, especially towards leak of fluids.

2.2.5. *Complete change of a portion of pipe*

This is the case when the damage is so extended that any tentative of repairing the original tube is inapplicable. Therefore, either the service of the pipeline is stopped or a bypass is installed, and change of portion is possible with the pipeline active.

2.3. ESTIMATION OF SERVICE LIFE AFTER REPAIRING

After a repairing technique is chosen, it is necessary to estimate the service life of that portion, as the geometry of the pipe could have been modified, and the material characteristics could have been altered by welding, for instance. Tools used in this section are mainly fatigue life, creep estimation or aging estimation. Other parameters should be considered, when social and environmental conditions are difficult.

2.4. COSTS ESTIMATION

Estimation of total costs must take into account all technical and economical costs, but also social costs.

2.4.1. *Economical costs*

This section includes:
- Losses due to stopping of pipeline
- Costs of environment rehabilitation
- Costs of the different repairing solutions considered
- Eventual delayed reparation costs

If there is no stopping of production, the first item vanishes or is minimized. Environment rehabilitation and penalties are more and more asked by governments, which should be taken into account in case a risk of fracture of the repairing is non-negligible.

2.4.2. *Social costs*

If the initial fracture has provoked loss of human lives or serious health problems, if there are problems linked to stopping delivery of vital products such as drinking water or energy, this has a cost which should be taken into account.

Moreover, if the staff that has to apply the repairing solution is exposed to health, climate or even political risks, it needs at list an important salary compensation, which should be added to the other costs.

2.5. CRITERIA FOR A CHOICE OF A REPAIRING SOLUTION

After the up-described analysis has been made, it is necessary to give tools for decision making to the manager. In the analysis, three main items have been considered: economy, politics and technique.

If there are several technical possibilities, they probably induce different costs. It could also exist some political constraints that may influence the technical choice on a hand and the economical costs, on the other hand. All these choices should be presented to the manager, who will make decision with respect to them.

Once he has made decision, a feedback control is necessary to ensure that all steps of it will be conveniently applied. This has also a cost that should be included in the total cost of the action.

2.6. SUMMARY OF THE METHODOLOGY

We have seen that repairing a damaged portion of a pipeline cannot be reduced to a simple technical operation. Taken into account the importance of the fluid transported in, technical decisions have economical and social consequences that should be considered in the costs.

Here the main objective of the engineer would be to design technical solutions that minimize the economical and social consequences.

3. Different technical solutions

3.1. SUMMARY OF DIFFERENT OPTIONS

It has been seen in the preceding section that there could be several options in front of a degraded situation of a pipeline. They vary from "do nothing" to "complete replacement of a portion of tube". This, of course depends on the actual situation of the pipe, but also on the available means of observation.

If a portion of pipe is completely destroyed, it is obvious that a complete change of the destroyed portion is needed. But, if the pipe is not completely destroyed, it is necessary to have some investigation on the dimensions, position, shape and nature of the defective part.

If it is a surface pipe, it is relatively easy to an investigation of the assumed damaged part with eye or camera, with eventual image analysis for helping diagnostic. But if the pipe is running underground, it is more difficult to have this investigation, because it is necessary to localize with some precision the damaged place. Then, an excavation could be made and visual observation then becomes possible.

In both cases, these observations could hardly give information about the state of the inner of the tube as well as state of welding cordon. Other techniques, such as US, gamma ray or insertion of robot vehicle in the inner tube should be envisaged. The objective is to collect a maximum of data on the damage for the most accurate evaluation of harmfulness.

3.2. TOOLS FOR HARMFULNESS ASSESSMENT

Once the damage is localized and the dimensions and shape of the defect is estimated, fracture mechanics tools will be useful for the engineer for fracture assessment.

Of course, classical fracture mechanics parameters such as stress intensity factor, COD, J-integral could give a first approach of the risk of fracture, but they are deterministic.

A further step using a probabilistic approach is to use a failure assessment diagram. This method includes a safety factor, which is related to the accepted probability of failure. Another way is to use the SINTAP method coupled with a FORM/SORM (First Order/Second Order Reliability Method) (Jallouf et al., 2005). Both methods lead to an estimation of the remaining life of the damaged part, provided the operator accepts a given level of risk.

3.3. REPAIRING SOLUTIONS

There are six main techniques for repairing a damaged tube.

1. Changing the damaged portion
2. Installing a bypass
3. Grinding
4. Metal deposition
5. Metallic sleeves
6. Composite sleeves

3.3.1. *Changing the damaged portion*

This solution should be used when the damage is so large that the integrity of the pipe is not asserted, even pipe is already broken. This method needs stopping tube service for a long time, as it is necessary to replace a portion of the tube, and, therefore to prepare and carefully realize proper welded joints. These welds should also be controlled.

3.3.2. *Installing a bypass*

This operation is also possible before the first one, so that some stopping service time could be saved, if the replacement operation should be delayed for any reason. But it also needs stopping the service of the pipe for the necessary time for bypass installation. However, a new technique of installing a derivation tape on the tube without stopping the service is proposed by Furmanite® (Furmanite, 2008).

3.3.3. *Grinding*

One of the most important characteristic of a geometrical defect (for instance a notch created by accidental indentation by a diving machine) is the stress concentration it generates. As tubes are often over designed, a controlled grinding of the damaged part could be a simple solution to lower stress concentration effect. Of course the thickness decrease that is associated with this grinding must not affect the strength of the tube.

3.3.4. *Metal deposition*

On the opposite, notches or other surface defects can be filled in by metal deposition. If material continuity is asserted, then the stress concentration due to notch effect vanishes. This solution is applicable on tubes made from steel with good weldability. Combination with grinding can help to obtain a smooth repair.

3.3.5. *Metallic sleeves*

Two types of metallic sleeves can be used, which are a function of the seriousness of the situation. If only local reinforcement is needed, then the so-called *type A* is applied. Two half sleeves, which inner diameter is exactly equal to the outer diameter of the damaged tube are placed around the area to be reinforced and, then welded together. If necessary, the shape of the sleeves can be adjusted to exactly match that of the tube. This solution is applicable when there is no leak and is independent on the weldability of the tube.

If local reinforcement is needed together with preventing any further leak, it is therefore addressed to use the *type B* sleeves. In this type, two half sleeves are also used (and, if necessary with their shape adjusted to the shape of the tube), but they also need to be welded to the tube, so that leaking is prevented. In this case, they have to support higher stresses and need very skill welding operators for proper realization.

3.3.6. *Composite sleeves*

The principle of composite sleeves is the same as for metallic sleeves, but the implementation procedure is different. Of course, there are also several techniques of implementation, but we will mainly consider here the wrapping technique. In that case, the product is a wide ribbon made from epoxy pre-impregnated fabrics, it is relatively easy to implement on-site. The ribbon is wrapped around the damaged zone so that the desired thickness is obtained. Moreover, a pre-tension can be applied, so that a part of the strength, which can be increased due to a notch effect and/or local thickness reduction is compensated. The sleeve is then cured to create a strong sleeve around the damaged zone.

4. Mechanical behavior of the repairing with a composite sleeve

As composite sleeves seem to present several advantages with respect to steel ones and as notch fracture mechanics is a useful tool for the engineer who is in charge to suggest a technical solution for repairing a damaged pipe, we have realized some finite elements computations for a better knowledge of stresses and notch fracture mechanics parameters in the repaired area.

4.1. LONGITUDINAL EXTERNAL NOTCH

This situation has been presented in details in a preceding work (Jodin, 2007). It has been shown that J-integral reduction highly depends on the thickness of the sleeve and on the quality of adhesion between sleeve and tube. This is illustrated in Table 1.

TABLE 1. Relative reduction of J-integral for an elliptical longitudinal surface defect in a tube with respect to sleeve thickness and width and type of joint with the tube.

	Sleeve thickness (mm)	Sleeve width (mm)		
		0	36	48
No sleeve	0	1.9336		
Non glued	5		−48.9%	−48.3%
	10		−80.5%	−78.9%
Glued	5		−77.4%	−77.2%
	10		−84.0%	−83.9%

4.2. TUBE WITH A HOLE DRILLED IN

This situation may occur if somebody drills a hole in the tube to make uncontrolled taking off. Fast repairing can be achieved by wrapping a composite fabrics, which should be glued to the tube for ensuring liquid or gas waterproofness. It is clear that the stresses distribution around the hole plays an important role in the quality of the repairing and for its duration.

The goal of this study was to evaluate the influence of the hole diameter and the composite thickness on the stress that is normal to the adhesive plane. Moreover, two kinds of fiberglass fabrics are considered: one is unidirectional, the second is bidirectional. Computations were made through Castem® finite elements programme, using simple prismatic or cubic elements. A portion of tube (elastic, isotropic), a layer of adhesive (1 mm thickness, elastic

396   P. JODIN

isotropic) and a layer of composite (elastic, orthotropic or isotropic) were modelized. The maximum normal stress in the adhesive layer is extracted, and stress gradient is represented. Results for maximum stresses are shown in Figures 1 and 2. It is clear that the maximum stress value is highly dependent on diameter of the hole. Concerning the dependence on the composite thickness, the maximum stress value decreases while thickness increases, but trends to vanish with high thicknesses.

Another point is the stress distribution in the adhesive along the border of the hole. Figure 3a shows the stress gradient when a uniaxial fabrics is used and Figure 3b when a biaxial one is used. It can be seen that the stress gradient is higher in the uniaxial case as in the biaxial case.

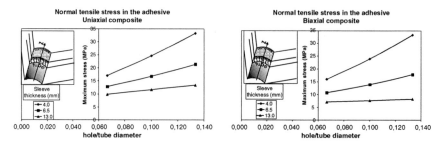

Figure 1. Variation of maximum stress in the adhesive on the border of the hole with hole diameter.

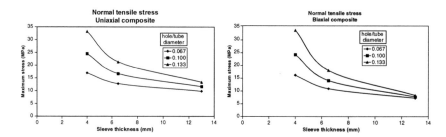

Figure 2. Variation of maximum stress in the adhesive on the border of the hole with composite thickness.

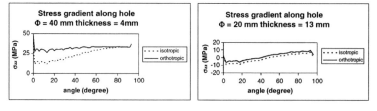

Figure 3. Variation of stress gradient along the hole in different cases.

4.3. CONCLUSION ON REPAIRING WITH FIBERGLASS COMPOSITE FABRICS

From a theoretical point of view, repairing a drilled pipe by wrapping fiberglass pre-impregnated fabrics composite is possible, provided the thickness is large enough to ensure hoop stress sustainment and pressure strength. A sensitive point is the stress distribution in the glue joint around the hole as, if it overcomes the strength of the adhesive layer, it could initiate a crack which should be able to propagate. It has been shown that the stress level is lowered when thickness of the composite increases and that the stress gradient is lowered when a biaxial composite (say quasi isotropic) is used.

From an engineering point of view, using a composite pre-impregnated ribbon is relatively easier than welding a steel sleeve, which should be machined to precisely fit the local shape of the pipe. This could be somewhat difficult, if the pipe is damaged and exhibits some local deformation.

However, it is necessary to take into account in the estimation of the repairing duration the loss of mechanical characteristics due to aging of fiberglass-epoxy composite.

5. Conclusion

It has been shown in this paper that the choice of method for repairing a damaged pipe depends on several factors:

- Nature and extension of damage
- Leak or pipe service is always possible
- Economical and social conditions
- Technical abilities of technicians that will proceed to repairing

In some cases, replacement of a portion of pipe is the unique and best solution, which is quite expensive a may take a lot of time, for providing the convenient replacement pipe. To restore pipe service, it is possible to install a bypass, but this increases the cost. An economical balance should be made before making decision.

In other cases, welding a steel sleeve partially or all around the tube may be a durable solution, but quite difficult to realize and rather expensive.

At last, wrapping a fiberglass-epoxy composite ribbon around the pipe is a relatively easy and cheap solution, but not so durable as the previous one.

In all cases, the engineer should realize an economical balance to choose the best adapted solution. This suppose him to have all necessary data about materials costs, labor costs, loss of production costs, and also social costs and environmental conditions.

At last, decision could also depend on politics, but this does not depend on the engineer range of responsibility.

References

Batisse, R., 2007, "Review of gas transmission pipeline repair methods", in: *Safety, Reliability and Risks Associated with Water, Oil and Gas Pipelines,* G. Pluvinage and M. H. Elwany, eds, Springer, Dordrecht, The Netherlands, pp. 335–349.

Furmanite, 2008, Richardson, Texas: http://www.furmanite.com/

Jallouf, S., Milović, Lj., Pluvinage, G., Carmasol, A. and Sedmak, S., 2005, "Determination of safety margin and reliability factor of boiler tube with surface crack", *Structural Integrity and Life,* **5**(3), 131–142.

Jodin, Ph., 2006, "Repairing of damaged pressure pipes with a composite sleeve", in: *Recent Developments in Advanced Materials and Processes – YUCOMAT VII,* Trans Tech Publications Ltd, Switzerland, pp. 531–536.

Jodin, Ph., 2007, "Fracture mechanics analysis of repairing a cracked pressure pipe with a composite sleeve", in: *Safety, Reliability and Risks Associated with Water, Oil and Gas Pipelines,* G. Pluvinage and M. H. Elwany, eds, Springer, Dordrecht, The Netherlands, pp. 325–333.

Pluvinage, G. and Elwany, M. H., (eds) 2007, "Preface", in: *Safety, Reliability and Risks Associated with Water, Oil and Gas Pipelines,* Publisher, Springer, Dordrecht, The Netherlands, pp. 9–11.

DAMAGE CONTROL AND REPAIR FOR SECURITY OF BUILDINGS

DRAGOSLAV ŠUMARAC
University of Belgrade, Faculty of Civil Engineering
Bul. kralja Aleksandra 73, 11000 Belgrade, Serbia

ZORAN PETRAŠKOVIĆ[*]
Earthquake Engineering Innovation Centre System DC 90
Vele Nigrinove 1, 11000 Belgrade, Serbia

Abstract Civil engineering objects, especially buildings, suffer the damage and failure under earthquake. Invented DC Damper System can increase the resistance and strengthen the construction, enabling tougher behaviour under seismic load. Experimental research and experience in repairing 350 damaged objects on four continents is the base for system development.

Keywords: Damage, seismic load, building, damper, structural integrity, retrofit

1. Introduction

Earthquakes are dangerous impacts on civil engineering structures. It is practically impossible to understand the behaviour of buildings subjected to the seismic loads without basic knowledge of the construction members' behaviour regarding low-cycle fatigue and the stability of the construction during all phases of earthquake activity, because masonry structures are sensitive to it. It is well known that these structures have large mass, due to bad cohesion between bricks (stones) and mortar they suffer damage when exposed to earthquakes, under non linear post elastic condition. The need to find an effective object protection and to realize tougher constructions resulted in a new, the DC 90 Construction System[1] and associate devices.

---

[*] E-mail: dc90@eunet.yu

2. Development and testing of the DC 90 Construction System

System DC 90 comprises a number of structural elements which make the walls more secure, increasing their ductility and toughness. They make floor slabs and ceilings stiff and capable to transmit the load in their own plane, and connect them by foundation collars. These elements make structure stronger to accept the horizontal and vertical loads. This invention is based on the construction system with dampers – absorbers, which makes the building structures more resistant and lets them withstand the highest values of earth tremors through elastic plastic work and plastic deformation control. The damper member is deformed in low cyclic fatigue in accordance with the values of accumulated dilatations so that can accept more than three or four high quakes. The construction is very effective at the masonry objects of historical value, at the modern nuclear power plants and other objects of any security importance during the structure life. The constructions capable to achieve higher level of post elastic non linear condition (that means higher ductility) are likely to survive the damage that seismic loading may cause to such structures. Analyzing the behaviour of different type of materials commonly used for construction of building (concrete, bricks, stone, wood, plastics) the outstanding ductility of steel elements in building constructions is doubtless. The need to provide deformation control that due to the unacceptable large deformation scale may cause the destruction of the elements or total structure collapse was particularly considered.

The aim of the new device (damper-absorber) is to provide an accurate and controlled elastic-plastic work.[2] The most important parameters that define the damper construction and its properties are shown in Figures 1 to 3.

Special contribution presents the innovative design of Damper DC 90. It is patented in the U.S.A.[3] and awarded by Gold Medal in Brussels. Damper is involved as a part of vertical stiffener elements, and thanks controlled fatigue defines position of plastic hinges and instant of its initiation, intensity of load and deformation, and their control, affecting structure behaviour in seismic condition. The DC 90 Construction System defines the locations for DC Damper installing and the parameters of low-cycle fatigue.

The hard work from inventing and innovating to testing and final implementation of the technology over four continents contributed to understand new technology. The permanent innovation of device design and manufacturing process, model testing and in-situ testing for technology transfer are of the great importance.

Peculiarity of damper design is the middle part shape ("dog bone"),[3] with finished surface to eliminate surface cracks and prevent undercuts. By three limit rings and one movable ring it controls displacement at predetermined length of controlled low cycle fatigue caused by variable seismic loading, and by reduction of cross section the load intensity is controlled. Special elements assure local buckling stability of pressed elements. In this way it is possible to locate and control the position of plastic hinge. The function is controlled by changing stiffness and dynamic characteristics.

As an illustration, in Figure 1 damper of type IRAN is shown, developed for retrofit of buildings in Asia, and in Figure 2 damper of type Mionica, applied for repairing of buildings in the region Kolubara in Serbia after a heavy earthquake.

The damper behaviour under cyclic activity through time, i.e. number of cycle, depends on the following factors: accumulated deformation, frequency, cycle number and damper properties, defined by model testing in the laboratory for dynamic testing.[1,2]

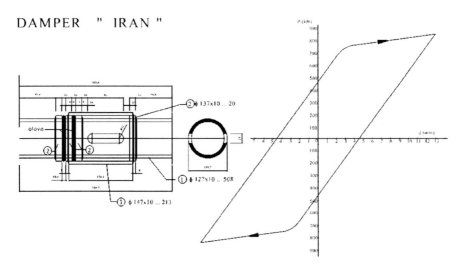

Figure 1. The design of IRAN type damper with basic hysteresis loop.[1]

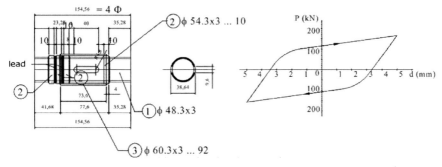

Figure 2. The design of MIONICA type damper with basic hysteresis loop.[1]

The principal feature of the hysteresis loop in diagram force vs. displacement (Figure 3) enables deformation control.

Diagram displacement vs. number of cycles (Figure 4) shows a good performance of tested damper even at high number of cycles. It is obvious that accumulated strain increases with decreasing number of cycles to collapse. This feature is used to determine the damper dimensions and application.

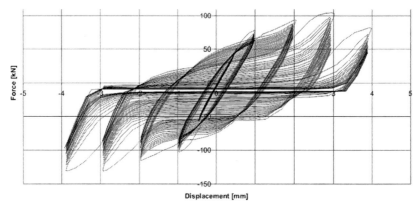

Figure 3. Hysteresis loops diagram force vs. displacement of DC 90 damper.

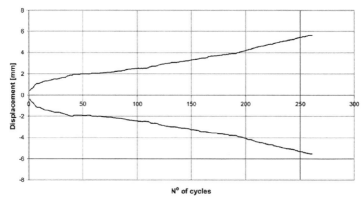

Figure 4. Diagram displacement vs. number of cycles.[4]

Energy increases with increasing accumulated strain, but decreases with number of cycles to collapse of damper, as shown in graph in Figure 5.[4]

Diagrams force–displacement analyses allow to conclude that low cycle fatigue is in question. Analyzing ring operation, in displacement control, permanent plastic deformations occur, that is with increasing number of cycles material weakens and maximum force reduces (material is weaker when force decreases for the same deflection). For smaller displacement, number of cycles needed for failure is greater.

Accepted approach for development of dampers is to test experimentally each new model in order to assure best performance for practical use. Data of one test are presented in Figures 6 to 8.

Dampers for DC 90 system (Figure 6) were tested by variable loading on MTS servo-hydraulic closed-loop machine (Figure 7) in Military Technical Institute (VTI), Belgrade.[5] Obtained hysteresis diagram is given in Figure 8.

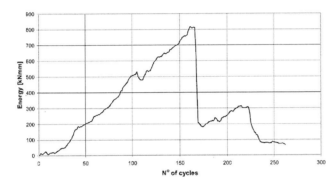

Figure 5. Energy decrease with number of cycles to collapse of damper.[4]

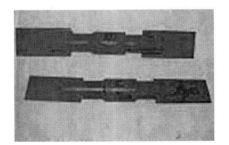 

Figure 6. Typical dampers of DC 90 system.   Figure 7. Damper testing.

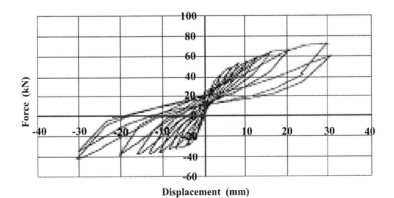

Figure 8. Hysteresis loop diagram obtained by testing of dampers.

Following experiments were performed, including system models:

1. Hysteresis diagram load vs. displacement of biaxial damper model X
2. Uniaxial testing of damper model F for masonry (hysteresis diagram)
3. Extended testing of damper MIONICA, with different frequencies in dilatation range 1% to 5% in dependence of loading mode
4. Testing of sensibility to buckling of 24 models in Civil Engineering Faculty in Ljubljana (Slovenia) in the scope of project PROHITECH[6]
5. Testing of two stores building wall model by quasidynamic loading, up to crack occurrence and final fracture, without and with damper, (hysteresis diagram force vs. displacement, strains measured in vertical and diagonal system elements, as well as model displacement vs. load relation, with continuous monitoring of crack initiation and growth)
6. Experimental research of models (1/10) of masonry walls strengthened by DC 90 system, performed on shaking table in the Dynamic Testing Laboratory of Institute of Earthquake Engineering and Engineering Seismology (IZIIS) in Skopje, Macedonia

Figure 9 presents testing by quasidynamic loading of experimental two store wall, made of hollow blocks 19 × 19 × 25, framed by girders, as classical solution (left), as strengthened according to DC 90 system (middle) and during testing after few cycles (right),[7] performed in the Institute for material testing (IMS), Belgrade.

Figure 9. Two-store building wall model, before retrofit (left), with dampers installed (middle) and during testing of strengthened wall (right).[7]

For sample top point displacement of 10 mm, large flaws (8–10 mm) opened in the first field from the point of force application. In the region of wall constraint, cracks typical for dominant bending stress appeared.[7] Measured initial shear stiffness was $E_{s\text{-}start}$ = force/displacement = F/d = 60/6 = 10 kN/mm; after few cycles, at deflection about 30 mm, it was decreased to $E_{s\text{-}end}$ = 9/6 = 1.5 kN/mm. This significant decrease of wall stiffness can be considered as a collapse of the wall and whole building.

Measured shear stiffness of strengthened – retrofitted wall was initially $E_s = 20/3.5 = 5.7$ kN/mm, and later it slightly decreased, mostly due to yield of lower anchorage tie, and then it increased again. That leads to the conclusion that anchorage must be done properly, and that quality control of performed works must be strict. Wall behaves much tougher at cyclic loading. For load of 60 kN, and displacement of 25 mm far less cracks occurred, opened only for 2–3 mm.

Tested stiffener for larger strains exhibited sufficient durability and probably can, as part of the vertical stiffening frame, save the building at increasing deflections from collapse, making the building more secure during earthquake, especially if building is built without girders.

Experimental research of models (1/10) of masonry walls strengthened by DC 90 system was performed on the shaking table (Figure 10) in the Dynamic Testing Laboratory of Institute of Earthquake Engineering and Engineering Seismology (IZIIS)[8] in Skopje, Macedonia. Three models have been constructed, 30 cm long, 25 cm high, 3.5 cm thick, made of gitter bricks, of plane burned bricks and plane dried bricks. Each type of model was made as conventional and as strengthened by DC 90 System.

Model 1. Hollow brick   Model 2. Burned brick   Model 3. Dried brick

Figure 10. Shaking table test of a wall model.

The single component shaking table has been used to test the models under harmonic excitation within the frequency range of 1–100 Hz and amplitude range (0–10)·g (g stands for gravity). The idea was to compare dynamic response of models produced by traditional and strengthening method of construction for the same brick type, and also to compare the effect of brick type on dynamic behaviour of the models.

The testing program consists of several phases:

– Definition of resonant frequencies
– Definition of elastic response of both models
– Determination of limit state and fracture mechanism

Results of testing are presented in Table 1. The calculated stiffness should be related to the scale factor 1/10 to obtain the actual stiffness of the real wall element. The results have clearly shown increasing of stiffness of walls by implementation of the elements of System DC 90.

TABLE 1. Calculated lateral stiffness of the models, kN/cm.

Type of the model	Strengthened model	Non-strengthened model
Model 1 – hollow brick	35.53	24.75
Model 2 – burned brick	19.95	10.49
Model 3 – dried brick	6.48	4.40

The implementation of DC 90 System after performed experimental testing was not a difficult task, because important data were collected, significant experience gained and acceptability of system has been proved. The orders came from all over the word (Europe, Asia, Africa and America), from different customers for different applications (residential buildings, important industrial object, historical monumental buildings). The most important examples will be presented in Section 4.

3. Numerical modelling of building and its verification

Earthquakes are dangerous impact on civil engineering structures, especially for masonry structures, what has to be considered in building design. In order to reduce the costs and to get better insight in the problem, numerical modelling should be applied, in addition to classical design methods.

3.1. DEVELOPMENT OF NUMERICAL MODEL

Additional contribution to the System DC 90 Damper design is obtained by finite elements (FE) modelling of building. This approach became inevitable for analysis before retrofitting by DC 90 technology of damaged building. Numerical modelling will be successful only if the behaviour of repaired and stiffened objects is tested. Testing of dampers, including vibro platform, was performed for verification of developed numerical models.

Typical earthquake damage of two store building made from bricks and mortar is presented in Figure 11, showing that cracks grow in direction of cross diagonals started from openings (windows, doors).

The damage was analysed applying convenient modelling approach. The analysis, performed for two directions of earthquake effect, shows that the largest tensile stress, responsible for cracking, occur between holes on the building (Figure 12). The earthquake is represented in model by convenient El Centro response.[7] Based on gathered information the building design was strengthened, and new model is presented in Figure 13.[9]

Figure 11. Damage of two store building produced by earthquake.

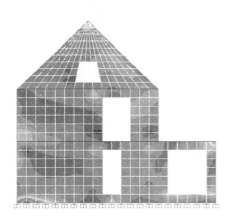 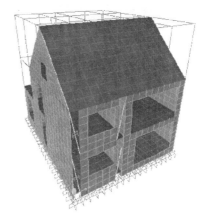

Figure 12. Largest tensile stresses found in corners of openings by model.

Figure 13. Finite elements model of redesigned building.[9]

Vertical elements – walls are strengthened by vertical stiffeners that connect all horizontal slabs and the foundation. Vertical stiffeners – trusses consist of the vertical ties, which are pre-stressed, and the other elements are diagonals with the seismic energy absorber, and horizontals as a part of stiff floor slabs. Walls strengthened in this way exhibited sufficient toughness and capacity to accept the alternative horizontal dynamic displacements. If horizontal elements are not stiff in their own plane, floor slabs and ceilings have to be strengthened by impregnation with a thin, slightly reinforced, concrete slab or incorporating horizontal bracings, linked to vertical stiffeners. The foundation structure is confined with a proper collar, connected by anchors, in which the vertical stiffening elements are anchored.

3.2. TESTING OF RETROFITED AND NON-RETROFITED OBJECTS

Experimental testing for verification of numerical model was performed on both, non-retrofitted (Figure 14) and retrofitted object (Figure 15) in the region of earthquake attack, Mionica, Serbia.[10]

Figure 14. Experimental testing of non-retrofitted building.

Figure 15. Experimental testing of retrofitted building.

Applied testing methods of ambient and forced vibrations by Institute for earthquake engineering and engineering seismology (IZIIS),[7] Skopje, have shown that dominant frequencies of built objects are within the range of 6–8 Hz, while after retrofit stiffness raises up to approximately 35%.

From numerical analysis performed during this study and from experimental results it could be concluded that System DC 90 is powerful tool for engineers to solve problems of retrofitting of damaged building structures. Since this system is cheap, fast and applicable in side, it could be applied properly and broadly for earthquake regions, as it is done in Kolubara region, Serbia, and in Algeria.

4. Examples of DC 90 System application

Thanks to its attractive performance, DC 90 System became very popular. It has been successfully applied on four continents: America, Africa, Asia and. Europe, Most interesting applications will be shortly presented.

4.1. WALL CONSTRUCTION IN HALL OF BORNUDA HYDROELECTRIC POWER PLANT, QUEBEC, CANADA

The wall construction of the engineering hall at Bornuda Hydroelectric Power Station in Quebec, Canada highlights some basic properties of applied "Mionica+" type damper. The designed damper construction was tested in VTI, Belgrade, and as presented wall model testing (Figure 10) in IZIIS, Skopje.[7] Testing of more than 50 specimens was conducted at Civil Engineering Faculty, Ljubljana, Slovenia, and VTI, Belgrade.

The estimated seismic loading of 0.20·g and frequent appearance of minor quakes in the area of large hydroelectric objects in Quebec region, Canada, motivated the "Hydro Quebec" company to analytically approach the problem of damage risk estimation of company's objects. Numerous detailed tests of materials and constructions, model testing and numerical analyses were conducted to cover the topic. The numerical analyses were made by Canadian, Britain and Indian expert teams, dynamic model testing was executed by means of the vibrating platform at IZIIS, Skopje, and dynamic testing of objects by ambient vibration method was made in Canada. The requested damper model testing in VTI, Belgrade, preceded to these actions.

4.1.1. *Numeric modelling*

Numeric modelling of damper performance designed for wall construction of engineering hall in "Beauharnais" Hydroelectric Power Plant in Canada is made based on following elements of stress–strain curve $(\sigma - \varepsilon)$[11]:

1. Initial modulus to linear elastic limit E_o = 20.000 kN/cm² (valid for strain values up to ε = 0.0011 and stress values up to σ = 22 kN/cm²).
2. Elasticity modulus at yielding E_1 = 4.333 kN/cm² (up to ε = 0.0060 and σ = 26 kN/cm²). As number of cycles and frequency increase (strain/s), the elasticity modulus decreases due to accumulated strain. For strain range from 0.0011 to 0.0060, stress values of initial phase vary from 22 to 26 kN/cm², it is about to yield limit. It can be presented by any approximate curve. However, for simplification it is done by means of COSMOS software as a bilinear curve.
3. Elasticity modulus (actually, slope of curve of damper operation in plastic regime) E_2 = 120 kN/cm² (for ε > 0.006, σ=26–32 kN/cm²). The elasticity modulus increases with number of cycles and frequency increase.

The values of E_1 and E_2 elasticity modules depend on:

– Cycle number
– Accumulated strain
– Values of accumulated strain through time

The Coffin–Manson relation is given as:

$$\Delta\varepsilon_p N_f^\alpha = C_1 \qquad (1)$$

where $\Delta\varepsilon_p$ is the cyclic plastic strain range (accumulated strain), N_f is the number of cycles to failure, C_1 and α are material constants. The exponent α usually varies between 0.5 to 0.7, and depends on whether torsion or axial loading is in question. The constant C_1 is defined from damper serial test.

The Palmgren–Miner cumulative damage rule proposes that:

$$\Sigma n/N_f = (n_1/N_{f1}) + (n_2/N_{f2}) + \ldots = 1.0 \qquad (2)$$

where n_1 is the number of cycles at the stress or strain range level 1, N_{f1} corresponding number of cycles for fracture, and so on. The problem of low-cycle fatigue and operation of damper in large strain range up to 5% can be presented by numerical methods, but bilinear relation is sufficient for this analysis.

4.1.2. General information on applied DC90 damper performance

The damper behaviour (approximate damper operation) can be presented by bilinear stress–strain curve, valid for the ascendant and E_2 curve. The slope of the ascendant curve decreases depending on the number of cycles and accumulated strain – Coffin–Manson law (Eq. (1)), the function of two parameters: the number of cycles, N_f and the value of accumulated strain, ε_p.

When it is necessary to limit the displacement, e.g. by 5 mm, that corresponds to the maximum strain of 5%, it is possible according to calculated average slope of the ascendant curve. The slope of E_2 curve for this damper type is 3–10% of the ascendant curve slope, obtained by experiments.

Consideration of frequency effect additionally complicates the diagram.

In plastic range DC 90 damper works in 100 mm weakening length. The pipe, sized 16 × 1 mm, should be weakening by two axial cuts 7 mm long.

The pipe cross section surface is $P = 0.50$ cm^2. The surface of the weakened cross-section is $P - 2 \times 0.7 \times 0.1 = 0.36$ cm^2. Damper design prevents lateral bending and local buckling by means of external and internal elements built into the pipe (micro reinforced polymer concrete and pipe covered by aluminium foil). The overall damper weakening length should be covered by plastics. It is about 100 mm. For simplification, one can consider the average values of strain and stress for total weakening length, i.e. the length situated in plastic zone. Besides, to sustain the fatigue (low cycle fatigue) the surface of the pressed part should be removed to exclude the possibility of any local or surface cracks (damages).

The damper is made of structural steel of high ductility produced by "US Steel" Company (Smederevo, Serbia). All the dampers should be made of the steel of same composition of guaranteed mechanical characteristics.

There is a variety of steel compositions with different tensile properties found, but all were inside the specified values.

The initial part of curve slope (E – elasticity modulus) should be at standard level, as well as yield limit. But in this case the E_2 curve, as well as total strain increases, that provides damper effects (behaviour) at low cycle fatigue. The deformation is controlled by distance control elements (rings). The lead addition affects the reduction of stroke and brittle fracture as well as damper behaviour and work.

Anyhow, the above described design is supposed to provide high strain (accumulated strain) during large earthquakes.

4.1.3. *Residency of Finland Ambassador in Algeria, Africa*

The detailed research of construction condition, as well as non linear dynamic analysis and seismic strengthening of the object were performed to meet requirements of Ministry of Foreign Affaires.[12] For that, an intensive technological research was necessary during realization of this object.

The realization of the object was aimed to investigate the technological process in aspects of humanization, economical realization and modifying or simplifying construction solutions of the system in this work.

4.1.4. *Principle features of the object*

Object dimensions: 27.00 × 17.75 × 13.02 m
Floors: basement, ground-floor, first floor, second floor, tower
Walls: made of stone, thickness d = 80 cm
Inter-storey structure: steel bracings I180 with bow of bricks (h = 40 cm)
Foundation: stone wall, d = 80 cm

The view of the object during repairing action is presented in Figure 16.

Figure 16. View of residency of Finland ambassador in Algeria under retrofitting.

4.1.5. *Technology and process features*

Weight of steel elements: 3,552 kg

Total number of dampers-absorbers 41 pieces, consisting of 18 pieces of type "Algeria", 19 of type "Mionica" and 4 pieces of $\Phi 30$

Total number of displacement compensators: 29

Total number of wall connectors: 59

Total time spent on cracks and splits repairing: 553 h

Time duration of cracks and splits repairing – from 21 to 28 November 2006, duration of object strengthening – from 06 to 30 December 2006.

Total working time spent on construction strengthening: 2,280 h

Working hours: 10 h/day with eventual breaks caused by rain

Engaged experts staff included two civil engineers, one interpreter, four metal structure experts, six experts in erection and assembling, one assistant, and the cook.

Equipment: pipe and movable aluminium scaffolding, diamond saw "Stihl", drilling hand tools "Bosch" and "Hilty", injecting pump, welding equipment, concrete mixers, dozers, digital weighing machines, control and measuring equipment and necessary small instruments and appliances.

4.1.6. *Technologies applied on the object*

Vertical wall bracings with dampers

Horizontal tension of inter-storey structure by means of displacement compensators – time deformations

Vertical tension of the tower by means of damper and displacement compensators through time

Wall connectors at the positions of wall conjunction to preserve wall integrity and avoid separation

4.1.7. *The conclusions made after technological research and analysis*

The technological problems of high noise and huge dust during wall cutting are solved applying hydraulic machines with diamond saws and dust aspirators. Moistening the surface during wall cutting was an alternative.

The problem in construction, significant wall destruction after cutting procedure, particularly of "Algeria" type of $\Phi 100$ mm diagonal bracings, were solved by applying all bracing members (vertical, horizontal, diagonal) of filled steel square or circular pipes, of size minimum 20 mm.

The technology of concrete pressing should be used. The concrete must have an adequate consistence (W/C).

4.1.8. *Design solutions*

The system should be based on the following assumptions:
1. All the members (vertical, horizontal, diagonal) have filled square or round cross-section. The construction members can be built inside or outside the walls with specially designed details which provide the connection between walls and construction elements.
2. The connection details should be typified and unified.
3. The life span of all steel elements is of great importance. The problem will be studied by the experts in technology and structural integrity and life.

4.2. AZERBAIDJAN PRESIDENT RESIDENCE, BAKU (ASIA)

"SERBAS" Company, Baku, applied the DC 90 System technology to provide seismic strengthening and protection of the Azerbaijan President Residence building in Baku.[13]

The conclusions and recommendations gathered in previous activity were used in the Azerbaijan President Residence rehabilitation project. They contributed much to technology promotion and to avoid unfavourable observations and comments during realization of project in Baku.

4.3. SOME EXAMPLES OF APPLICATION OF DC 90 SYSTEM IN EUROPE

In Serbia, Montenegro and in Slovenia many retrofitting projects have been realized on different objects, mostly on the residential houses (Figure 15). In January 2001 DC 90 System Innovation Centre accepted an order for experimental strengthening of the six different objects damaged by earthquake in Kolubara region (Figure 11) in Ljig place.

After detailed experimental verification the technology was applied to repair 350 objects in trussed Kolubara region in Serbia. The experimental reconstruction of house (property Lazić) and object strengthening by applying DC 90 System technology are presented in Figure 17.[14]

Figure 17. Experimentally reconstructed house, Lazić property (left) and strengthen object (right).

5. Project "PROHITECH"

The project "PROHITECH" is an Euro-Mediterranean project aimed to provide an earthquake protection of historical heritage objects.[6] It is leaded by Professor F. M. Macovani from Naples University, who invited the Innovation Centre for Seismic Engineering, Belgrade, to take part in the project. The participants of the project are from Italy, Greece, Portugal, Morocco, Romania, Macedonia, Belgium, Slovenia, Turkey, Israel, Egypt and Algeria. The DC 90 System technology was analyzed and tested in six FW projects "PROHITECH". As the research is under way some finding are presented here without any analysis or comment.

6. Discussion

It is well known for that people always wish to preserve their buildings from destructions of any kind, including earthquake. The process of creation, innovation and implementation of the invention in how to save or retrofit the integrity of building enabled to apply new damper system all over the world (America, Africa, Asia and Europe).

It is reasonable here to cite the reflexion about his invention topic of great scientist Nikola Tesla, probably the most brilliant inventor born. "Invention is a crown of intellect. The development of human kind is substantially dependent on the invention, as most important product of the creative brain. Final goal of mankind is to master the nature by intellect and to exploit its power for mankind needs". Nikola Tesla recognised well that science can't be realised only by theory and mathematics, since the possibility for demonstration by symbolised process are minor and of less significance compared to great truth gathered by experience in practice.

It is not possible to imagine development of mechanics without basic knowledge of material and physical fields (electrical, magnetic, gravity), not only on macro, but also at micro and nano level, and also beyond. "Beyond" is always present. It is difficult to explain how at the end of infinity there is also next point, and how it is possible to divide infinitesimal value on two, so it is impossible to predict development of mechanics and its unforeseeable capacity. In that sense we are prone to think that presented simple innovation is a contribution in mastering the nature. In same category belongs the development and implementation of new materials and structures, in this moment at the level of nano structures.

New design of dampers will ask for new, tougher material, and in the scope of science and invention, this topic is steadily present.

7. Conclusion

Performed researches and investigation made it clear that the system for seismic strengthening of object needs to be innovated continuously, covering new systems design, technology, and numerical modelling as well.

The analysis at levels of crystals disarrangement and interatomic connections weakening, and even beyond that, determines the materials behaviour in non-elastic zone of major importance. It can stimulate the discovery of different materials or structural members reacting properly at seismic loads.

As far as bridge construction is concerned the investigations should be directed to numerical analysis that can provide necessary damping performances (relations between stresses, strains, forces, displacements).

For special objects (nuclear power stations) the investigation should be made in the same direction according to the design schemes, necessary damper performances should be obtained by the numerical analysis.

It is undoubtedly that further innovations will give numerous new and better solutions of constructions and damper devices designed for different building structures. The technological research of the process conducted at experimental and exhibited objects is one of the strategies of the "DC 90 System" Innovation Centre for Seismic Engineering.

The loss of fracture ductility during the low-cycle fatigue process was investigated regarding small surface cracks. The crucial cause for the loss of fracture ductility was elucidated on the basis of microscopic observations. The results are summarised as follows:

1. The low-cycle fatigue process in an annealed medium carbon steel (0.46% C steel) was almost 100% dominated by the growth process of a single crack. In an extreme case, microcrack initiation was observed on the surface of a plain specimen during the first stress cycle.
2. If the surface of fatigued specimen is removed by machining and thereafter by electro-polishing excluded the possibility of surface crack, there is no sign of the fracture ductility loss. The conclusion has been derived that the fracture ductility loss during low cycle fatigue is caused by the existence of fatigue cracks on the specimen surface.
3. After the material constants C_1 and α are determined the "DC 90 System" damper behaviour obeys satisfactory both, the Coffin–Manson law and the Palmgren–Miner rule.

References

1. Z. Petraškovic, Seismic strengthening and protection of objects, Monograph System DC 90, Innovation Centre Belgrade for Earthquake Engineering, Belgrade, 2005.
2. Z. Petraskovic, D Šumarac, M. Anđelković, S. Miladinović, M.Trajković, Retrofitting damaged masonry structures by Technology DC 90, Structural integrity and life (IVK), Belgrade, 2005, Vol. 5, 2, pp. 59–71.
3. Patent USA No. 10/555,131 from 31 October 2005, patent Australia No. AU 2003254327A1 from 23 November 2004.
4. Z. Petrašković, S. Miladinović, D. Šumarac, Technology of seismic strengthening of masonry structures by applying vertical ties and diagonals with seismic energy absorber "System DC 90", International Conference on Earthquake Engineering, Parallell Session, Topic: Retrofit of structures, pp. T6–9, August–September 2005.
5. D. Šumarac, Z. Petraskovic, M. Maksimović, S. Miladinoić, J. Petrašković, Structure retrofit for residental house of Finland's Ambassador in Algier, International Scientific Meeting, Žabljak, Montenegro, 2006, pp. 367–373.
6. F. Mazzolani, Z. Petraskovich, Sixth Framework Program, Priority FP6-2002-INCO-MPC-1, Earthquake Protection of Historical Buildings by Reversible Mixed Technologies PROHITECH, WP6, Naples, 2004–2007.
7. Lj. Tashkov, M. Manic, Shaking table test of a brick-masonry models in scale 1/10, strengthened by DC 90 System, Institute of Earthquake Engineering and Engineering Seismology, University "Ss. Cyril and Methodius", Skopje, Republic of Macedonia, Skopje, May 2004.
8. Lj. Tashkov, M. Manic, Z. Petrashkovich, Vibroplatform testing of brick- masonry models strengthened by System DC 90 in 1:10 ratio, JGDK Symposion, Vrnyachka Banya, 29 September–01 October 2004.
9. D. Šumarac, Z. Petraskovic, M. Maksimović, S. Miladinoić, I.Džuklevcki, N. Trišovič, Retrofit of masonry structures applying vertical braces with dampers System DC 90 and newly designed wall buildings, International Scientific Meeting, Žabljak, Montenegro, 2006, pp. 373–381.
10. Lj. Tashkov, M. Manic, Z. Petrashkovich, R. Folich, B. Bulajich, Experimental verification of dynamic behavior of "System DC 90" under seismic conditions, Belgrade 2003.
11. Z. Petrasković, D. Šumarac, S. Miladinoić, M. Trajkovic, M. Andjelković, N. Trišovič, Absorbers of seismic energy for damaged masonry structures, Alexandropoulos, ECF 16, European Structural Integrity Society (ESIS), 2006.
12. Lj. Taškov, L. Krstavska, Z. Petraskovic, Experimental testing and strengthening of president palace in Alzir by DC 90 System, Institute of Earthquake Engineering and Engineering Seismology, University "Ss. Cyril and Methodius", Skopje, Republic of Macedonia, Skopje, May 2005.
13. Lj. Taškov, L. Krstavska, Z. Petraskovic, Experimental testing and strengthening of president palace in Baku by DC 90 System, International Scientific Meeting, Žabljak, Montenegro, 2008, pp. 475–481.
14. La recherche des vibrations ambiantes, Institut IZIIS, Skopje, Macedonie – CGS, Algérie, La recherche des vibrations forcées – le séisme artificiel, Institut IZIIS, Skopje, Macedonie-CGS, Algérie, Le rapport du contrôle d'entreprise des travaux, CTC, Ain Defla, Algérie.

INDEX

A

accident 265
aluminum alloy 26
application 224

B

bi-material interface 139
blowdown 253
bootstrap method 26
building, damper 399

C

cohesion 378
corrosion 78
corrosion current density 78
corrosion damaging 75
costs
   economical 397
   social 397
crack driving force 91
crack parameters 209
crack path 139

D

damage 399, 413
deformation 360
delamination 337
design tools 175
development 209
discrete process 360

E

elasto-plastic fracture mechanics 1
energy fatigue-function 281
equivalent initial flaw 26

F

failure 26
failure assessment diagram 26, 175
fatigue 399
fatigue crack growth 26
fatigue modeling 281
fatigue-crack growth 281
fluid dynamic forces 253
fractals 360
fracture 360
fracture mechanics 301, 387
fracture parameters 139

G

grain size 337

H

High Cycle fatigue design 325
high strength low alloy steel 91

I

Integrity of structure 91

J

J-integral 157, 288

L

load separation 157

M

macro 360
meso 360
Monte Carlo simulation 26

multiaxial cyclic loading 325
multiscale 360

N

nano 360
non-standard curved specimen 157
natural fatigue-tendency of materials 281
notched components 325
notch fracture toughness 157

P

petrochemistry 267
pipelines 91, 253, 387
POD 26
potentiodynamic polarisation curves 75
pressure equipment 231
probabilistic fracture mechanics 26
probability of failure 26

Q

quantitative NDI 26

R

reliability 267
repairing 399
residual strength 1
retrofit 399

S

safety 301, 387
Safety 267

safety assessment 1
safety factors 175
security 253
seismic load 399
square-root area parameter 281
steel 231
steel bridges 301
strain 337
strain based design 175
strength 337
stress intensity factor 91, 281
structural integrity 209, 231, 399
structural steel 91
substrate 337
surface crack 1
sustainable development 337

T

tensile 337
testing 224, 337

U

uncertainty 26

V

verification 301
volumetric method 157

W

welded joint 231
welded joints 75

Y

yield strength 337
Young's modulus 337

Breinigsville, PA USA
20 August 2009
222668BV00003B/30/P